Polypropylene Based Blends and Composites

Polypropylene Based Blends and Composites

Dr. Vikas Mittal
Editor and Lead Author

CWP
Central West Publishing

This edition has been published by Central West Publishing, Australia

For more information about the books published by Central West Publishing, please visit https://centralwestpublishing.com

Disclaimer

Every effort has been made by the publisher, editor and authors while preparing this book, however, no warranties are made regarding the accuracy and completeness of the content. The publisher, editor and authors disclaim without any limitation all warranties as well as any implied warranties about sales, along with fitness of the content for a particular purpose. Citation of any website and other information sources does not mean any endorsement from the publisher and authors. For ascertaining the suitability of the contents contained herein for a particular lab or commercial use, consultation with the subject expert is needed. In addition, while using the information and methods contained herein, the practitioners and researchers need to be mindful for their own safety, along with the safety of others, including the professional parties and premises for whom they have professional responsibility. To the fullest extent of law, the publisher, editor and authors are not liable in all circumstances (special, incidental, and consequential) for any injury and/or damage to persons and property, along with any potential loss of profit and other commercial damages due to the use of any methods, products, guidelines, procedures contained in the material herein.

NATIONAL LIBRARY OF AUSTRALIA

A catalogue record for this book is available from the National Library of Australia

ISBN (print): 978-1-922617-12-5

Contents

Preface

Polymer blends and composites represent functional materials generated for combining useful properties of their components. Effective mixing of the components in these materials usually results in synergistic enhancements in mechanical, thermal, flow, barrier and flammability performances. This tuning and control on the material properties leads to the applications of these functional blends and composites in many applications. A great degree of advancements in these materials have been achieved in the recent years on all aspects related to material combinations, synthesis routes, organic-inorganic phase compatibilization methodologies, etc. This text aims to present a number of such blend and composite systems based on polypropylene (PP). For this reason, a majority of the chapters presented in this text focus on experimental details as well as accurate material characterizations to present deep insights into controlling structure-property correlations.

In Chapter 1, the effect of amine functionalized polyolefin compatibilizer and graphite oxide (GO) nanofiller on the mechanical as well as thermal properties of polypropylene has been investigated. In Chapter 2, polypropylene nanocomposites were generated using binary filler system containing graphene and surface modified sepiolite. The influence of the mixed filler system on the mechanical, thermal and morphological properties of the composites was studied. The objective of Chapter 3 was to investigate the effect of different copolymers (ethylene vinyl acetate-g-maleic anhydride (EVA-MA), ethylene butyl acrylate copolymer (EBA) and ethylene-α- olefin copolymer (EAO)) on the mechanical, thermal and rheological properties of different grades of PP, which included isotactic PP (iPP) as well as impact propylene copolymer (IPC). In Chapter 4, the approach of generating blend nanocomposites using reactive compatibilization method has been further explored to achieve significant enhancement in the composite properties, especially at a low filler content. In Chapter 5, the blend nanocomposites of polypropylene and polystyrene have been further explored by using graphene as filler. Two different blend compositions were employed to study the effect on composite morphology and properties. In Chapter 6, the effect of hygrothermally induced ageing on the properties of neat PP and PP/copolymer blend samples was studied. On the other hand, the behavior of three PP grades was studied under the influence of hygrothermal environment in Chapter 7. In Chapter 8, iPP and two IPC grades were analyzed for the effect of accelerated UV weathering conditions. In Chapter 9, PP nanocomposites have been developed with graphene as filler for achieving enhancement in mechanical and thermal properties, along with thermal and electrical conductivities. In Chapter 10, cyclohexylethylene-co-ethylene (CHEE) copolymer was used to develop PP blends and graphene reinforced PP blend nanocomposites using direct melt blending technique. Chapter 11 presents further efforts in the direction of achieving good toughness-stiffness balance of PP. In this respect, PP has been incorporated with different weight fractions of a variety of copolymers. Chapter 12, an amine functionalized PP compatibilizer (PP-g-DA6) has been explored with GO nanofiller to improve the morphological, mechanical, thermal and rheological properties of PP.

The support of all the chapter contributors is greatly acknowledged towards the successful accomplishment of the book. The book is dedicated to my family for their constant support.

Vikas MITTAL

1

PP/GO Nanocomposites with Amine Functionalized PE-g-MA Compatibilizer

1.1 Introduction

Polypropylene (PP) is a versatile semi-crystalline thermoplastic polymer with a wide range of commercial applications. It presents several benefits like low cost, easy processing, low density, thermal stability, moisture resistance and high melting point [1,2]. In spite of its functional advantages, poor impact resistance of polypropylene, especially under intense conditions such as low temperatures or high train rates, hinders its utilization as an engineering plastic [3]. Further, due to the non-polar nature of the resin, it is challenging to obtain the homogenous dispersion of the polar fillers in the PP matrix [4-6].

Various methods have been proposed to enhance the impact performance of PP, mainly based on the incorporation of the soft rubbery phase elastomers, however, these materials impact the stiffness of the polymer negatively [7]. The crystallinity of the polymer matrix is also negatively impacted. The chain modification during synthesis can also enhance the impact strength, however, the process is generally cumbersome. The incorporation of the amphiphilic materials, consisting of both polar as well as non-polar groups, is a promising approach, which results in the simultaneous enhancement of the toughness and stiffness, owing to the nanoscale dispersion of the filler phase in the matrix [8].

Among the different nanofillers, graphene, a two dimensional nanomaterial, offers opportunities for generating advanced nanocomposites having optimal micro-mechanical behavior [9]. Incorporation of a small extent of graphene nano-sheets in the polymer matrices has been observed to remarkably enhance the physical properties of the host polymers. The possibilities to functionalize its surface further enhance the potential of achieving simultaneous enhancements in stiffness, toughness and other properties. In recent years, different studies have reported the preparation and structure-property

Haleema Saleem, M. R. Vengatesan and Vikas Mittal
Khalifa University of Science and Technology, Abu Dhabi, UAE

relationships of PP/graphene nanocomposites [10]. Song *et al.* [11] reported the generation of PP/graphene nanocomposites with enhanced thermal as well as mechanical properties. It was observed that the incorporation of about 0.42 vol% graphene resulted in 74% increment in the polymer stiffness, whereas the yield strength was enhanced by 75%. In addition, the thermal oxidative stability of the polymer was also enhanced. In another study, Varghese *et al.* [12] studied the effect of thermally reduced graphene oxide (TRG) on the properties of polypropylene/maleic anhydride-graft-ethylene vinyl acetate (PP/EVA-g-MA) blends. The addition of up to 3 wt% TRG in the PP/EVA-g-MA blend enhanced the tensile strength and modulus, while retaining the impact strength. PP/EVA-g-MA/TRG nanocomposites also exhibited higher electrical and thermal conductivities as compared to PP or PP/EVA-g-MA blends.

In spite of the enhancements in the mechanical properties of the nanocomposites, aggregation and stacking of graphene are common issues hindering the optimal enhancement in performance. These phenomena result in stress concentration and hinder the load transfer from the PP matrix [13,14]. For reducing the cohesive forces between the graphene sheets and for developing specific interactions with the host polymer matrix, the functionalization of graphene is, thus, needed.

Graphite oxide (GO), where the carboxyl and hydroxyl functional groups are covalently bonded to the carbon edges and basal planes, has been employed as filler in various polymer matrices [15,16]. It is generated by the chemical oxidation of graphite. However, it is difficult to achieve the homogenous dispersion of GO in non-polar polymers such as PP, polyethylene (PE) and polystyrene. Further, it also exhibits poor electrical properties due to the disconnected sp^2 network and oxygenated functional groups [17]. To overcome these challenges, surface modification of GO is carried out using the oxygenated functional groups present on the surface [18,19]. For achieving the homogenous dispersion of GO in the PP matrix, *in-situ* reduction in the polymer also presents a viable strategy.

Furthermore, as mentioned earlier, the addition of a compatibilizer also serves to bridge the non-polar PP and polar graphene phases. Shin *et al.* [20] used octyl-triethoxysilane (OTES) as a compatibilizer for PP/GO nanocomposites. It was noted that the mechanical performance of the composites increased due to the stronger interaction between the PP and GO phases. For instance, the fracture toughness increased by 117% by the incorporation of OTES and 1.0

wt% GO. To enhance the compatibility of PP with the polar fillers, PP-graft-maleic anhydride (PP-g-MA) or PE-g-MA have been widely employed as compatibilizers. However, literature studies have also confirmed better adhesion of the amine-functionalized PP/PE than the maleated PP/PE compatibilizer. Kobayashi *et al.* [21] conducted comparative study of the influence of PE-g-MA and amino-functionalized PE for enhancing the bonding between polyurethane and polyolefins. The amine functionalized polymer was observed to exhibit greater adhesion than the PE-g-MA compatibilizer.

Several studies have proved that the blending of PP with PE can improve the impact strength of PP at lower temperature [22,23]. Jose *et al.* [24] examined the mechanical properties, crystallization behavior and phase morphology of high density polyethylene (HDPE)/isotactic polypropylene (i-PP) blends. It was observed that the properties of both polymers related closely to each other. Fel *et al.* [25] observed that the incorporation of a small amount of PE (typically < 20 wt%) can increase the impact strength of the PP blends. In another study, Sanchez-Valdes *et al.* [26] studied the compatibilization effect of PE-g-MA and different NH_2-functionalized PEs on the PE based nanocomposites. The NH_2-functionalized PEs were generated by the chemical reaction of PE-g-MA with a tertiary amine 2-[2-(dimethylamino)ethoxy]ethanol (DMAE) and two primary amines, 2-aminoethanol and 1,12-aminododecane to form the corresponding PE-g-DMAE, PE-g-EA, and PE-g-D12 compatibilizers. The PE-g-DMAE compatibilizer was observed to provide an optimally exfoliated morphology and an agreeable balance between optical, mechanical and thermal properties.

In the current study, the effect of amine functionalized polyolefin compatibilizer and GO nanofiller on the mechanical as well as thermal properties of PP has been investigated. In this approach, an amphiphilic compatibilizer, PE-alt-DA6, is added to the PP matrix, along with GO (Figure 1.1). The amine groups present in the PE-alt-DA6 structure react with the carboxyl as well as epoxy groups present on GO, thus, causing the covalent binding of the filler platelets with the compatibilizer. During this process, GO also gets chemically reduced, thus, generating reduced GO (r-GO). PE-alt-DA6 was prepared by melt mixing hexamethylenediamine (HMDA) and polyethylene-alt-maleic anhydride (PE-alt-MA). The mechanical, thermal, rheological and morphological properties of the nanocomposites were examined by keeping the compatibilizer content constant, while varying the amount of GO in the polymer matrix (0.5, 1.0, 3.0 and 5.0 wt %).

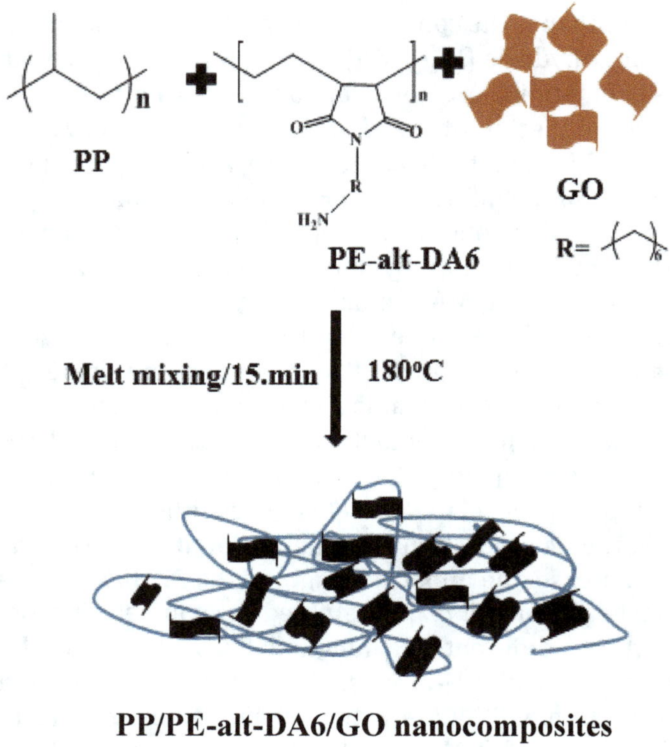

Figure 1.1 Generation of PP/GO nanocomposites from PP, PE-alt-DA6 and GO.

1.2 Experimental

1.2.1 Materials

The isotactic PP used in this study was supplied by Abu Dhabi Polymers Company (Borouge), with melt flow rate (MFR) of 9 g/min (200 °C, 2.16 kg). HMDA, dimethyl formamide (DMF) and PE-alt-MA were purchased from Sigma Aldrich, Germany. PE-alt-MA consisted of ≤0.2 wt% maleic anhydride and ≤1.5 wt% water. It had an average molecular weight in the range of 100,000-500,000 g/mol, viscosity in the range of 0.75- 1.05 poise and glass transition temperature of about 235 °C. HMDA had a melting point in the range of 42-45 °C. GO nanoplatelets were generated by the oxidation of graphite, as reported earlier [27]. The polymers and reagents were used as received without any further purification.

1.2.2 Generation of PE-alt-DA6

To prepare PE-alt-DA6 compatibilizer, 50 g PE-alt-MA and 10 g HMDA were taken in a round bottomed flask, followed by the immersion of the flask in a heating oil bath at 170 °C under reflux and stirring. At the top of the reflex condenser, a guard tube was placed to prevent the moisture entering the flask. The mixture turned yellow after the completion of the reaction. After cooling to the room temperature, distilled water (200 mL) was added to the flask. After overnight stirring, the mixture was added to 1.5 L methanol and stirred for 2 h. Subsequently, the filtered residue was added to 500 mL of 8% $NaHCO_3$ solution and re-filtered. Further, the residue was washed with water, followed by washing with ethanol. The product was finally dried at of 50 °C for 12 h. The chemical reaction describing the generation of PE-alt-DA6 is shown in Figure 1.2.

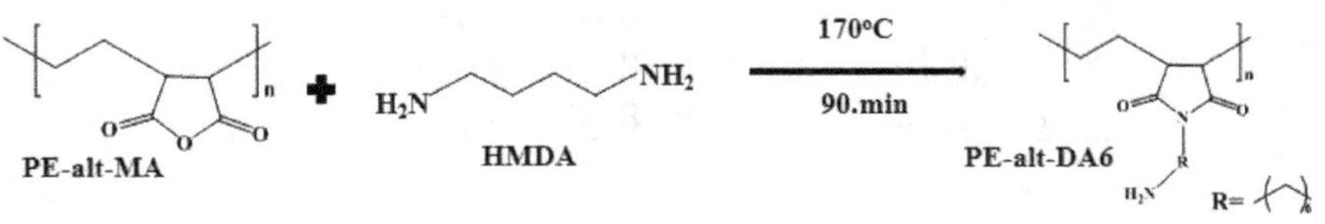

Figure 1.2 Generation of PE-alt-DA6.

1.2.3 Generation of Compatibilizer/Filler Hybrid

For identifying the chemical reaction between PE-alt-DA6 and GO, their hybrid was prepared without PP. Here, GO (0.2 g) was added to 100 mL DMF solvent, followed by ultra-sonication to achieve effective GO dispersion. PE-alt-DA6 (0.2 g) was subsequently added to the dispersion and stirred for 5 h at 110 °C. Later, centrifugation and separation of the solvent were carried out. Finally, the hybrid PE-alt-DA6/GO was obtained, which was examined using Fourier transform infrared spectroscopy (FTIR) spectroscopy to analyze the chemical bonding between PE-alt-DA6 and GO.

1.2.4 Preparation of Nanocomposites

The PP nanocomposites were generated by melt mixing PP, GO and PE-alt-DA6, utilizing a twin screw extruder Thermo Scientific Haake Minilab II. The extruder was operated in co-rotating mode at 100 rpm and 180 °C. The materials were melt mixed for about 15 min. Table 1.1 also depicts the composite formulations. In all combinations, the amount of PE-alt-DA6 compatibilizer was fixed at 10 wt%. After extrusion, the test specimens were injection molded using Thermo Scientific Haake Minijet Pro molding machine. The mold and cylinder temperatures were 135 °C and 190 °C, respectively. Also, the injection and post pressures were 430 and 500 bar, respectively, applied for 10 s.

Table 1.1 Composition of PP nanocomposites (in wt%)

Polymer/ Nanocomposites	PP (%)	GO (%)	PE-alt-DA6 (%)
Pure PP	100	0.0	0
PP/PE-alt-DA6	90.0	0.0	10
PP/PE-alt-DA6/0.5% GO	89.5	0.5	10
PP/PE-alt-DA6/1.0% GO	89.0	1.0	10
PP/PE-alt-DA6/3.0% GO	87.0	3.0	10
PP/PE-alt-DA6/5.0% GO	85.0	5.0	10

1.2.5 Characterization

FTIR was performed on a Nicolet iS10 spectrometer having SmartiTR diamond ATR accessory, DTGS KBr detector and KBr beam splitter. The analysis was carried out in transmission mode. Thermogravimetric analysis (TGA) was performed on a Discovery thermogravimetric analyzer from TA Instruments over the temperature range of 35 °C to 700 °C, in an inert nitrogen atmosphere, using a heating rate of 10 °C/min. Differential scanning calorimetry (DSC) analysis of the nanocomposites was carried out using a Discovery DSC from TA Instruments. Sample weight was approx.. 5-8 mg, and nitrogen was employed as the carrier gas with a constant flow rate. The samples were subjected to two cycles of heating and cooling at a rate of 10 °C/min. The specimens were heated from room temperature (25 °C) to 200 °C, followed by cooling to -60 °C. Subsequently, the samples were heated to 200 °C and then cooled down to 25 °C.

Tensile strength, tensile modulus and elongation at break of the samples were measured using Instron 5567 UTM at the strain rate of 10 mm/min. The tests were performed at room temperature in accordance with ASTM D638. The Izod impact strength analysis was performed on rectangular bar shaped specimens according to ASTM

D256 at room temperature. An average value from five test specimens was recorded. The melt rheological performance of pure PP and its nanocomposites was analyzed using a AR2000 813901 rheometer from TA Instruments. The analysis was performed using an angular frequency range of 0.1-500 rad/sec, at 2% strain and 180 °C. The Raman spectroscopic analysis was performed using Horiba LabRAM HR, equipped with 633 nm laser source.

Microscopy analysis of the nanocomposite samples was carried out using FEI, TECNAI transmission electron microscope (TEM) at an acceleration voltage of 200 kV. The sample sections were microtomed using PowerTome microtome, equipped with a diamond knife. For image processing, digital micrograph software (Gatan, USA) was employed. Panalytical X'Pert Pro diffractometer was used to analyze the wide angle X-ray diffraction (WAXRD) of materials in reflection mode, and the scanning angle range was $2\theta = 5\text{-}60°$. Laser flash analysis (LFA) was carried out using Netzsch LFA 447 to determine the thermal conductivity of PP nanocomposites.

1.3 Results and Discussion

FTIR analysis was performed to determine the nature of interactions between GO, PP and PE-alt-DA6 compatibilizer. Figure 1.3 presents the FTIR spectra of pure PP, GO, PE-alt-DA6, PE-alt-DA6/GO hybrid and PP based nanocomposites. Pure PP exhibited distinctive peaks at 1167 cm^{-1} (C-C stretching of C-CH_3 group), 1375 cm^{-1} (symmetric C-H bending vibration of –CH_3 and –CH_2 groups), 1455 cm^{-1} (asymmetric C-H bending vibration of –CH_3 and –CH_2 groups), 2837 cm^{-1} (C-H symmetric stretching of –CH_2 group), 2866 cm^{-1} (C-H symmetric stretching of –CH_3 group), 2916 cm^{-1} (C-H asymmetric stretching of –CH_2 group) and 2949 cm^{-1} (C-H asymmetric stretching of –CH_3 group). For pristine GO, the FTIR spectrum confirmed the presence of characteristic oxygen containing groups. Specifically, the peaks at 1225 cm^{-1} (C-O-C stretching vibration), 1393 cm^{-1} (O-H deformation vibration), 1625 cm^{-1} (C=C stretching vibration), 1733 cm^{-1} (C=O stretching vibration) and 3222 cm^{-1} (O-H stretching vibration) were observed [28]. For PE-alt-DA6 compatibilizer, specific peaks were observed at 1686 cm^{-1} (1° amide C=O stretching), 1438 cm^{-1} (N-H secondary amine stretching) and 1402 cm^{-1} (amide carbonyl stretching), among others. Further, in the case of PE-alt-DA6/GO hybrid, a shift of the aromatic C=C stretching vibration was observed from 1625 cm^{-1} to 1655 cm^{-1}, due to r-GO formation. Thus, the FTIR analysis confirmed the effective bond formation between PE-alt-DA6 and GO, to form r-GO.

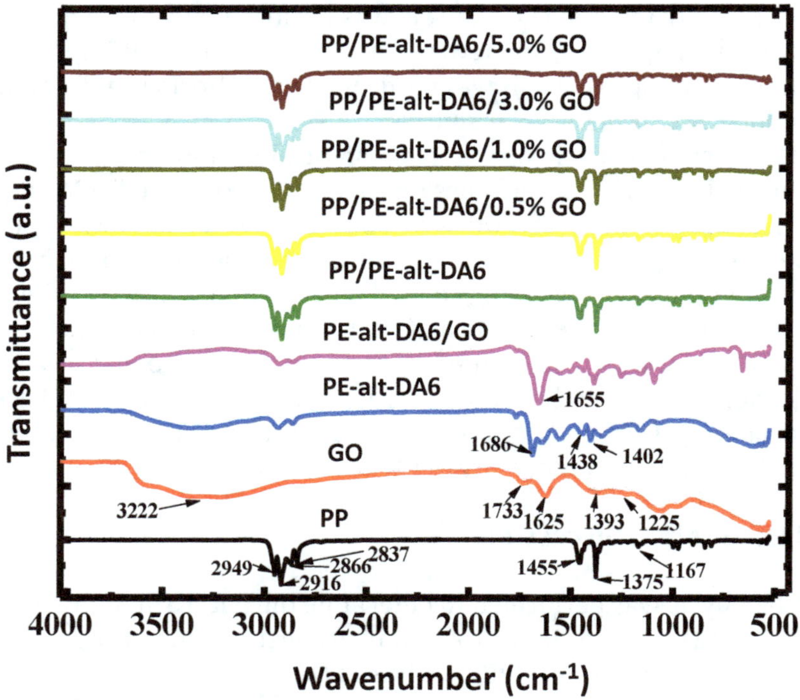

Figure 1.3 FTIR spectra of pristine PP, GO, PE-alt-DA6, PE-alt-DA6/GO hybrid and PP/PE-alt-DA6/GO nanocomposites.

Raman spectroscopic analysis was performed to gain further insights about the interaction between the GO and PE-alt-DA6 compatibilizer phases (Figure 1.4). The dominant feature of the GO spectrum is the prominent D and G peaks observed at almost 1352 and 1577 cm^{-1}, respectively [29], where the G peak indicates the graphitic carbon structure [30]. Using the G band, the thickness of graphene layers has also been estimated. The D peak provides information about the graphene disorder arising because of the structural edge effects [31,32]. From the Raman spectra of GO and PE-alt-DA6/GO hybrid, the D band of the specimens was observed to remain unchanged. However, due to the covalent bonding between PE-alt-DA6 and GO, the G band of the hybrid shifted from 1577 cm^{-1} to 1595 cm^{-1}. This indicated the reduction of GO to form r-GO. The evaluation of I_D/I_G ratio (I_D: Raman intensity of D peak and I_G: Raman intensity of G peak) of GO and PE-alt-DA6/GO hybrid is also presented in Table 1.2 [33]. When compared to GO, the I_D/I_G ratio of the hybrid PE-alt-DA6/GO was enhanced by 10%. The removal of the oxygen containing groups in GO nanofiller would have resulted in the observed enhancement. Additionally, it lowered the average size of the sp^2 domain.

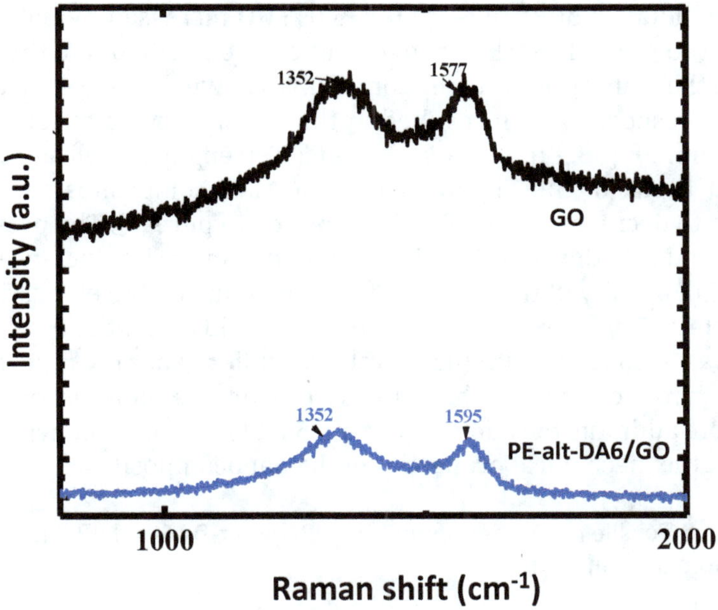

Figure 1.4 Raman spectra of pristine GO and PE-alt-DA6/GO hybrid.

Table 1.2 Raman intensity ratio (I_D/I_G) of pristine GO and PE-alt-DA6/GO hybrid

Material	D band (cm^{-1})	G band(cm^{-1})	Raman intensity ratio (I_D/I_G)
GO	1352	1577	1.00
PE-alt-DA6/GO hybrid	1352	1595	1.10

It has been observed that the mechanical properties of the nanocomposites depend on filler dispersion state, percentage crystallinity and polymer-filler interactions [34-36]. Wypych [37] also stated that the polymer-filler interactions are a critical factor influencing the tensile properties of a polymer-filler system. To gain qualitative insights about this aspect, Table 1.3 illustrates the tensile strength, modulus and elongation at break of pure PP, PP/PE-alt-DA6 hybrid and PP/PE-alt-DA6/GO nanocomposites. Figure 1.5 also presents the tensile modulus of the PP/PE-alt-DA6/GO nanocomposites as a function of GO content. A significant increase in the tensile modulus was observed in the composites as compared to pure PP. For instance, the tensile modulus of PP/PE-alt-DA6/5.0% GO increased by 88% with respect to pure PP. Such improvement can be attributed to the reaction of the amine groups in the compatibilizer with the oxy-functional graphene sheets, which resulted in the homogenous distribution of GO within PP [38,39]. Further, the tensile strength as well as elongation at break of the PP/PE-alt-DA6/GO nanocomposites were observed to decrease marginally as compared to pure PP. The presence of alt groups in the PP/PE-alt-DA6/GO nanocomposites increased the crosslink density of the samples, thereby, reducing the elongation at break [40]. The marginal reduction in the yield strength and elongation was in contrast with the literature studies generally reporting a significant reduction in these properties on the addition of nano-fillers. In addition, the addition of even 5 wt% filler did not negatively impact the mechanical properties of the nanocomposites.

Table 1.3 Tensile properties of pure PP, PP/PE-alt-DA6 and PP/PE-alt-DA6/GO nanocomposites

Material	Elongation at break (%)	Yield stress (MPa)	Tensile modulus (MPa)
Pure PP	7	38	879
PP/PE-alt-DA6	5	34	1447
PP/PE-alt-DA6/0.5% GO	5	34	1509
PP/PE-alt-DA6/1.0% GO	5	35	1547
PP/PE-alt-DA6/3.0% GO	5	35	1632
PP/PE-alt-DA6/5.0% GO	5	34	1654

The fracture toughness of a material represents the amount of energy absorbed for fracturing the sample. The improvement in the toughness of PP due to the addition of the nanoparticles is mainly because of the micro-cracking mechanism [41]. The addition of fillers in the polymer results in the establishment of a heterogeneous system [42]. Hence, an external load can generate stress concentration in the polymer composites, thereby influencing the fracture behavior of the material [43]. The fracture in the polymer starts with the plastic deformation before the initial crack. Thus, the debonding of particles occurs, and the zones are stretched and ruptured [44,45].

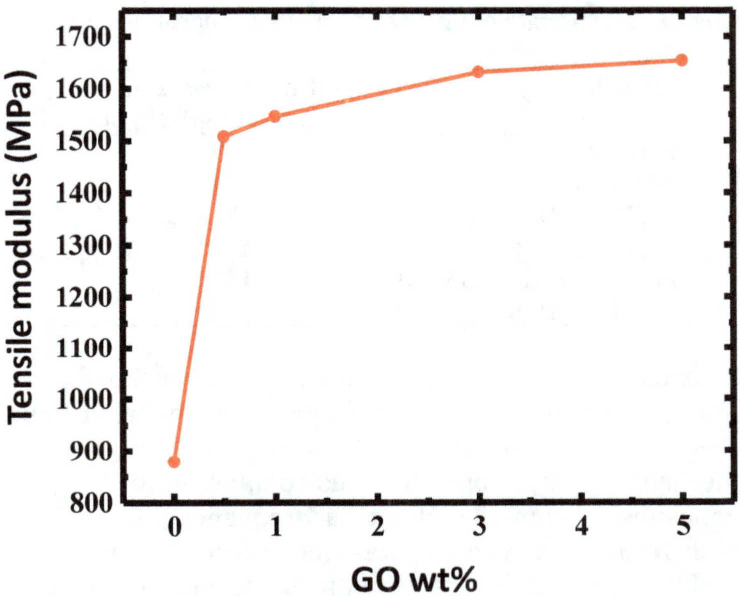

Figure 1.5 Tensile modulus of PP/PE-alt-DA6/GO nanocomposites as a function of GO content.

The findings from the impact analysis of pristine PP, PP/PE-alt-DA6 and PP/PE-alt-DA6/GO nanocomposites are presented in Table 1.4. The impact strength of pure PP was observed to be 7.96 kJ/m^2. The impact strength of PP/PE-alt-DA6 was observed to increase by 18% as compared to pure PP. For PP/PE-alt-DA6/GO nanocomposites, the addition of 0.5 wt%, 1.0 wt% and 3.0 wt% GO enhanced the impact strength to 11.06 kJ/m^2 (39% increase), 11.48 kJ/m^2 (44% increase) and 13.52 kJ/m^2 (70% increase) respectively. Further increase in GO content resulted in a reduction in the impact strength of the composite. This was due to the greater stacking of the graphene sheets within the PP matrix [46,47], caused due to van der Waals forces. The results also indicated that the PE-alt-DA6 compatibilizer led to substantial bonding between the GO and PP phases, which delayed the crack initiation as well as propagation in the PP nanocomposites. The enhancement in the crack growth initiation energy, thus, caused a significant increase in the toughness of the PP nanocomposites [46]. Overall, a remarkable increase in the impact strength as well as a good toughness-stiffness balance was observed for PP/PE-alt-DA6/3.0% GO.

Table 1.4 Impact strength of PP/PE-alt-DA6/GO nanocomposites

Material	Unnotched Izod impact strength (kJ/m^2)
Pure PP	7.96
PP/PE-alt-DA6	9.41
PP/PE-alt-DA6/0.5% GO	11.06
PP/PE-alt-DA6/1.0% GO	11.48
PP/PE-alt-DA6/3.0% GO	13.52
PP/PE-alt-DA6/5.0% GO	10.34

The rheological analysis is performed to evaluate the filler dispersion, internal structure and processing properties of nanocomposites [48]. Figures 1.6-1.9 demonstrate the storage (G') and loss modulus (G"), mechanical loss factor (tan δ) and complex viscosity (η*) of the PP composites as a function of angular frequency (ω). The G' of the PP/PE-alt-DA6/GO nanocomposites was observed to be lower than pristine PP. The reduced G' value of the nanocomposites could be attributed to the matrix plasticization [49]. Due to the low surface friction of GO, the PP/GO interlayer slipping occurs. Therefore, the PP nanocomposite melt behaves like a viscous PP liquid [50]. Additionally, the G' curves were observed to converge with each other at higher frequencies. This clearly indicated the concentration independence behavior of the materials at such frequencies. G" of pristine PP was higher than G' at lower angular frequency values indicating the dominance of the viscous behavior.

Tan δ of PP/PE-alt-DA6/GO nanocomposites was observed to be higher than pristine PP. It was noticed from the Figure 1.8 that the pristine PP and PP/PE-alt-DA6/GO nanocomposites showed the viscoelastic nature at tan δ value of about 0.2 rad/sec, and glassy nature at almost 100 rad/sec [51,52].

Newtonian fluid is represented by a horizontal line in the η* vs. ω plot. At the same time, the shear thinning of the sample is indicated by a reduction in the complex viscosity with increase in ω or shear rate [53]. In PP nanocomposites too, η* was observed to decrease with increase in ω (Figure 1.9), revealing the shear thinning nature. The shear thinning behavior at higher ω is related to the disentaglement mechanism. In addition, η* of PP nanocomposites was observed to be lower than pristine PP, especially at lower frequencies.

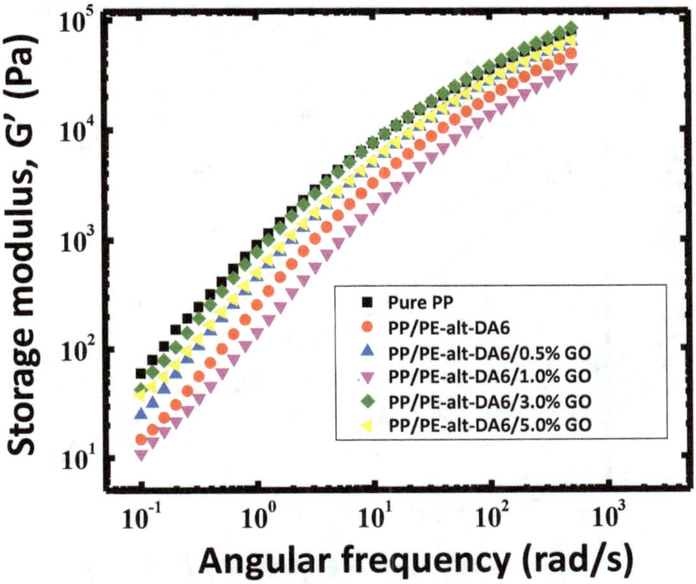

Figure 1.6 Storage modulus (G') vs. angular frequency for pristine PP, PP/PE-alt-DA6 blend and PP/PE-alt-DA6/GO nanocomposites.

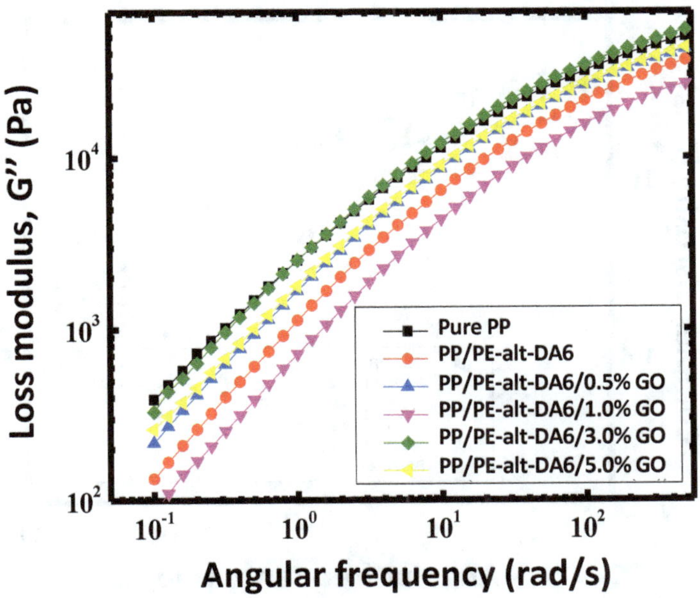

Figure 1.7 Loss modulus (G") vs. angular frequency for pristine PP, PP/PE-alt-DA6 blend and PP/PE-alt-DA6/GO nanocomposites.

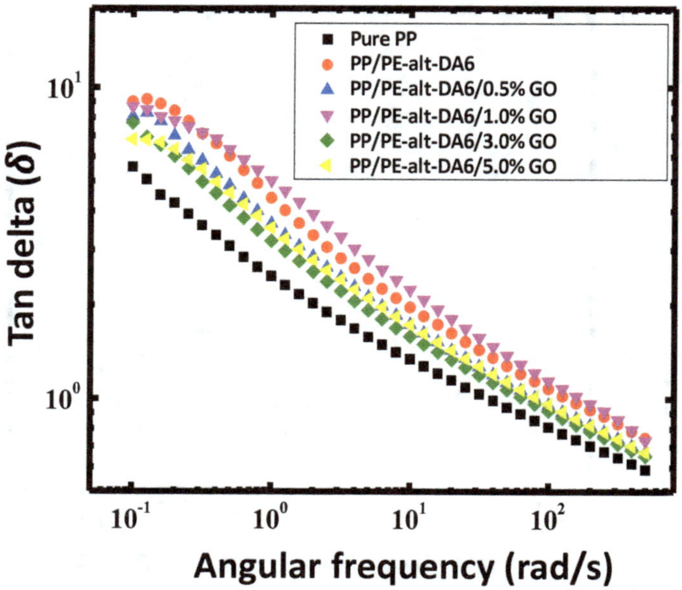

Figure 1.8 Tan delta vs. angular frequency for pristine PP, PP/PE-alt-DA6 blend and PP/PE-alt-DA6/GO nanocomposites.

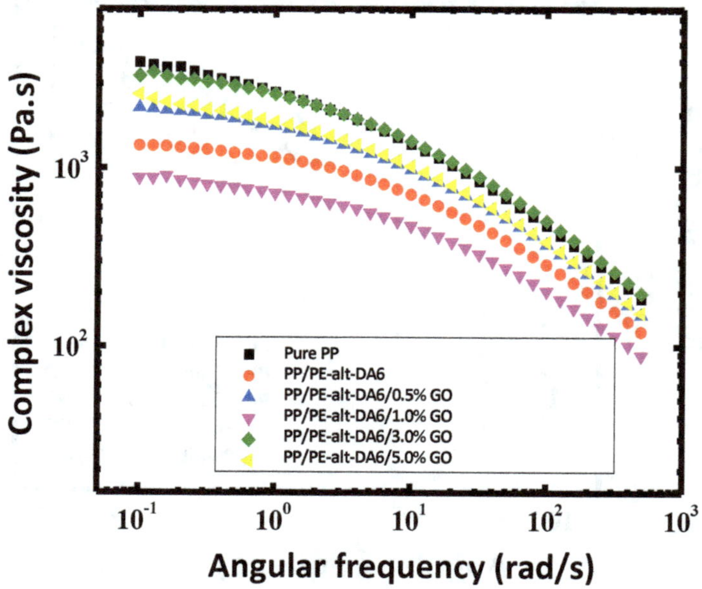

Figure 1.9 Complex viscosity (η*) vs. angular frequency for pristine PP, PP/PE-alt-DA6 blend and PP/PE-alt-DA6/GO nanocomposites.

The calorimetric properties of pristine PP and PP/PE-alt-DA6/GO nanocomposites are presented in Table 1.5. The Figures 1.10 and 1.11 also illustrate the melting and cooling thermograms of the materials. The degree of crystallinity was calculated using the enthalpy of fusion of 100% crystalline PP (209 J/g) [54].

Table 1.5 Calorimetric properties of pristine PP and PP/PE-alt-DA6/GO nanocomposites

	Heating			Cooling			X_c
	ΔH_m (cal/g)	ΔH_m (J/g)	T_m (°C)	ΔH_c (cal/g)	ΔH_c (J/g)	T_c (°C)	
Pure PP	20.8	86.9	167	22.7	95.1	128	42
PP/PE-alt-DA6/0.5% GO	21.3	89.1	167	22.0	92.2	127	43
PP/PE-alt-DA6/1.0% GO	21.3	89.1	168	21.7	90.9	126	43
PP/PE-alt-DA6/3.0% GO	21.5	89.9	169	21.3	89.2	126	44
PP/PE-alt-DA6/5.0% GO	21.2	88.7	168	20.3	84.8	126	45

A single endothermic peak was observed in the melting thermograms of all samples at temperature >160 °C. This indicated the presence of the α- crystals solely, with no β- crystals [55]. The incorporation of PE-alt-DA6 and GO to the PP matrix marginally enhanced the crystallinity, which was also a function of GO fraction. The observed improvement in the crystallinity can be related to the heterogeneous nucleation effect in the PP matrix. Additionally, no remarkable change in the T_m and T_c values of PP was observed on the incorporation of PE-alt-DA6 and GO. Several literature studies have reported that the grafting of long chain alkylamines, such as decylamine (A10), hexadecylamine (A16) and octadecylamine (A18), on the 2-D nanosheets resulted in the oriented crystalline arrangement of the alkylamines [56]. However, the compatibilizer used in the study was largely amorphous in nature. Thus, the observed changes in crystallinity in the nanocomposites resulted from the combined effects of the compatibilizer and GO.

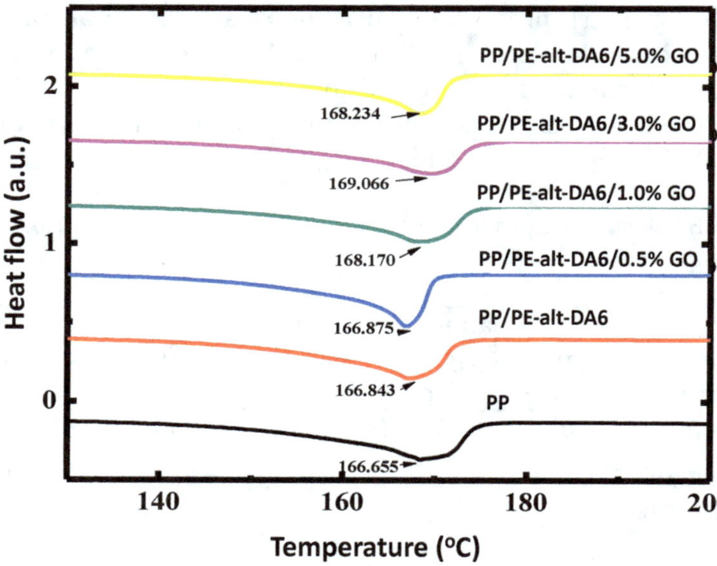

Figure 1.10 Melting endotherms of pristine PP, PP/PE-alt-DA6 blend and PP/PE-alt-DA6/GO nanocomposites.

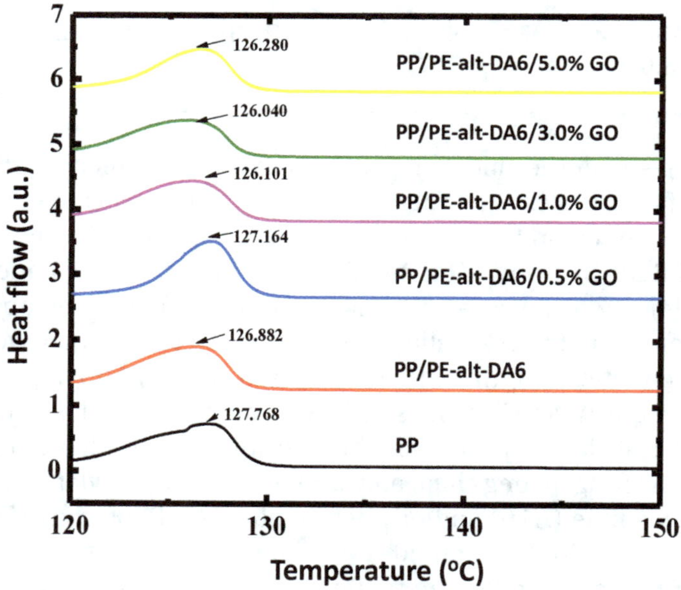

Figure 1.11 Crystallization isotherms of pristine PP, PP/PE-alt-DA6 blend and PP/PE-alt-DA6/GO nanocomposites.

Figure 1.12 presents the TGA thermograms of pristine PP, PP/PE-alt-DA6 blend and PP/PE-alt-DA6/GO nanocomposites. A single step degradation mechanism was observed for all the materials. The T_{ini} (5% weight loss temperature), $T_{40\%}$ (40% weight loss temperature) and char yield (mass remaining at 700 °C) are also compiled in Table 1.6.

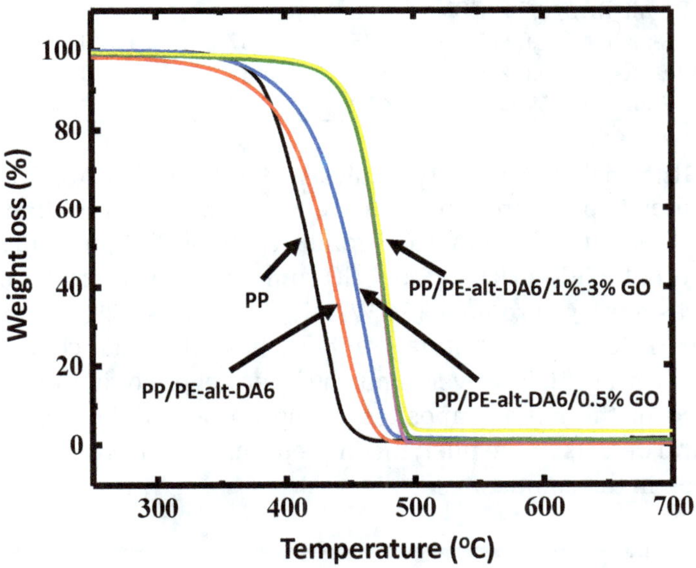

Figure 1.12 TGA thermograms of pristine PP, PP/PE-alt-DA6 blend and PP/PE-alt-DA6/GO nanocomposites

The addition of PE-alt-DA6 and GO nanofiller enhanced the thermal stability to a large extent (except the initiation of degradation temperature for the PP/PE-alt-DA6 blend). The thermal degradation temperatures for 5% as well as 40% weight loss increased with the GO loading until 1.0 wt%. The degradation temperatures were observed to be similar for the PP nanocomposites having GO content higher than 1.0 wt%. The observed thermal behavior revealed high thermal stability of the PP/PE-alt-DA6/GO nanocomposites, as compared to pristine PP [57]. The char yield of PP/PE-alt-DA6/GO nanocomposites also increased with GO content, as compared to pure PP. Overall, the enhancement in the thermal stability of the PP nanocomposites was due to the stronger interactions between the GO and PP phases, which enhanced the heat transfer at the interface, thus, increasing the thermal degradation activation energy [58].

Table 1.6 Thermal degradation behavior of pristine PP, PP/PE-alt-DA6 blend and PP/PE-alt-DA6/GO nanocomposites

	$T_{ini\%}$ (°C)	$T_{40\%}$ (°C)	Char yield at 700 °C (wt %)
Pure PP	375	414	1.26
PP/PE-alt-DA6	347	424	0.00
PP/PE-alt-DA6/0.5% GO	375	441	0.53
PP/PE-alt-DA6/1.0% GO	425	467	0.58
PP/PE-alt-DA6/3.0% GO	428	467	0.92
PP/PE-alt-DA6/5.0% GO	430	470	3.04

The thermal conductivity analysis of PP/PE-alt-DA6/GO nanocomposites is presented in in Figure 1.13. The incorporation of reduced GO enhanced the thermal conductivity of PP significantly. For instance, in PP/PE-alt-DA6/5.0% GO composite, the thermal conductivity was observed to be 0.336 W/mK, which was 68% higher than pristine PP [59]. The pristine graphene has a thermal conductivity >5000 W/mK [60], however, a much lower degree of increment was observed in the nanocomposites owing to various factors such as structural defects in the filler, incomplete reduction of GO and GO aggregation in the composites.

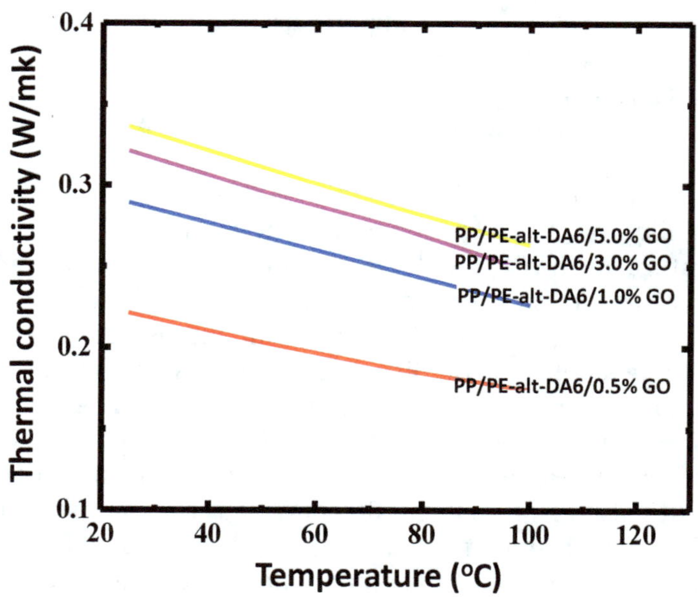

Figure 1.13 Thermal conductivity of PP/PE-alt-DA6/GO nanocomposites.

The physical and chemical properties of the nanocomposites are mainly driven by the polymer-filler interactions as well as filler dispersion [61]. Figure 1.14 presents the morphological analysis of the PP/PE-alt-DA6/5.0% GO composite to gain insights about the extent of filler dispersion. The composite exhibited well dispersed and exfoliated morphology, thus, also indicating optimal polymer-filler interactions [34].

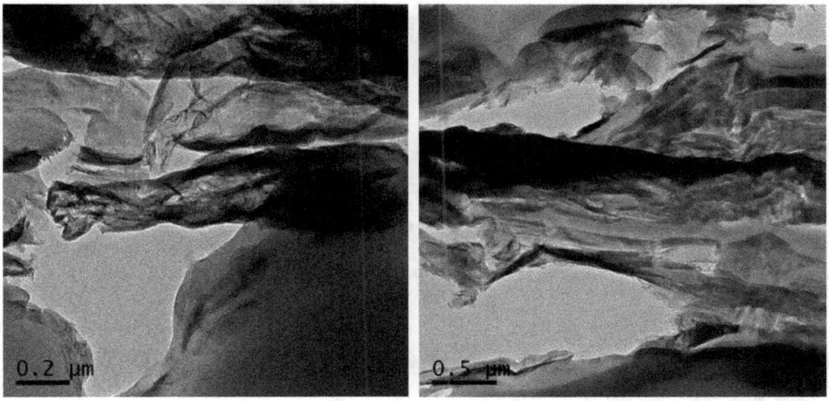

Figure 1.14 TEM images of PP/PE-alt-DA6/5.0% GO nanocomposite.

Figure 1.15 presents the WAXRD patterns of pristine PP, GO, PE-alt-DA6, PP/PE-alt-DA6 blend and PP/PE-alt-DA6/GO nanocomposites. In the case of pristine GO, a strong reflection peak was observed at $2\theta = 9.6°$. The diffraction peaks of pure PP were observed at 14.5°, 17.2°, 18.8° as well as 22.1°, which corresponded to the (110), (040), (130) and (111) planes of the PP crystal, respectively [62]. The XRD pattern of PE-alt-DA6 exhibited no sharp peak, suggesting the amorphous nature. Broad and weak diffraction peaks were noticed in the diffraction patterns of PP/PE-alt-DA6/GO nanocomposites at 2θ range 25-30°, due to the reduction of GO [63]. As previously identified in the DSC analysis, no β and γ phases were noticed in the materials.

1.4 Conclusion

Melt mixing method was used to generate PP/PE-alt-DA6/GO nanocomposites, by varying the content of GO (0.5, 1.0, 3.0 and 5.0 wt%) and keeping the concentration of PE-alt-DA6 compatibilizer constant (10.0 wt%). The microscopy analysis confirmed the homogenous dispersion of GO in the PP matrix owing to the presence of PE-alt-DA6

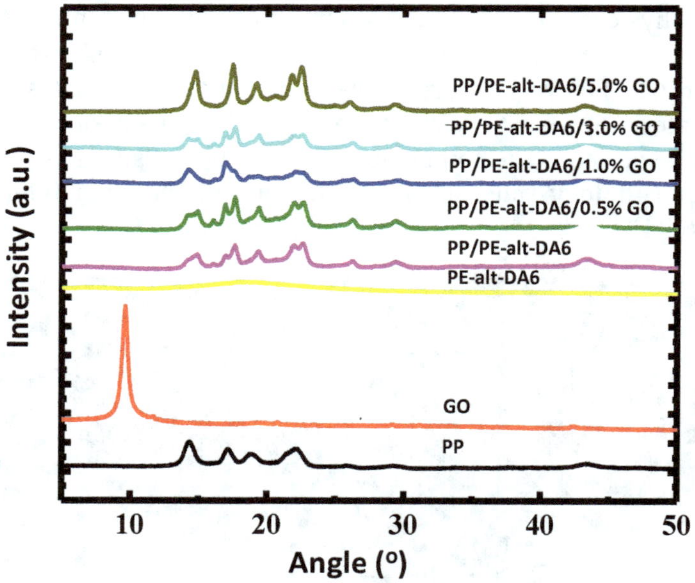

Figure 1.15 XRD patterns of pure PP, GO, PE-alt-DA6, PP/PE-alt-DA6 and PP/PE-alt-DA6/GO nanocomposites.

compatibilizer. Additionally, the formation of a bond between GO and PE-alt-DA6, forming r-GO in the PP nanocomposites, was confirmed by Raman, FTIR and XRD analyses. The effective polymer-filler interactions resulted in an enhancement in the mechanical and thermal properties of the resultant PP nanocomposites. These also delayed the crack initiation and propagation in the PP/PE-alt-DA6/GO nanocomposites, thereby increasing the toughness of the nanocomposites. For instance, PP/PE-alt-DA6/3.0% GO exhibited a 70% increase in toughness as compared to pure PP. Shear thinning nature of the samples was noted at higher frequencies in the rheological analysis. The DSC analysis indicated that the crystallinity of PP enhanced to a small extent in the nanocomposites. The thermal conductivity of PP was also observed to increase by the addition of GO. A 68% increase in the thermal conductivity of PP/PE-alt-DA6/5.0% GO was observed, as compared to pristine PP.

References

1. Pasquini, N., and Sgarzi, P. (2005) Polypropylene - The market. In: *Polypropylene Handbook*, Pasquini, N. (ed.), 2nd edition, Carl Hanser

Verlag, Germany.
2. Kissel, W. J., Han, J. H., Meyer, J. A. (2003) Polypropylene: Structure, properties, manufacturing and applications. In: *Handbook of Polypropylene*, Karian, H. (ed.), Marcel Dekker AG, Switzerland, pp. 10-27.
3. Lin, Y., Chen, H., Chan, C. M., Wu, J. (2008) High impact toughness polypropylene/$CaCO_3$ nanocomposites and the toughening mechanism. *Macromolecules*, **41**, 9204-9213.
4. Lopez-Quintanilla, L. M., Sanchez-Valdes, S., Ramos-de-Valle, L. F., and Miranda, R. G. (2006) Preparation and mechanical properties of PP/PP-g-MA/Org-MMT nanocomposites with different MA content. *Polymer Bulletin*, **57**, 385-393.
5. *Advances in Polyolefin Nanocomposites*, Mittal, V. (ed.), CRC Press, USA (2011).
6. *Polymer-Graphene Nanocomposites*, Mittal, V. (ed.), Royal Society of Chemistry, UK (2012).
7. Jancar, J., and Di Benedetto, A. T. (1994) The mechanical properties of ternary composites of polypropylene with inorganic fillers and elastomer inclusions. *Journal of Material Science*, **29**, 4651-4658.
8. Arencon, D., Velasco, J. I., Realinho, V., Sanchez-Soto, M. A., and Gordillo, A. (2007) Fracture toughness of glass microsphere filled polypropylene and polypropylene/poly(ethylene terephthalate-coisophthalate) blend-matrix composites. *Journal of Material Science*, **42**, 19-29.
9. Hong, J.-Y., Shin, K.-Y., Kwon, O. S., Kang, H., and Jang, J. (2011) A strategy for fabricating single layer graphene sheets based on a layer-by-layer self assembly. *Chemical Communications*, **47**, 7182-7184.
10. Kuilla, T., Bhadra, S., Yao, D., Kim, N. H., Bose, S., and Lee, J. H. (2010) Recent advances in graphene based polymer composites. *Progress in Polymer Science*, **35**(11), 1350-1375.
11. Song, P., Cao, Z., Cai, Y., Zhao, L., Fang, Z., and Fu, S. (2011) Fabrication of exfoliated graphene-based polypropylene nanocomposites with enhanced mechanical and thermal properties. *Polymer*, **52**(18), 4001-4010.
12. Varghese, A. M., Rangaraj, V. M., Mun, S. K., Macosko, C. W., and Mittal, V. (2018) Effect of graphene on polypropylene/maleic anhydride-graft-ethylene-vinyl acetate (PP/EVA-g-MA) blend: Mechanical, thermal, morphological, and rheological properties. *Industrial & Engineering Chemistry Research*, **57**, 7834-7845.
13. Li, D., Muller, M. B., Gilje, S., Kaner, R. B., and Wallace, G. G. (2008) Processable aqueous dispersions of graphene nanosheets. *Nature Nanotechnology*, **3**, 101-105.
14. Layek, R. K., and Nandi, A. K. (2013) A review on synthesis and properties of polymer functionalized graphene. *Polymer*, **54**(19), 5087-

5103.

15. Cao, Y., Feng, J., and Wu, P. (2012) Polypropylene-grafted graphene oxide sheets as multifunctional compatibilizers for polyolefin-based polymer blends. *Journal of Materials Chemistry*, **22**, 14997-15005.
16. Szabo, T., Berkesi, O., and Dekany, I. (2005) DRIFT study of deuterium-exchanged graphite oxide. *Carbon*, **43**(15), 3186-3189.
17. Eda, G., and Chhowalla, M. (2010) Chemically derived graphene oxide: Towards large-area thin-film electronics and optoelectronics. *Advanced Materials*, **22**, 2392-2415.
18. Stankovich, S., Piner, R. D., Nguyen, S. T., and Ruoff, R. S. (2006) Synthesis and exfoliation of isocuanate-treated graphene oxide nanoplastelets. *Carbon*, **44**(15), 3342-3347.
19. Park, S., Dikin, D. A., Nguyen, S. T., and Ruoff, R. S. (2009) Graphene oxide sheets chemically cross-linked by polyallylamine. *Journal of Physical Chemistry C*, **113**(36), 15801-15804.
20. Shin, K.-Y., Hong, J.-Y., Lee, S. and Jang, J. (2012) Evaluation of anti-scratch properties of graphene oxide/polypropylene nanocomposites. *Journal of Materials Chemistry*, **22**, 7871-7879.
21. Kobayashi, S., Song, J., Silvis, H. C., Macosko, C. W., and Hillmyer, M. A. (2011) Amino-functionalized polyethylene for enhancing the adhesion between polyolefins and polyurethanes. *Industrial and Engineering Chemistry Research*, **50**(6), 3274-3279.
22. Zhou, X.-Q., and Hay, J. N. (1993) Structure property relationships in annealed blends of linear low density polyethylene with isotactic polypropylene. *Polymer*, **34**(22), 4710-4716.
23. Hill, M. J., Oiarzabal, L., and Higgins, J. S. (1994) Preliminary studies of polypropylene/ linear low density polyethylene blends by transmission electron microscopy. *Polymer*, **35**(15), 3332-3337.
24. Jose, S., Aprem, A. S., Francis, B., Chandy, M. C., Werner, P., Alstaedt, V., and Thomas, S. (2004) Phase morphology, crystallisation behaviour and mechanical properties of isotactic polypropylene/high density polyethylene blends. *European Polymer Journal*, **40**(9), 2105-2115.
25. Fel, E., Khrouz, L., Massardier, V., Cassagnau, P., and Bonneviot, L. (2016) Comparative study of gamma-irradiated PP and PE polyolefins part 2: Properties of PP/PE blends obtained by reactive processing with radicals obtained by high shear or gamma-irradiation. *Polymer*, **82**, 217-227.
26. Sánchez-Valdes, S., Méndez-Nonell, J., Medellín-Rodríguez, F. J., Ramírez-Vargas, E., Martínez-Colunga, J. G., de Valle, L. F. R., Mondragón-Chaparro, M., López-Quintanilla, M. L., García-Salazar, M. L. (2010) Evaluation of different amine-functionalized polyethylenes as compatibilizers for polyethylene film nanocomposites. *Polymer International*, **59**(5), 704-711.
27. Chaudhry, A. U., and Mittal, V. (2013) High density polyethylene

nanocomposites using master batches of chlorinated polyethylene/Graphene oxide. *Polymer Engineering and Science*, **53**(1), 78-88.

28. Tien, H. N., Luan, V. H., Lee, T. K., Kong, B. S., Chung, J. S., Kim, E. J., and Hur, S. H. (2012) Enhanced solvothermal reduction of graphene oxide in a mixed solution of sulfuric acid and organic solvent. *Chemical Engineering Journal*, **211-212**, 97-103.
29. Elias, D.C. Nair, R.R. Mohiuddin, T.M.G. Morozov, S.V. Blake, P. Halsall, M.P. Ferrari, A. C., Boukhvalov, D. W., Katsnelson, M. I., Geim, A. K., and Novoselov., K. S. (2009) Control of graphene's properties by reversible hydrogenation: Evidence for graphene. *Science*, **323**(5914), 610-613.
30. Lonkar, S. P., Pillai, V. V., Stephen, S., Abdala, A., and Mittal, V. (2016) Facile in situ fabrication of nanostructured graphene–CuO hybrid with hydrogen sulfide removal capacity. *Nano-Micro Letters*, **8**(4), 312-319.
31. Rao, C. N. R., Sood, A. K., Subrahmanyam, K. S., and Govindaraj, A. (2009) Graphene: the new two-dimensional nanomaterial. *Angewandte Chemie, International Edition*, **48** 7752-7777.
32. Xu, Z., Bando, Y., Liu, L., Wang, W., Bai, X., Golberg, D. (2011) Electrical conductivity, chemistry, and bonding alternations under graphene oxide to graphene transition as revealed by in situ TEM. *ACS Nano*, **5**(6), 4401-4406.
33. Jiao, L., Wang, X., Diankov, G., Wang, H., and Dai, H. (2010) Facile synthesis of high-quality graphene nanoribbons, *Nature Nanotechnology*, **5**, 321-325.
34. Ryu, S. H., and Shanmugharaj, A. M. (2014) Influence of long-chain alkylamine-modified graphene oxide on the crystallization, mechanical and electrical properties of isotactic polypropylene nanocomposites. *Chemical Engineering Journal*, **244**, 552-560.
35. Shen, B., Zhai, W., Tao, M., Lu, D., and Zheng, W. (2013) Chemical functionalization of graphene oxide toward the tailoring of the interface in polymer composites. *Composites Science and Technology*, **77**, 87-94.
36. Yuan, B., Bao, C., Song, L., Hong, N., Liew, K. M., and Hu, Y. (2014) Preparation of functionalized graphene oxide/polypropylene nanocomposites with significantly improved thermal stability and studies on the crystallization behavior and mechanical properties. *Chemical Engineering Journal*, **237**, 411-420.
37. Wypych G. (2016) *Handbook of Fillers*, 4th edition, Chem Tec Publishing, Canada.
38. Mandhakini, M., Chandramohan, A., Vengatesan, M. R., and Alagar, M. (2011) Synthesis and characterization of linseed vinyl ester fatty amide modified epoxy layered silicate Nanocomposites. *High Performance Polymers*, **23**(5), 403-412.

39. Qian, D., Dickey, E. C., Andrews, R., and Rantell, T. (2000) Load transfer and deformation mechanisms in carbon nanotube polystyrene composites. *Applied Physical Letters,* **76**, 2868.
40. Tamboli, S. M., Mhaske, S. T., and Kale, D. D. (2004) Crosslinked polyethylene. *Indian Journal of Chemical Technology,* **11**, 853-864.
41. *Polymer Toughening,* Arends, C. B. (ed.), Marcel Dekker Inc., USA (1996).
42. Kowalewski, T., Galeski, A., and Kryszewski, M. (1984) The structure and tensile properties of cold drawn modified chalk filled polypropylene. In: *Polymer Blends. Processing, Morphology and Properties,* Kryszewski, M., Galeski, A., and Martuscelli, E. (eds.), Plenum Press, USA, pp. 223-241.
43. Arencon, D., and Velasco, J. I., (2009) Fracture toughness of polypropylene-based particulate composites. *Materials,* **2**, 2046-2094.
44. Pukanszky, B., Vanes, M., Maurer, F. H. J., and Voros, G. (1994) Micromechanical deformations in particulate-filled thermoplastics: Volume strain measurements. *Journal of Material Science,* **29**, 2350-2358
45. Zhuk, A. V., Knunyants, N. N., Oshmyan, V. G., Topolkaraev, V. A., and Berlin, A. A. (1993) Debonding microprocesses and interfacial strength in particle-filled polymer materials. *Journal of Material Science,* **28**, 4995-5606.
46. Albdiry, M. T., Yousif, B. F., Ku, H., and Lau, K. T. (2012) A critical review on the manufacturing processes in relation to the properties of nanoclay/ polymer composites. *Journal of Composite Materials,* **47**(9), 1093-1115.
47. de Villoria, R. G., and Miravete, A. (2007) Mechanical model to evaluate the effect of the dispersion in nanocomposites. *Acta Materialia,* **55**, 3025-3031.
48. Woo, D. K., Kim, B. C. and Lee, S. J. (2009) Preparation and rheological behavior of polystyrene/multi-walled carbon nanotube composites by latex technology. *Korea-Australia Rheology Journal,* **21**(3), 185-191.
49. Chaudhry, A. U., and Mittal, V. (2014) Blends of high-density polyethylene with chlorinated polyethylene: Morphology, thermal, rheological, and mechanical properties. *Polymer Engineering and Science,* **54**, 85-95.
50. Ferry, J. D. (1980) *Viscoelastic Properties of Polymers,* 3rd edition, Wiley, New York.
51. Li, Y., Zhu, J., Wei, S., Ryu, J., Sun, L., and Guo, Z. (2011) Poly(propylene)/graphene nanoplatelet nanocomposites: melt rheological behavior and thermal, electrical, and electronic properties. *Journal of Macromolecular Chemistry and Physics,* **212**(18), 1951-1959
52. a) Chen, X., Wei, S., Yadav, A., Patil, R., Zhu, J., Ximenes, R., Sun, L., and Guo, Z. (2011) Poly(propylene)/carbon nanofiber nanocomposites:

Ex situ solvent-assisted preparation and analysis of electrical and electronic properties. *Macromolecular Materials and Engineering*, **296**, 434-443; b) Wei, S., Patil, R., Sun, L., Haldolaarachchige, N., Chen, X., Young, D. P., and Guo, Z. (2011) Ex situ solvent-assisted preparation of magnetic poly(propylene) 8nocomposites filled with Fe@ FeO Nanoparticles. *Macromolecular Materials and Engineering*, **296**(9), 850-857.

53. Sugimoto, M., Tanaka, T., Masubuchi, Y., Takimoto, J. I., and Koyama, K. (1999) Effect of chain structure on the melt rheology of modified polypropylene. *Journal of Applied Polymer Science*, **73**, 1493-1500.
54. *Polymer Handbook*, Brandup, J., and Immergut, E. H. (eds.), Wiley, USA (1988).
55. Bao, R.-Y., Cao, J., Liu, Z.-Y., Yang, W., Xie, B.-H., and Yang, M.-B. (2014) Towards balanced strength and toughness improvement of isotactic polypropylene nanocomposites by surface functionalized-graphene oxide. *Journal of Materials Chemistry A*, **2**(9), 3190-3199.
56. Lin, Z., Liu, Y., and Wong, C. P. (2010) Facile fabrication of superhydrophobic octadecylamine-functionalized graphite oxide film. *Langmuir*, **26**, 16110-16114.
57. Vengatesan, M. R., Singh, S., Pillai, V. V., and Mittal, V. (2016) Crystallization, mechanical, and fracture behavior of mullite fiber-reinforced polypropylene nanocomposites. *Journal of Applied Polymer Science*, **133**(30), doi: 10.1002/app.43725.
58. Li, B., and Zhong, W.-H. (2011) Review on polymer/graphite nanoplatelet nanocomposites. *Journal of Material Science*, **46**, 5595-5614.
59. Kalaitzidou, K., Fukushima, H., and Drzal, L. T. (2007) Multifunctional polypropylene composites produced by incorporation of exfoliated graphite nanoplatelets, *Carbon*, **45**, 1446-1452.
60. Prasher, R. (2010) Graphene spreads the heat. *Science*, **328**, 185-186.
61. Yang, L., Phua, S. L., Teo, J. K., Toh, C. L., Lau, S. K., Ma, J., and Lu, X. (2011) A biomimetic approach to enhancing interfacial interactions: polydopamine-coated clay as reinforcement for epoxy resin. *ACS Applied Materials & Interfaces*, **3**, 3026-3032.
62. Nishino, T., Matsumoto, T., and Nakamae, K. (2000) Surface structure of isotactic polypropylene by X-ray diffraction. *Polymer Engineering and Science*, **40**, 336-343.
63. Cao, N., and Zhang, Y. (2015) Study of reduced graphene oxide preparation by hummers' method and related characterization. *Journal of Nanomaterials*, **2015**, Article ID 168125.

2

Polypropylene-Sepiolite/Graphene Hybrid Composites: Thermal, Mechanical and Morphological Properties

2.1 Introduction

Layered silicates (clay) are one of the commonly used low cost nanofillers for polymer nanocomposites due to high surface area and aspect ratio, resulting in remarkable mechanical performance and thermal stability owing to the large number of filler-matrix interactions at the interface leading to effective stress and heat transfer from the matrix to the filler phase [1-4]. Different structural varieties of layered silicates, such as montmorillonite (MMT), bentonite, hectorite and sepiolite, have been used as nanofillers for the development of polymer nanocomposites [5-8]. More specifically, sepiolite is one of the naturally occurring clays and exhibits good textural properties with fibrous morphology. The crystalline structure of sepiolite clay consists of an octahedral sheet of magnesium hydroxide sandwiched between two tetrahedral sheets of silica [9]. Similar to other silicates, sepiolite exhibits limited dispersion in the polymer matrices due to its hydrophilic nature [10]. In order to overcome this limitation, surface modification with organic molecules such as surfactants, polymers and silane coupling agents has been reported [11-13]. The surface of the sepiolite clay contains large number of silanol (Si-OH) groups, which readily react with organic coupling agents [14]. In a recent study, Chen *et al.* [15] developed sepiolite-polyurethane nanocomposites and reported the tensile strength and elongation at break of the nanocomposites to increase with increasing amount of nanoclay. In a similar study, Bilotti *et al.* [16] observed the sepiolite clay to exhibit higher reinforcing efficiency in polyamide composites compared to MMT clay.

Graphene being an allotrope of carbon has drawn significant research attention as filler for polymer nanocomposites because of its attributes like high surface area, large charge carrier mobility, chemical stability, excellent mechanical properties, as well as high thermal

S. Singh, M. R. Vengatesan[a], Sevim Isci[b] and Vikas Mittal[a]
[a]Khalifa University of Science and Technology, Abu Dhabi, UAE; [b]Istanbul Technical University, Istanbul, Turkey

and electrical conductivity [17-20]. Graphene has also been reported to exhibit better properties than other carbon fillers in polymer nanocomposites [21-23]. However, the tendency of graphene platelets to aggregate significantly limits the efficient dispersion in polymer matrices [24].

More recently, research has focused on exploring the synergistic effects of mixed filler systems on the properties of the polymer matrices [25-28]. Graphene has been widely used as one of the components for the preparation of mixed nanofiller systems by physically mixing with a variety of one dimensional (1D) nanomaterials [29-32]. Such mixing reduces the sheet staking of graphene and generates a synergistic effect on the resultant nanocomposite properties [29]. Noh *et al.* [30] generated cyclic butylene terephthalate nanocomposites using pitch based carbon fibers and graphene sheets through physical mixing method and observed a synergistic effect of the fillers on the isotropic thermal conductivity with 82% improved efficiency as compared to the individual fillers. Cho *et al.* [31] developed polypropylene/polyaniline/graphene nanocomposite by a facile physical mixing of polyaniline fibers and graphene sheets with polypropylene in organic solvent and observed that 10 vol% of co-fillers in the PP matrix resulted in a high dielectric constant of $\varepsilon' \approx 51.8$ with small dielectric loss $\varepsilon'' \approx 9.3 \times 10^{-3}$. Luan *et al.* [32] developed conductive epoxy composites using graphene and silver nanowires *via* simple physical blending method and the nanocomposites exhibited a significant enhancement in the electrical conductivity due to the decreased tunneling resistance (Rt) between the silver nanowires in the presence of the conductive 2-D chemically reduced graphene. Less number of studies have been reported for mixed fibrous clay and graphene platelets for the generation of polymer nanocomposites. Kwon *et al.* [25] compared the properties of polyimide nanocomposites generated with organoclay, functionalized graphene and organoclay/functionalized graphene complex. The authors concluded that the clay/graphene complex composite exhibited superior properties compared to other nanocomposites due to the synergistic effect of the binary nanofiller system. Due to potential of achieving enhanced composite properties using such mixed filler systems, further work involving such systems is required, especially using polyolefins like polypropylene.

In this study, polypropylene nanocomposites were generated using binary filler system containing graphene and surface modified sepiolite. The influence of the mixed filler system on the mechanical, thermal and morphological properties of the composites was studied. Amine functionalized polypropylene (PP-*g*-DA8) was synthesized and used as compatibilizer to obtain chemical interactions with the filler surface.

2.2 Experimental

2.2.1 Materials

Polypropylene (PP) with a trade name of HD51CF was received from Abu Dhabi Polymers (Borouge), UAE. Sepiolite (S) clay was received from Sigma Aldrich. It has a characteristic diffraction peak (d_{001}) at 12.12 Å. The main characteristic bands of sepiolite were reported at 3684 cm^{-1}, 3550 cm^{-1}, 1693 cm^{-1}, 1658 cm^{-1}, 1204 cm^{-1}, 1022 cm^{-1}, 780 cm^{-1}, 691 cm^{-1} and 648 cm^{-1}, which are ascribed to bending of Mg-OH, -OH stretching, OH bending, Si-O-Si bands, bending of Mg-Fe-OH and Mg-OH bending. 1,8-diaminooctane, polypropylene-graft-maleic anhydride (PP-*g*-MA) (M_w ~9,100, M_n ~3,900 and maleic anhydride grafting percentage of 8-10%), acetone, methanol, *o*-xylene, sodium bicarbonate ($NaHCO_3$) and polyurethane (PU) liquid were received from Sigma Aldrich and were used without further purification. Graphene (G) with a trade name of N002-PDR was purchased from Angstron Materials, USA (specific surface area of 400-800 m^2/g, average plane dimension = 1.0-1.2 nm). Ethylene vinyl acetate-*graft*-maleic anhydride (EVA-*g*-MA) copolymer (density 0.962 g/cm^3 and melt flow rate (190 °C/2.16 kg) 1.4 g/10 min) with a trade name of Fusabond C190 was obtained from DuPont.

2.2.2 Surface Modification of Sepiolite

The surface of the sepiolite fibers was modified with PU and EVA-*g*-MA copolymer. For modification with PU, PU was dispersed in deionized (DI) water at a concentration of 1 mg/ml at 70 °C. Sepiolite (2% w/w) was added to the polymer dispersion and shaken for 2 h and ultra-sonicated for 5 min. The content was further stirred overnight. The sample was washed with DI water followed by acetone. The sample was finally dried at 50 °C. The final product was termed as PU-S clay.

Same procedure was used to prepare the EVA-*g*-MA modified sepiolite filler (denoted as EVA*g*MA-S) in toluene solvent. The modified was dried at 100 °C overnight under vacuum to remove the solvent.

2.2.3 Preparation of Amine Functionalized Polypropylene (PP-*g*-DA8) Compatibilizer

The compatibilizer was prepared as per the reported literature [33]. In brief, 10 g of PP-g-MA and 40 g of 1, 8-diaminooctane were added together in N_2 atmosphere. The mixture was heated to 130 °C and stirred for 30 min. Subsequently, it was allowed to cool at room temperature. The content was washed with 5 wt% $NaHCO_3$ solution followed by water and acetone. The material was dried under vacuum for 12 h at 50 °C.

2.2.4 Generation of PP Nanocomposites

The nanocomposites were prepared using solution blending of PP and PP-*g*-DA8 with the fillers. Typically, 400 mL of xylene was placed in a round bottom flask and 0.5 g of EVA*g*MA-S was added to it, followed by 1 h ultra-sonication. 1.0 g of PP-*g*-DA8 and 8.5 g of PP were subsequently added to the flask and stirred at 130 °C for 2 h with a reflux condenser. The solvent was then evaporated at 100 °C under vacuum (Scheme 2.1a). The prepared composite, named as P/5%EVA*g*MA-S, was molded using injection molding (Thermo Scientific) at 190 °C with 400 bar for 10 s to obtain test specimens. The remaining nanocomposites were similarly prepared using 10 wt% EVA*g*MA-S, 5 wt% PU-S, 10 wt% PU-S, 2.5 wt% EVA*g*MA-S + 2.5 wt% G, 2.5 wt% PU-S + 2.5 wt% G, 5 wt% EVA*g*MA-S + 5 wt% G and 5 wt% PU-S + 5 wt% G, with PP-*g*-DA8 concentration fixed at 10 wt%. The nanocomposites were respectively named as P/10%EVA*g*MA-S, P/5%PU-S, P/10%PU-S, P/2.5%EVA*g*MA-S/2.5%G, P/2.5%PU-S/2.5%G, P/5%EVA*g*MA-S/5%G and P/5%PU-S/5%G (Scheme 2.1b). In order to confirm the covalent interaction between the nanofiller and compatibilizer, hybrid of PP-*g*-DA8 and modified sepiolite was also generated. For this, 0.5 g of modified sepiolite was dispersed in 50 mL of dimethyl formamide using an ultra-sonication bath at 30 °C for 1 h. Subsequently, 0.5 g of PP-*g*-DA8 was added to the dispersion and the mixture was stirred for 2 h at 110 °C. The hybrid was obtained by evaporation of the solvent under vacuum at 100 °C.

2.2.5 Characterization

IR spectra of the samples were collected on Nicolet iS10 spectrometer equipped with SmartiTR diamond ATR accessory (angle of incidence

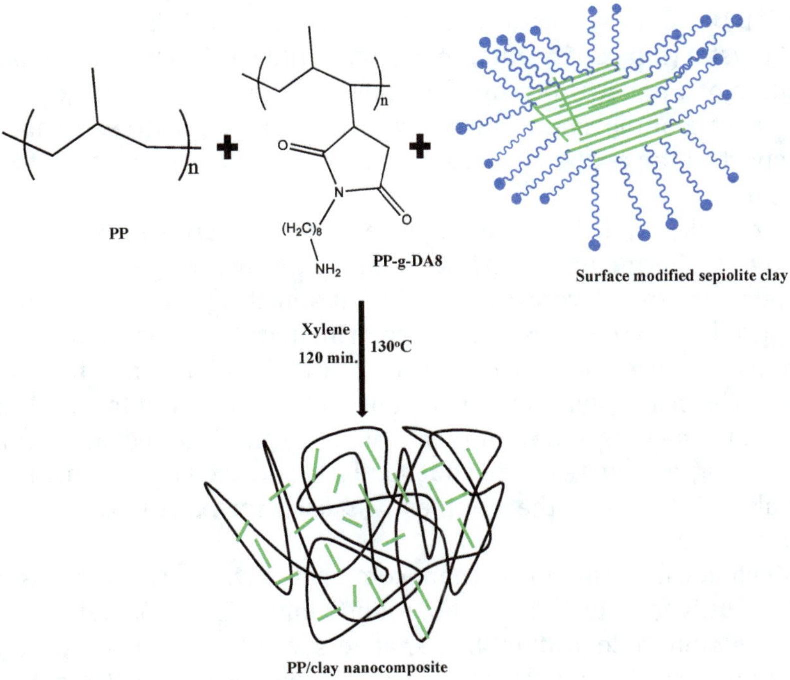

(a)

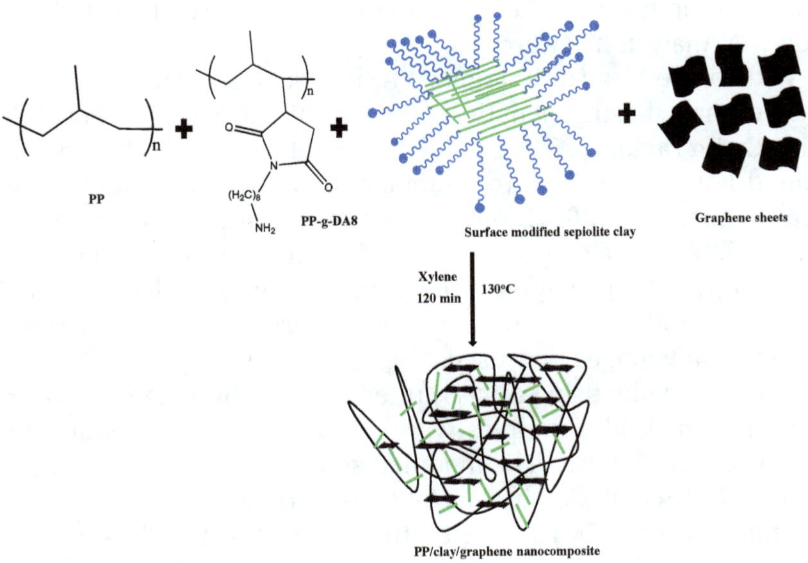

(b)

Scheme 2.1 Schematic of the generation of (a) polypropylene/clay and (b) polypropylene/clay/graphene nanocomposites.

of 45°), DTGS KBr detector and KBr beam splitter. It had a diamond ATR crystal (index of refraction = 2.4 at 1000 cm^{-1}) and a depth permeation of 2 μm at 1000 cm^{-1} for sample with refractive index of 1.5. The spectra of the samples were recorded using OMNIC software in the spectral range of 4000-600 cm^{-1} with a resolution of 4 cm^{-1} from 32 scans.

Thermal properties of the nanocomposites were recorded using Discovery Thermogravimetric Analyzer (TGA) in nitrogen medium. A temperature range from 35 to 700 °C at a heating rate of 10 °C/min was used. Differential scanning calorimetric thermograms of the composites were recorded on a Discovery DSC under nitrogen atmosphere. The scans were obtained from 35 to 200 °C and from 200 to 35 °C using heating and cooling rate of 10 °C/min. The second heating and cooling runs were carried out similarly and used to calculate melt enthalpy. 5-10 mg of the samples was used for both DSC and TGA analysis.

Mechanical testing of the nanocomposites was carried out using Instron universal testing machine following ASTM D 638 standard. The injection molded dumbbell-shaped samples were used, and an average of five specimens was recorded. A loading rate of 10 mm/min was used, and the tests were carried out at room temperature. The tensile strength and modulus of the composites were calculated using Blue-Hill Analysis software.

Wide-angle X-ray diffraction (WAXRD) analysis of the composites was performed using analytical powder diffractometer (X'Pert PRO) using Cu Kα radiation (λ = 1.5406 Å) in reflection mode. A zero-background holder was used to minimize the noise. The samples were step-scanned from 2θ = 5-60° at room temperature using a step size of 2θ = 0.02° and a step time of 10 s. For the transmission electron microscopy (TEM) analysis of the ultra-thin film samples, Philips CM 20 (Philips/FEI, Eindhoven) electron microscope at 120 and 200 kV accelerating voltages was used. The samples were prepared by microtoming the film using a glass cutter followed by a diamond cutter. 70-90 nm thick film samples were micro toned and supported on TEM grids for analysis. The modified sepiolite fillers were analyzed in Hitachi field emission in-lens S-900 high resolution scanning electron microscope (SEM) at accelerating voltages of 10-20 kV.

2.3 Results and Discussion

Figure 2.1 demonstrates the SEM images of the pristine and modified

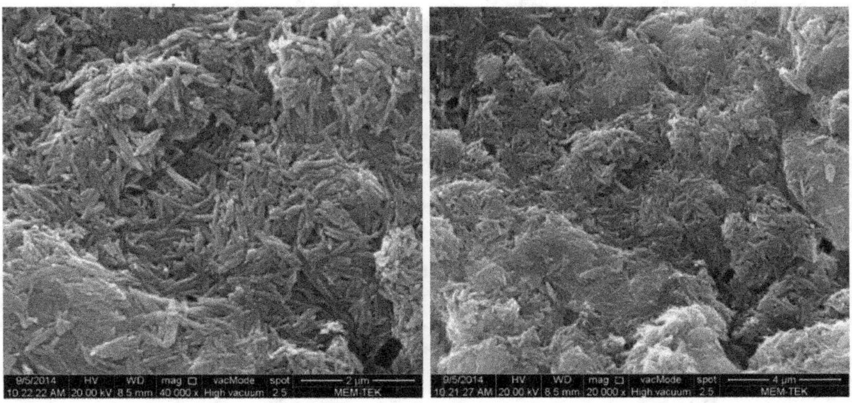

(a)

(b)

(c)

Figure 2.1 SEM images of (a) pristine sepiolite, (b) PU-S and (c) S-EVA*g*MA-S at two different magnifications.

sepiolite. The pristine particles were observed to be uniformly dispersed as primary particles without forming any large sized aggregates. On the other hand, the filler particles were observed to form large aggregates in these modified sepiolite materials. However, it should be noted that the observed aggregates may be only loosely held aggregated structures, which may be easily broken when blended with polymers.

Figure 2.2a shows the FT-IR spectra of PP-g-DA8, EVA*g*MA-S and

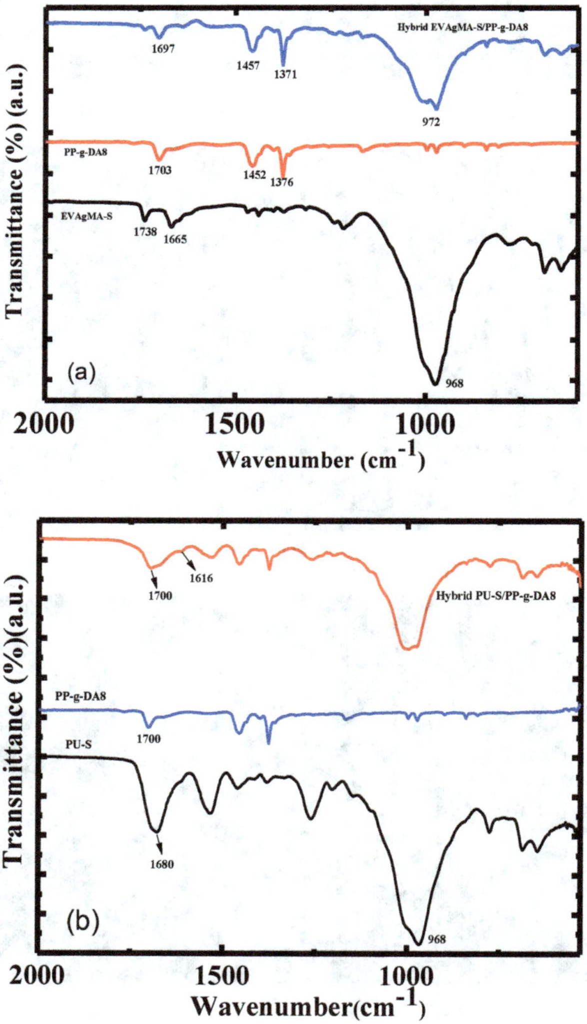

Figure 2.2 FTIR spectra of a) EVA*g*MA-S, PP-*g*-DA8 and hybrid EVA*g*MA-S/PP-*g*-DA8, b) PU-S, PP-*g*-DA8 and hybrid PU-S/PP-*g*-DA8.

PP-*g*-DA8/EVA*g*MA-S hybrid. The spectrum of EVA*g*MA-S exhibited maleic anhydride group with characteristic absorption bands at 1738 and 1665 cm^{-1} associated with the carbonyl C=O symmetric and asymmetric stretching bands. PP-*g*-DA8 exhibited an absorption band at 1703 cm^{-1} attributed to primary amide C=O stretching and another absorption band at 1457 cm^{-1} associated with N–H secondary amine stretching. In the hybrid of EVA-*g*-MA and PP-*g*-DA8, most of the carbonyl C=O symmetric and asymmetric stretching bands at 1738 cm^{-1} vanished indicating the reaction between the EVA*g*MA-S and the compatibilizer [33,34]. In addition, a broad intense peak appeared at 972 cm^{-1} representing the magnesium silicate of sepiolite. Figure 2.2b represents the FT-IR spectra of PP-*g*-DA8, PU-S and PP-*g*-DA8/PU-S hybrid. The hybrid of PU-S with PP-*g*-DA8 did not exhibit any chemical interaction and only a small peak was observed at 1616 cm^{-1} due to the weak hydrogen bonding between the C=O of PU-S and N-H of PP-*g*-DA8 [35], thus, representing a physical blend. Figure 2.3a shows the FT-IR spectra of the 10% PU-S and EVA*g*MA-S filled PP nanocomposites. Same reaction between EVA*g*MA-S and PP-g-DA8 was observed to occur when PP was added. In addition, the incorporation of graphene resulted a characteristic hump like peak at 1583 cm^{-1}, which corresponded to the C=C stretching frequency of graphene (Figure 2.3b) [36].

Table 2.1 presents the melting and crystallization properties of pure PP, PP/PP-*g*-DA8 blend and nanocomposites. The degree of crystallinity (X_c) was calculated using the melting enthalpy (ΔH_m) of the 100 % crystalline PP (209 J/m) [37]. The incorporation of 10 wt% PP-*g*-DA8 reduced the melt enthalpy and the degree of crystallinity of the polymer. These results suggested that the polar amine-PP restricted the ability of the pure PP chains to form crystals and the amine-PP acted as a plasticizer in the blend [33]. The crystallinity of nanocomposites increased with increasing percentage of sepiolite. Both surface modifications for sepiolite resulted in a significant increase in the % crystallinity of the polymer in the composites, thus, confirming the nucleation effect. In the case of hybrid filler composites, addition of clay and graphene in the PP matrix resulted in slightly enhanced % crystallinity as compared to clay composites due to enhanced nucleating effect in the presence of graphene.

The thermal degradation temperature for 10 wt% and 40 wt% loss as well as char yield value at 700 °C are reported in Table 2.2. It was observed that the addition of compatibilizer resulted in an increase in the degradation temperature of the PP matrix. For example,

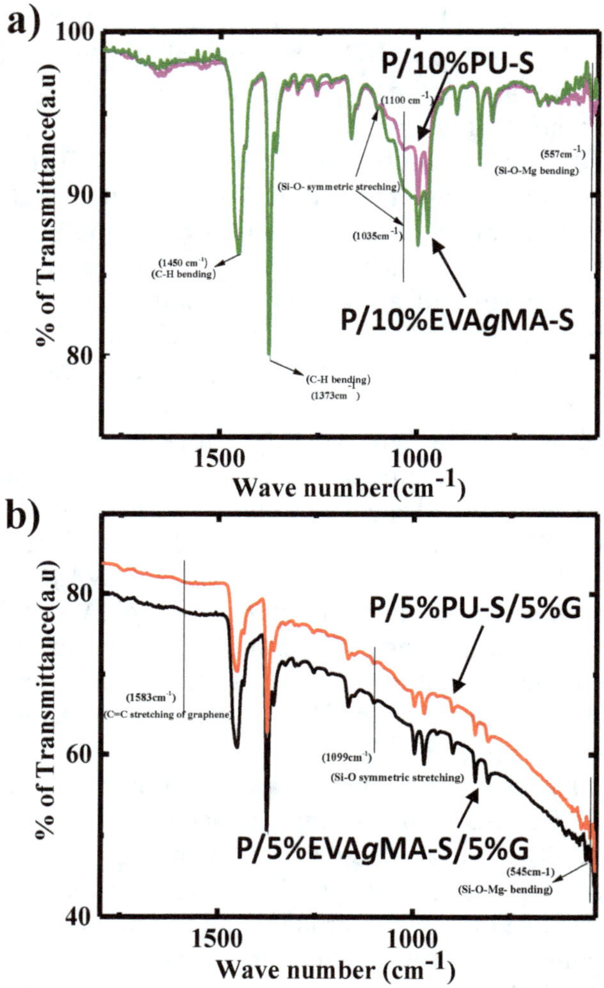

Figure 2.3 FTIR spectra of a) PP/clay nanocomposites, b) PP/clay/graphene nanocomposites.

the 10 wt% loss temperature for pure PP was observed to be 374 °C in nitrogen atmosphere, which was 43 °C less than the PP/PP-*g*-DA8 matrix. As the compatibilizer had both imide linkage and terminal amine group, it was expected to impart enhanced thermal stability. In addition, clay and hybrid clay-graphene filled nanocomposites exhibited almost similar 10% and 40% weight loss temperatures. The char yield of the nanocomposites increased with increase in the nanoclay fraction as compared to pure PP and PP/PP-*g*-DA8 blend, which can be attributed to the role of nanoclay as thermal shield in the polymer nanocomposites, thus, increasing the char yield. The hybrid clay/graphene nanofillers did not exhibit any significant synergistic effect on the char yield of the PP/clay/graphene nanocomposites.

Table 2.1 Calorimetric properties of nanocomposites

Sample	T_c (°C)	T_m (°C)	ΔH_m (J/g)	ΔH_c (J/g)	*Xc* (%)
PP	128	167	17	24	35
PP/PP-*g*-DA8	128	168	11	23	26
P/5%EVAgMA-S	128	167	17	22	41
P/5%PU-S	128	166	18	22	42
P/10%EVAgMA-S	129	167	17	21	43
P/10%PU-S	127	167	18	20	45
P/2.5%EVAgMA-S/2.5%G	127	169	19	20	45
P/2.5%PU-S/2.5%G	130	167	20	21	47
P/5%EVAgMA-S/5%G	130	168	17	19	43
P/5%PU-S/5%G	130	169	19	18	47

Table 2.2 Thermal stability and char yield of PP/clay and PP/clay/graphene nanocomposites

Sample	T_d^{10} (°C)	T_d^{40} (°C)	Char yield @ 700 °C
PP	386	414	1.3
PP/PP-*g*-DA8	429	456	1.4
P/5%EVAgMA-S	431	457	3.3
P/5%PU-S	432	458	2.2
P/10%EVAgMA-S	433	458	7.0
P/10%PU-S	431	460	5.0
P/2.5%EVAgMA-S/2.5%G	429	458	3.4
P/2.5%PU-S/2.5%G	429	458	3.2
P/5%EVAgMA-S/5%G	439	460	9.2
P/5%PU-S/5%G	440	462	8.0

The mechanical properties of nanocomposites are summarized in Table 2.3. Incorporation of 10 wt% compatibilizer to the PP matrix resulted in a significant decrease in the tensile modulus and strength due to the plasticization effect. Subsequently, the addition of EVA*g*MA-S to the polymer resulted in an increasing trend for both modulus and strength, with increase in clay content. For instance, at 10% EVA*g*MA-S content, the modulus and strength were enhanced

Table 2.3 Mechanical properties of PP/clay and PP/clay/graphene hybrid nanocomposites

Composites	Max. load	UTS [MPa]	Modulus [MPa]	Strain at break (%)
PP	212± 2	36 ± 2	891 ± 76	13 ± 2
PP/PP-*g*-DA8	194 ± 16	32 ± 3	784 ± 11	8 ± 3
P/5%EVA*g*MA-S	201 ± 4	33 ± 2	852 ± 54	6 ± 2
P/5%PU-S	152 ± 27	25 ± 4	838 ± 34	4 ± 1
P/10%EVAgMA-S	215 ± 3	36 ± 2	917 ± 32	6 ± 2
P/10%PU-S	190 ± 7	32 ± 2	911 ± 37	5± 1
P/2.5%EVAgMA-S/2.5%G	325 ± 12	41± 2	944 ± 54	7± 1
P/2.5%PU-S/2.5%G	322 ± 17	41 ± 2	935 ± 49	8 ± 2
P/5%EVAgMA-S/5%G	328 ± 5	43 ± 2	872 ± 41	8 ± 2
P/5%PU-S/5%G	290 ± 11	37 ± 2	867 ± 65	5 ± 1

by 16% and 11% respectively as compared to the PP/PP-*g*-DA8 blend. The reaction of the maleic anhydride functional group on the filler surface with the amine functionalized compatibilizer, thus, forming strong imide linkage between the compatibilizer and the filler phase, would have contributed to the enhanced mechanical performance of the composites. This was further confirmed from the mechanical performance of the nanocomposites with PU-S filler. The nanocomposites with PU-S exhibited significant decrease in the tensile strength, and the modulus was enhanced to a lesser extent as compared to EVA*g*MA-S. In comparison with the chemical linkage of EVA*g*MA-S with the compatibilizer, intermolecular H-bonding between the PU-S and PP-*g*-DA8 along with the plasticization effect of the urethane linkage in PU would have resulted in such performance. In the case of hybrid filler nanocomposites, a remarkable synergistic effect on both modulus and strength was observed for P/2.5%EVA*g*MA-S/2.5%G and P/2.5%PU-S/2.5%G. However, P/5%EVA*g*MA-S/5%G and P/5%PU-S/5%G composites exhibited a decrease in the tensile modulus probably due to filler aggregation. The tensile strength for P/5%PU-S/5%G was also observed to decrease, though the values were still higher than the corresponding sepiolite composites. It is well known that the nanofillers have a threshold limit of the loading amount in the polymer matrix. It is probable that the sepiolite phase failed to reduce the stacking of graphene sheets at higher concentration, as the filler materials have a tendency to aggregate due to van der Waals forces at higher contents. Thus, for the mixed filler system, the higher filler loading induced the aggregation due to weaker interfacial interaction between the fillers.

XRD spectra of pure PP, PP/PP-*g*-DA8 blend and nanocomposites (Figure 2.4) contained six major reflections of α -phase at 14.6°, 17.7°,

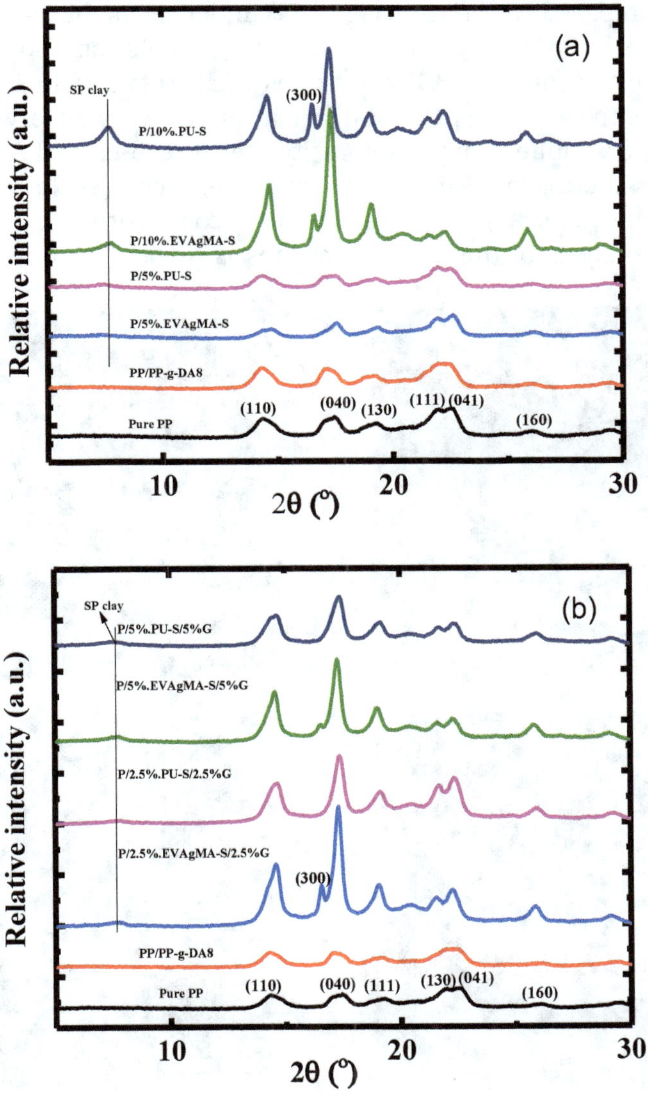

Figure 2.4 XRD spectra of a) PP/clay nanocomposites, b) PP/clay/graphene nanocomposites.

19.0°, 21.9°, 22.6° and 26.0° corresponding to planes (110), (040), (130), (111), (041) and (060) respectively. The (110) α and (040) α peaks were compared taking (110) α as a reference peak. The increase in the ratio of intensities between the two peaks confirmed an increase in the degree of crystallinity, which also supported the DSC results. The addition of 10 wt% sepiolite induced the β-crystallinity of PP at the plane (300). The β-crystalline phase is thermally less stable as compared to the α-phase of PP under normal crystallization conditions [39]. The presence of nucleating agents can also assist the formation of β-form in PP. However, in the case of sepiolite/graphene hybrid composites, the β- crystal formation was only observed in the EVA*g*MA-S/graphene nanocomposites. As the β-crystal formation improves the mechanical properties of PP, these results coincided with the mechanical properties of PP nanocomposites.

TEM analysis of the nanocomposites is presented in Figure 2.5.

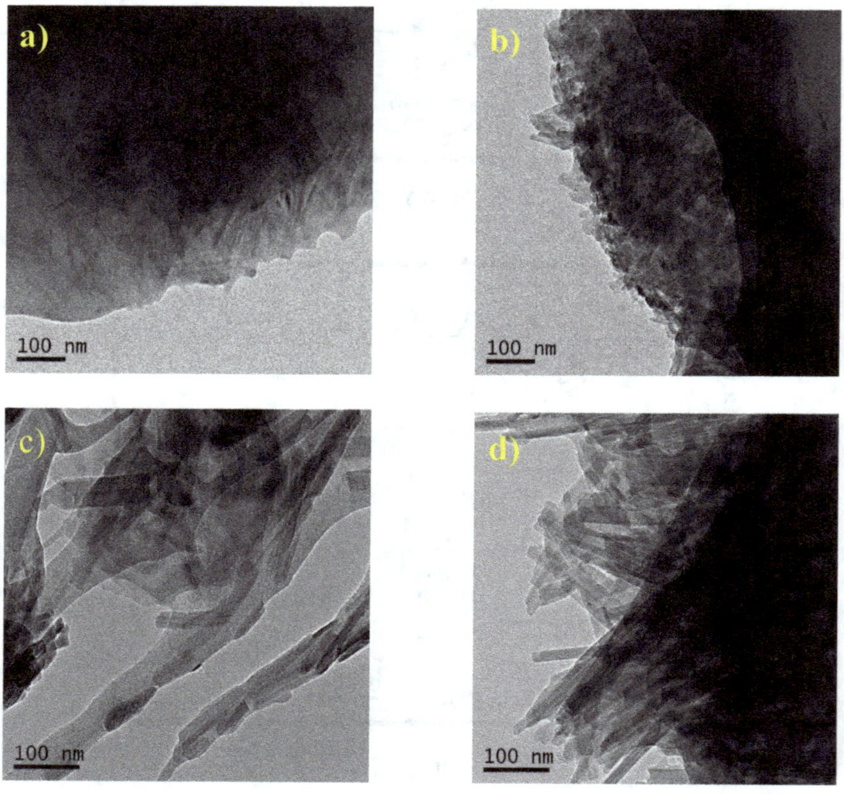

Figure 2.5 TEM images of (a) P/5%PU-S, (b) P/5%EVAgMA-S, (c) P/2.5%PU-S/2.5%G and (d) P/5%EVAgMA-S/5%G nanocomposites.

Figure 2.5a, corresponding to P/5%PU-S nanocomposite, exhibited uniform distribution of filler in the PP matrix, however, intercalated clay layers was also observed. Thus, though the compatibilizer led to secondary H-bonding with sepiolite and resulted in overcoming the energy barrier to form randomized delamination of sepiolite, complete delamination was, however, not achieved. Figure 2.5b shows the TEM image of P/5%EVA*g*MA-S nanocomposite. The sepiolite phase was observed to be uniformly distributed in the PP matrix and exhibited an exfoliated morphology. Chemical interaction of the compatibilizer with the filler surface resulted in an enhanced degree of filler delamination, which also translated into mechanical performance of the composites. Figure 2.5c and 2.5d show the images for the nanocomposites with 2.5wt%EVA*g*MA-S+2.5%G and 5%EVA*g*MA-S+5%G respectively. No filler aggregation was observed, and the particles were distributed at the nanoscale in the polymer matrix. Some degree of filler aggregation was also observed in the composites with higher filler content.

2.4 Conclusions

In the present study, PP nanocomposites were developed using organo-modified sepiolite clay (EVA*g*MA-S and PU-S) as well as graphene-modified sepiolite clay hybrid in the presence of amine functionalized PP as a compatibilizer.

EVA*g*MA modified sepiolite was observed to chemically interact with the compatibilizer. TEM images displayed the uniform dispersion of nanofillers, especially EVA*g*MA-S, in the PP nanocomposites at lower filler content. An improved thermal stability was observed in the composites along with an increase in the char yield. DSC studies showed that the nanofillers acted as weak nucleating agents, thus, resulting in an increase in the crystallinity of the polymer matrix. Further from XRD, an increase in the degree of crystallinity was confirmed when the ratio of (110) α and (040) α peaks was compared, with (110) α as the reference peak. An improvement in Young's modulus was observed for the nanocomposites, especially with EVA*g*MA-S clay containing composites, due to the covalent interaction of the filler surface with the compatibilizer, thus, resulting in better filler dispersion.

Addition of graphene further enhanced the tensile modulus, thus, confirming the synergistic effect of the mixed filler system on the polymer properties.

Acknowledgment

The authors sincerely thank the PI Research Centre, The Petroleum Institute for the financial support.

References

1. Kawasumi, M., Hasegawa, N., Kato, M., Usuki, A., and Okada, A. (1997) Preparation and mechanical properties of polypropylene-clay hybrids. *Macromolecules,* **30,** 6333-6338.
2. Garcia-Lopez, D., Picazo, O., Merino, J. C., and Pastor, J. M. (2003) Polypropylene-clay nanocomposites: effect of compatibilizing agents on clay dispersion. *European Polymer Journal,* **39**, 945-950.
3. Liu, X., and Wu, Q. (2001) PP/clay nanocomposites prepared by grafting melt-intercalation. *Polymer,* **42**, 10013-10019.
4. Vengatesan M. R., and Mittal, V. (2015) Surface modification of nanomaterials for application in polymer nanocomposite: An overview. In: *Surface Modification of Nanoparticle and Natural Fiber Fillers,* Wiley-VCH, Germany, pp. 1-28.
5. Saravanan, S., Gowda, K. M. A., Varman, K. A., Ramamurthy, P. C., and Madras, G. (2015) In-situ synthesized poly (vinyl butyral)/MMT-clay nanocomposites: The role of degree of acetalization and clay content on thermal, mechanical and permeability properties of PVB matrix. *Composites Science and Technology,* **117**, 417-427.
6. Sapalidis, A. A., Katsaros, F. K., Steriotis, T. A., and Kanellopoulos, N. K. (2012) Properties of poly (vinyl alcohol)- Bentonite clay nanocomposite films in relation to polymer-clay interactions. *Journal of Applied Polymer Science,* **123**, 1812-1821.
7. Zheng, S., Wang, T., Liu, D., Liu, X., Wang, C., and Tong, Z. (2013) Fast deswelling and highly extensible poly(N-isopropylacrylamide)-hectorite clay nanocomposite cryogels prepared by freezing polymerization. *Polymer,* **54**, 1846-1852.
8. Mittal, V., and Al Zaabi, K. (2013) Biodegradable polyester nanocomposites: Phase miscibility and properties. *Journal of Applied Polymer Science,* **130**(1), 516-525.
9. Garcia-Lopez, D., Fernandez, J. F., Merino, J. C., Santaren, J., and pastor, J. M. (2010) Effect of organic modification of sepiolite for PA6 polymer/organoclay nanocomposites. *Composites Science and Technology,* **70**, 1429-1436.
10. Bergaya, F., Then, B. K. G., and Lagaly, G. (2006) *Handbook of Clay Science,* Elsevier, Netherlands.
11. Samakande, A., Hartmann, P. C., Cloete, V., and Sanderson, R. D. (2007) Use of acrylic based surfmers for the preparation of exfoliated polystyrene clay nanocomposites. *Polymer,* **48**, 1490-1499.

12. Mansoori, Y., Atghia, S. V., Zamanloo, M. R., Imanzadeh, Gh., and Sirousazar, M. (2010) Polymer-clay nanocomposites: Free-radical grafting of polyacrylamide onto organophilic montmorillonite. *European Polymer Journal*, **46**, 1844-1853.
13. Choi, Y. Y., Lee, S. H., and Ryu, S. H. (2009) Effect of silane functionalization of montmorillonite on epoxy/montmorillonite nanocomposite. *Polymer Bulletin*, **63**, 47-55.
14. Grim, R. E. (1962) *Applied Clay Mineralogy*, McGraw-Hill, USA.
15. Chen, H., Zheng, M., Sun, H., and Jia, Q. (2007) Characterization and properties of sepiolite/polyurethane nanocomposites. *Materials Science and Engineering A*, **445-446**, 725-730.
16. Bilotti, E., Zhang, R., Denge, H., Quero, F., Fischer, H. R., and Peijs, T. (2009) Sepiolite needle like clay for PA6 nanocomosites: An alternative to layered silicates?. *Composites Science and Technology*, **69**, 2587-2595.
17. Novoselov, K. S., Geim, A. K., Morozov, S. V., Jiang, D., Katsnelson, M. I., Grigorieva, I. V., Dubon, S. V., and Firsov, A. A. (2005) Two-dimensional gas of massless Dirac fermions in graphene. *Nature*, **438**, 197-200.
18. Novoselov, K. S., Jiang, Z., Zhang, Y., Morozov, S. V., Stormer, H. L., Zeitler, U., Mann, J. C., Boebinger, G. S., Kim, P., and Geim, A. K. (2007) Room-temperature quantum hall effect in graphene. *Science*, **315**, 1379.
19. Wang, G., Yang, J., Park, J., Gou, X., Wang, B., and Liu, H. (2008) Facile synthesis and characterization of graphene nanosheets. *Journal of Physical Chemistry C*, **112**, 8192-8195.
20. Wang, G., Shen, X., Wang, B., Yao, J., and Park, J. (2009) Synthesis and characterisation of hydrophilic and organophilic graphene nanosheets. *Carbon*, **47**, 1359-1364.
21. Li, X., Wang, X., Zhang, L., Lee, S., and Dai, H. (2008) Chemically derived, ultrasmooth graphene nanoribbon semiconductors. *Science*, **319**, 1229-1231.
22. Eda, G., and Chhowalla, M. (2009) Graphene-based Composite Thin Films for Electronics. *Nano Letters*, **9**, 814-818.
23. Liang, J., Xu, Y., Huang, Y., Zhang, L., Wang, Y., and Ma, Y. (2009) Infrared-Triggered Actuators from Graphene-Based Nanocomposites. *Journal of Physical Chemistry C*, **113**, 9921-9927.
24. Kim, H., and Macosko, C. W. (2009) Processing-property relationships of polycarbonate/graphene composites. *Polymer*, **50**, 3797-3809.
25. Kwon, K., and Chang, J.-H. (2015) Comparison of the properties of polyimide nanocomposites containing three different nanofillers: organoclay, functionalized graphene, and organoclay/ functionalized graphene complex. *Journal of Composite Materials*, **49**, 3031-3044.

26. Paszkiewicz, S., Kwiatkowska, M., Rosłaniec, Z., Szymczyk, A., Jotko, M., and Lisiecki, S. (2015) The influence of different shaped nanofillers (1D, 2D) on barrier and mechanical properties of polymer hybrid nanocomposites based on PET prepared by in situ polymerization. *Polymer Composites,* **37**(7), 1949-1959.
27. Pradhan, B., Roy, S., Srivastava, S. K., and Saxena, A. (2015) Synergistic effect of carbon nanotubes and clay platelets in reinforcing properties of silicone rubber nanocomposites. *Journal of Applied Polymer Science,* **132**, 41818.
28. Kim, H., Abdala, A. A., and Macosko, C. W. (2010) Graphene/Polymer Nanocomposites. *Macromolecules,* **43**, 6515-6530.
29. Tang, Z., Wei, Q., Lin, T., Guo, B., and Jia, D. (2013) The use of a hybrid consisting of tubular clay and graphene as a reinforcement for elastomers. *RSC Advances,* **3**, 17057-17064.
30. Noh, Y. J., and Kim, S. Y. (2015) Synergistic improvement of thermal conductivity in polymer composites filled with pitch based carbon fiber and graphene nanoplatelets. *Polymer Testing,* **45**, 132-138.
31. Cho, S. Kim, M. Lee, J. S., and Jang, J. (2 015) Polypropylene/polyaniline nanofiber/reduced graphene oxide nanocomposite with enhanced electrical, dielectric, and ferroelectric properties for a high energy density capacitor. *ACS Applied Materials & Interfaces,* **7**, 22301-22314.
32. Luan, V. H., Tien, H. N., Cuong, T. V., Kong, B.-S., Chung, J. S., Kim, E. J., and Hur, S. H. (2012) Novel conductive epoxy composites composed of 2-D chemically reduced graphene and 1-D silver nanowire hybrid fillers. *Journal of Materials Chemistry,* **22**, 8649-8653.
33. Sanchez-Valdes, S., Mendez-Nonell, J., Medellin-Rodriguez, F. J., Ramirez- Vargas, E., Martinez-Colunga, J. G., Ramos de Valle, L.F., Mondragon-Chaparro, M., Lopez-Quintanilla, M. L., and Garcia-Salazar, M. L. (2010) Evaluation of different amine- functionalized polyethylene as compatibilizer for polyethylene film nanocomposites. *Polymer International,* **59**, 704-711.
34. Cui, L., and Paul, D. R. (2007), Evaluation of amine functionalized polypropylene as compatibilizers for polypropylene nanocomposites, *Polymer,* **48**, 1632-1640.
35. Deka, H., and Karak, N. (2009) Vegetable oil-based hyperbranched thermosetting polyurethane/clay nanocomposites. *Nanoscale Research Letters,* **4**, 758-765.
36. Vengatesan, M. R., Shen, T. Z., Alagar, M., and Song, J.-K. (2016) A facile chemical reduction of graphene-oxide using p-toluene sulfonic acid and fabrication of reduced graphene-oxide film. *Journal of Nanoscience and Nanotechnology,* **16**, 327-332.
37. Pedrazzoli, D., Khumalo, V. M., Kocsis, J. K., and Pegoretti, A. (2014) Thermal, viscoelastic and mechanical behavior of polypropylene with synthetic boehmite alumina nanoparticles. *Polymer Testing,*

35, 92-100.

38. Zaman, I., Manshoor, B., Khalid, A., Meng, Q., and Araby, S. (2014) Interface modification of clay and graphene platelets reinforced epoxy nanocomposites: A comparative study. *Journal of Materials Science*, **49**, 5856-5865.
39. Bao, S. P., and Tjong, S. C. (2 007) Impact essential work of fracture of polypropylene/montmorillonite nanocomposites toughened with SEBS-g-MA elastomer. *Composites Part A*, **38**, 378-38727.

3

Influence of Copolymers on the Mechanical, Thermal and Rheological Properties of Isotactic Polypropylene (iPP) and Impact Polypropylene Copolymers (IPCs): A Comparative Study

3.1 Introduction

Polyolefins is the largest group of industrial thermoplastics in terms of production and has superior rank among commodity plastics owing to their use in a large number of applications. The number of individual polyolefin members is limited and the number of applications based on such materials is constantly increasing, along with an increment in the service and environmental requirements posed on these materials. Thus, various polyolefin-based blends and composites are required to meet the increasing demand of materials with required properties and processing suitable for specific applications [1-3]. Polyolefin blends have received considerable acceptance due to the generation of materials with enhanced properties as compared to parent polymers, while having no impact on the existing processing technologies. The polymer blends also contribute significantly towards the overall economical, ecological and processing perspectives [4,5]. For instance, Wang *et al.* [6] reported the generation of the polyolefin based blends with styrene polymers and engineering plastics and developed effective strategy to compatibilize the blends by using a polyolefin based multiphase compatibilizer. Chlorinated polyethylene (CPE) blends with different polymers have also been commonly reported for many purposes such as improvements in the toughening properties of host polymers [7]. In a recent study, Mittal *et al.* [8] also reported blends of polyethylene (PE) with small fractions of chlorinated polyethylene (CPE) of two different chlorination levels (25% and 35%, thus named as CPE25 and CPE35). Figure 3.1 demonstrates the morphological analysis of the blends (with 5% CPE content) using transmission electron (TEM) and atomic force microscopy (AFM). In AFM images, PE+CPE25 polymer phase was observed to maintain the semi-crystalline nature of pure PE, however,

Anish M. Varghese, M. R. Vengatesan and Vikas Mittal
Khalifa University of Science and Technology, Abu Dhabi, UAE

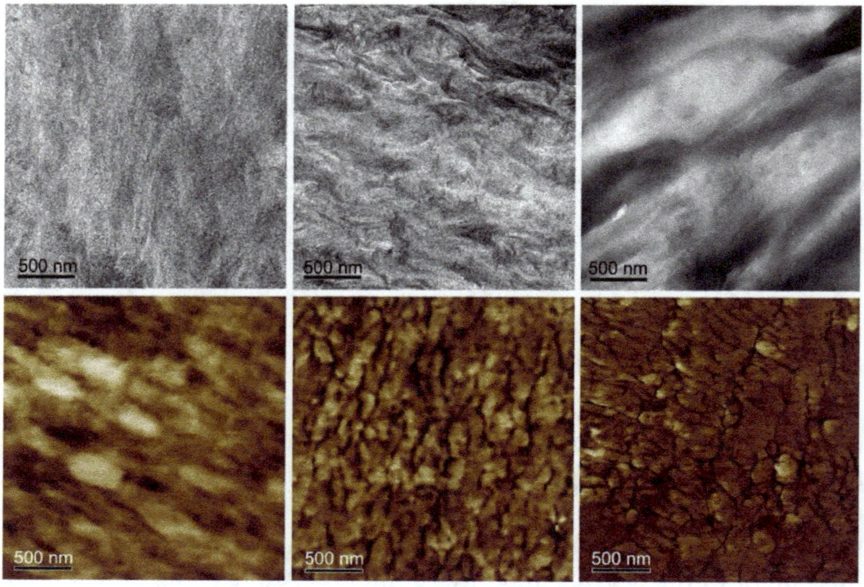

Figure 3.1 TEM and AFM images of (left) PE, (middle) PE+CPE25, and (right) PE+CPE35 samples. Reproduced from Reference 8 with permission from Wiley.

its morphology was quite different than PE as the chains seemed to be shorter in length. Also, the addition of CPE25 generated polycrystallinity in the sample. PE+CPE35 blend had more significant effect on the polymer microstructure. The crystalline phases were still present, but the order or periodicity was disturbed. Also, the crystalline phases were randomly distributed due to the hindrance of CPE35 phase to PE crystallization. However, the system was still miscible. As a result, CPE35 blend also exhibited higher extent of modulus reduction as compared to the blend with CPE25 [9]. In another recent study, Mittal *et al.* [10] reported polyethylene based functional blends with a variety of copolymers such as polyethylene-*g*-maleic anhydride copolymer (PE-MA), ethylene methacrylic acid copolymer with zinc ion (EMAZ) and ethylene vinyl acetate-*g*-maleic anhydride (EVA-MA) using the copolymer weight fractions of 5%, 10% and 15%. The EVA-MA and EMAZ based blends exhibited phase immiscibility due to the presence of separate melt transition of these copolymers in the DSC thermograms of the blend. As seen in Figure 3.2, the morphological analysis of the blends confirmed different degrees of phase immiscibility in EVA-MA and EMAZ based blends. On the other hand, the PE-MA based blends exhibited uniform morphology. As a result, the

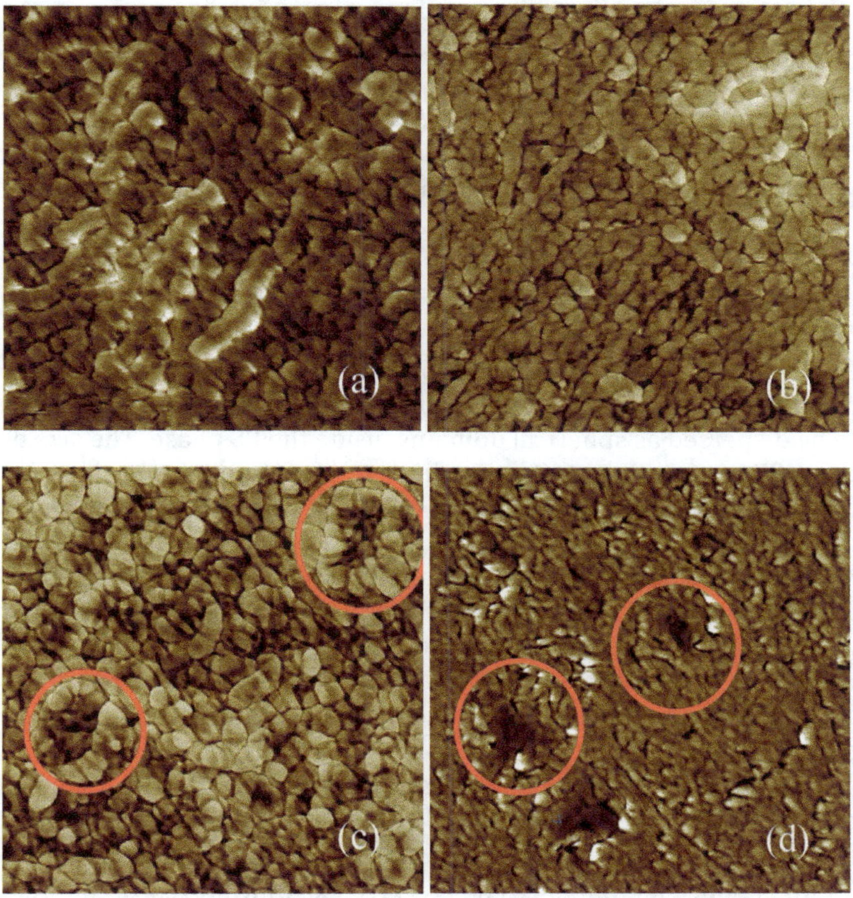

Figure 3.2 AFM images of the PE/15PE-MA1, PE/15%PE-MA2, PE/15%EMAZ and PE/15%EVA-MA blends. The phase image had the width of 3 μm. Fine dispersion of mixture of hard (bright phase representing crystalline region) and soft (dark phase representing amorphous region) regions is clearly visible in (a) and (b). On the other hand, the dispersion of the mixture of hard and soft regions was disturbed in (c) and (d), thus, representing immiscibility. In addition, the presence of "eye-like" structures with soft central part and dense borders was observed in (c) and (d). Reproduced from Reference 10.

mechanical properties were affected accordingly. The copolymers were also observed to have varying effect on thermal and rheological properties due to combined effects of plasticization and degree of phase miscibility. Thus, these studies underline the importance of optimizing the microstructure for obtaining optimal properties.

Polypropylene (PP) is a well-accepted thermoplastic commodity polyolefin and is an ideal choice for a wide variety of applications because of its low cost and versatile properties. The properties of PP can be further improved by blending with different polymeric materials [11-14]. For instance, a recent study by Mittal *et al.* [15] reported the blends of PP and polystyrene to enhance the properties of both polymers. The blends were also subsequently employed to generate nanocomposites with graphene. Figure 3.3 demonstrates the TEM micrographs of the PP/PS 70/30 and 30/70 blends. There was a difference in the phase morphology of the two blends. In the case of PP/PS 70/30 blend, though PP formed the continuous phase surrounding the large sized PS domains, however, it was also observed to be dispersed as spherical domains inside the PS phase. The size of these PP domains was 1-2 μm in diameter, indicating some phase intermixing between the PP and PS phases. As there is no thermodynamic interaction between the two polymers, therefore, the observed morphology would have solely resulted from the kinetic mixing in the compounder. In the case of PP/PS 30/70 blend, the polymer phases were more elongated in nature, though some degree of mixing of PP within the PS phase was still visible. The observed morphology of the blends correspondingly affected the mechanical and thermal properties.

The high performance engineering applications of isotactic polypropylene (iPP) are also occasionally restricted by its low impact performance, especially at low temperatures. Thus, the modification of iPP to enhance its toughness has received significant research interest [16,17]. Specifically, the low impact toughness of iPP is enhanced by applying the classical polymer melt blending with number of elastomers such as ethylene-propylene rubber (EPR) [18,19], ethylene-propylene-diene terpolymer (EPDM) [20,21], styrene-butadiene-styrene block copolymer (SBS) [22,23], styrene-ethylene-butylene-styrene triblock copolymer (SEBS) [24,25], ethylene-octene copolymer (POE) [26,27], etc. The other methods also include the addition of micro/nano-particulate fillers in the PP matrix [28-31]. In addition, the use of *in-situ* copolymerization of PP with ethylene in the polymerization reactor also results in the incorporation of toughness in the polymer main chain [32,33]. The product obtained from the *in-situ* copolymerization of PP and ethylene in the reactor is technically known as impact polypropylene copolymer (IPC) or heterophasic copolymer. The *in-situ* generated copolymers have gained significant attention due to their enhanced impact properties. The IPC synthesis includes two polymerization stages: preparation of iPP using bulk polymerization, followed by gas phase copolymerization of ethylene and propylene using different methods [34-36]. The existence of multiphase copolymeric system in the IPC was underlined by different

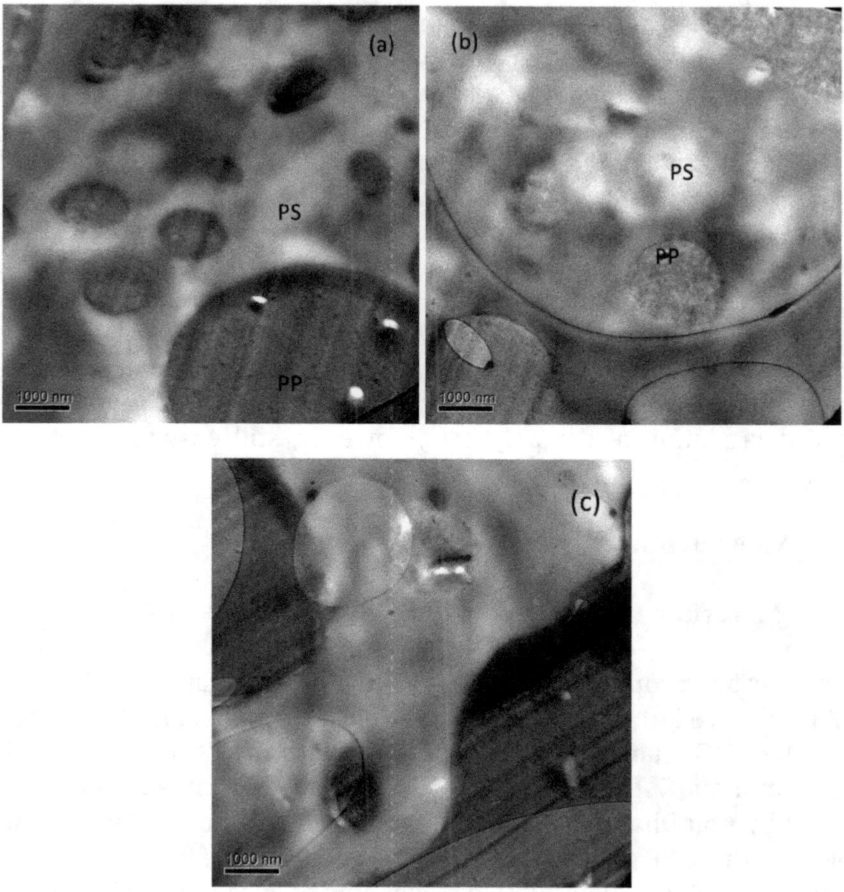

Figure 3.3 TEM images of the (a,b) PP/PS (70/30) and (c) PP/PS (30/70) blend. Reproduced from Reference 15.

literature studies and it was suggested that IPC included polypropylene homopolymer (iPP), amorphous ethylene-propylene random copolymers (EPR) and semi-crystalline ethylene-propylene block copolymers (E*b*P). It was also reported that the E*b*P phase in IPC acts as a compatibilizer and improves the interfacial interaction between the iPP matrix and dispersed EPR impact modifier phase [14,37]. During the application of high impact force, the dispersed EPR phase in IPC absorbs most of the sudden force and, thereby, experiences crazing and shear yielding in the iPP matrix. This results in the formation of microvoids and large plastic deformation, thus, preventing the crack growth. Therefore, the toughening behavior of IPC is decided by the size, shape and arrangement of dispersed EPR phase and the compatibility between the iPP matrix and dispersed EPR phases [38,39]. Blending of IPC with other elastomers, polymers or copolymers is also emerging as a suitable strategy for further enhancing the properties.

The objective of the current work was to investigate the effect of different copolymers (ethylene vinyl acetate-g-maleic anhydride (EVA-MA), ethylene butyl acrylate copolymer (EBA) and ethylene-α-olefin copolymer (EAO)) on the mechanical, thermal and rheological properties of different grades of PP, which included isotactic PP as well as IPC. The amount of copolymers in the blends was kept at a constant fraction of 20 wt%, and the blends were generated by using melt mixing by optimizing the compounding conditions to avoid thermal degradation of the polymers, however, still ensuring uniform mixing.

3.2 Experimental

3.2.1 Materials

Isotactic polypropylene (HD915CF) with melt flow rate of 8 g/10 min (230 °C/2.16 kg) as well as PP heterophasic copolymers (BH975MO & BD265MO, named as IPC1 and IPC2) with melt flow rates of 38 g/10 min and 7 g/10 min (230 °C/2.16 kg) respectively were provided by Abu Dhabi Polymers Company Limited (Borouge), UAE. All of these three PP grades had a density in the range of 900-910 kg/m^3. The EVA-MA copolymer with a tradename of Fusabond C190 having melt flow rate of 16 g/10 min (190 °C/2.16 kg), density of 0.950 g/cm^3 and peak melting point of 71 °C was provided by DuPont. EBA with tradename of Elvaloy AC 34035 had 35% by weight butyl acrylate co-monomer content. It had a melt flow rate of 40 g/10 min (190 °C/2.16 kg), density of 0.93 g/cm^3 and peak melting point of 90 °C was also procured from DuPont. EAO copolymer with tradename of Affinity EG8150G having melt flow rate of 0.5 g/10 min (2.16 kg, 190 °C, density of 0.868 g/cm^3 and peak melting point of 56 °C was obtained from Dow. The polymers and copolymers were used as received, without any subsequent treatment.

3.2.2 Preparation of PP/Copolymer Blends

PP/copolymer blends were prepared by melt blending of 20 wt% of each copolymer with different PP grades (iPP, IPC1, IPC2) using mini twin screw conical extruder (HAAKE MiniLab, Thermo Scientific) at a compounding temperature of 180 °C and 100 rpm for 15 minutes. The polymers were dried under vacuum at 70 °C overnight to remove surface bound moisture. The test specimens of the blended samples were prepared using mini injection molding machine (HAAKE MiniJet PRO, Thermo Scientific) with a cylinder temperature of 180 °C, the mold temperature of 125 °C, injection pressure of 430 bar for 10 s and post pressure of 500 bar for 6 s.

3.2.3 Materials Characterization

Differential scanning calorimetric (DSC) analysis of PP/copolymer blends was carried out under nitrogen atmosphere using Discovery Series DSC of TA Instruments. 3-8 mg of blend samples was scanned at a rate of 10 °C/min from -60 to 200 °C and then 200 to 25 °C. The second heating and cooling represented the melting and crystallization thermograms respectively. The thermal stability and decomposition characteristics of PP/copolymer blends were analyzed using thermogravimetric analyzer (TGA) from Discovery Series of TA Instruments under nitrogen atmosphere. 3-8 mg of the blend samples was heated from 25 to 700 °C at a rate of 10 °C /min and the mass loss was recorded as a function of temperature.

Tensile properties of the blend samples were performed at room temperature using universal testing machine Instron 3345 with a crosshead speed of 10 mm/min according to ASTM D638. The results represented the average of 5 dumbbell shaped specimens. Unnotched impact strength of PP/copolymer blended samples was measured using Resil Impactor of Ceast with hammer energy of 4 J and 3.46 m/s speed, according to ASTM D256 at room temperature. The Izod impact strength of each blend represented the average of five rectangular bar shaped specimens of 80 mm length, 10 mm width and 4 mm thickness.

Melt rheological properties such as complex viscosity (η^*), storage modulus (G'), loss modulus (G") and tan δ of blends were measured on AR 2000ex Rheometer from TA Instruments in a shear mode, according to ASTM D4440. Dynamic frequency sweep curves were obtained from 0.1 to 100 rad/s at 2% strain using disc shaped samples of 25 mm diameter and 1.5 mm thickness. The analysis was carried out at 180 °C in parallel plate geometry of 25 mm diameter with a geometric gap of 1.4 mm.

X-ray diffraction measurements were performed on a Panalytical Powder Diffractometer (X'Pert PRO) with Cu-Kα radiation having wavelength (λ) of 1.5406 Å, operated at 45 kV and 40 mA. The X-ray diffraction data were recorded with a step size of 0.0170°/s from 5 to 60° 2θ scale.

3.3 Results and Discussion

3.3.1 Calorimetric Properties and Phase Miscibility of PP/Copolymer Blends

Melting and crystallization thermograms of PP/copolymer blends are shown in Figures 3.4 and 3.5. The parameters such as peak temperature of melting (T_m) and crystallization (T_c), enthalpies of melting (ΔH_f) and crystallization (ΔH_c), as well as % crystallinity (X_c) of PP in the blends were obtained from the DSC curves and are presented in Table 3.1. The % crystallinity of PP in the PP/copolymer blends was measured using the below equation [11,40]:

$$X_c = \frac{\Delta H}{\Delta H_0 w_{pp}} \times 100$$

where ΔH_0 is the enthalpy of fusion for 100% crystalline isotactic PP (207.1 J/g [11]), ΔH is the enthalpy of fusion measured from DSC and w_{pp} is the fraction of PP in the PP/copolymer blend. From Table 3.1, it can be observed that the addition of copolymers did not adversely affect the T_m (169°C) and T_c (128°C) of iPP. The addition of both EVA-MA and EAO into iPP resulted in an increased extent of crystallinity from 44% to 49% and 47% in iPP/EVA-MA and iPP/EAO blends. No increment in X_c of iPP was observed in the iPP/EBA blend. It indicated the heterogeneous nucleating effect of EVA-MA and EAO in the iPP matrix [41]. In the case of IPC1/copolymers blends, the incorporation of copolymers did not show a significant variation in the T_m of pure IPC1 (167 °C), however a small reduction was observed in the T_c. The extent of crystallinity was also decreased from 40% for IPC1 to 37% in the blends containing EVA-MA and EAO, whereas the blend containing EBA did not influence the extent of crystallinity of IPC1. Thus, in contrast with iPP, both EVA-MA and EAO had a proscribing effect

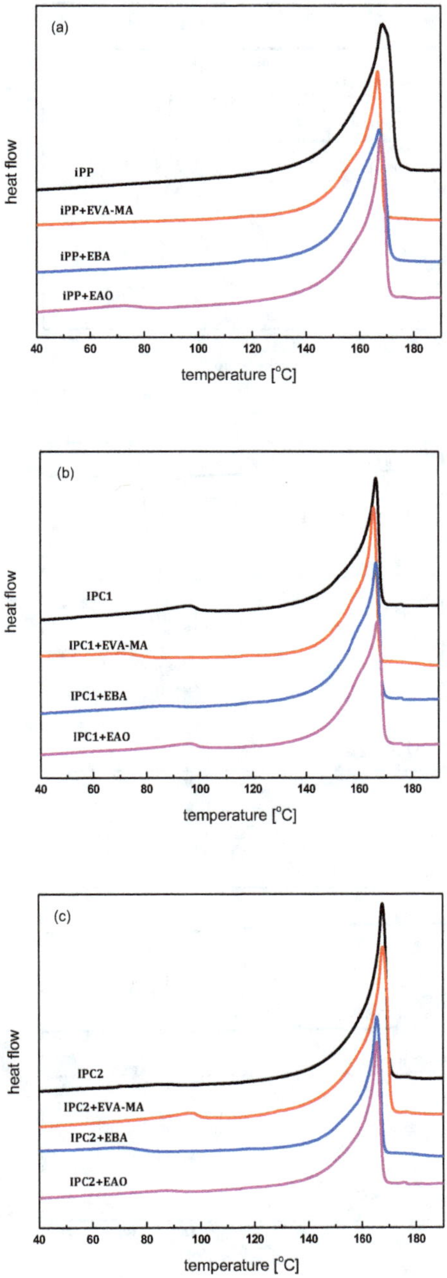

Figure 3.4 DSC melting thermograms of the PP blends with EVA-MA, EBA and EAO: (a) iPP, (b) IPC1 and (c) IPC2.

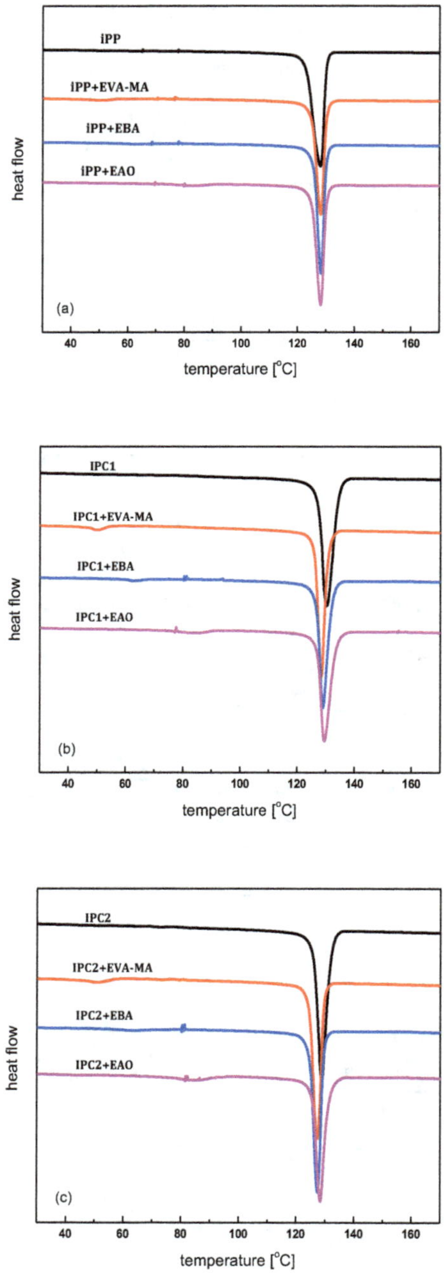

Figure 3.5 DSC crystallization thermograms of the PP blends with EVA-MA, EBA and EAO: (a) iPP, (b) IPC1 and (c) IPC2.

Table 3.1 Calorimetric properties of PP blends with EVA-MA, EBA and EAO

Sample	T_c (°C)	ΔH_c (J/g)	T_m (°C)	ΔH_f (J/g)	X_c (%)
*i*PP	128	99	169	91	44
*i*PP+EVA-MA	128	81	168	81	49
*i*PP+EBA	128	81	168	79	44
*i*PP+EAO	128	83	168	79	47
IPC1	131	84	167	82	40
IPC1+EVA-MA	128	65	166	62	37
IPC1+EBA	129	67	166	67	40
IPC1+EAO	130	66	167	62	37
IPC2	129	74	167	72	35
IPC2+EVA-MA	127	58	166	56	34
IPC2+EBA	127	58	165	53	32
IPC2+EAO	128	59	166	57	34

on the crystal formation of IPC1 in the blend. The incorporation of copolymers was also observed to have insignificant effect on both T_m and T_c of IPC2. However, the addition of EBA was observed to result in a decrease in % crystallinity from 35 to 32%, indicating hindrance to the crystal formation of IPC2 phase in the presence of EBA. Though the melting and crystallization thermograms did not exhibit significant individual transitions owing to the copolymer phase in the blends, however, weak signals corresponding to immiscible copolymer phase were nonetheless still observed in some cases. For instance, EVA-MA was observed to exhibit weak transitions (both melting and crystallization) in the blends, especially IPC1 and IPC2, indicating some degree of immiscibility with the PP matrices. Similarly, EBA was also observed to exhibit its melting transition in the blend with IPC1 and IPC2, as well as EAO in blend with iPP. In order to gain further insights into phase miscibility of blends, the rheological behavior of the blends was also analyzed using criteria reported for compatibility of polymer blends by Han and Chuang [42,43], as shown in Figures 3.6(a,c,e). On plotting G' vs. G", such analysis generates composition independent correlation for compatible blends, whereas the correlation is composition dependent for incompatible blends. As also observed in the DSC analysis, different extents of deviation from the common curve was exhibited by EVA-MA in the blends, especially at lower modulus values (or lower angular frequency). Similarly, EBA was also observed to exhibit significant deviation in the blend with IPC2. Figures 3.6(b,d,f) also demonstrate the

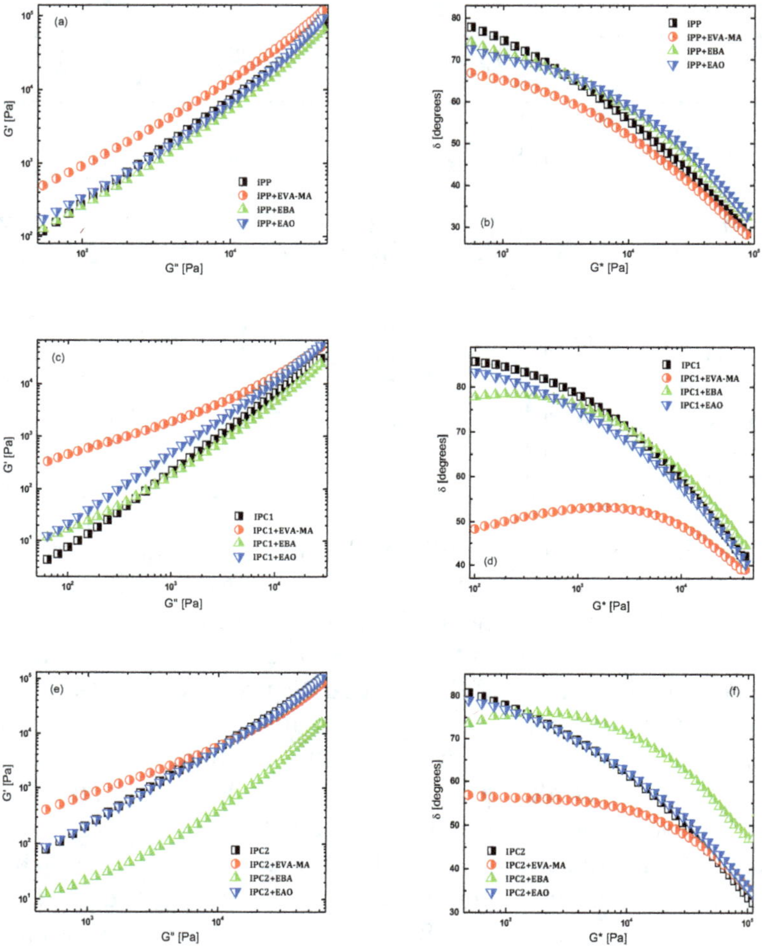

Figure 3.6 (a,c,e) Han-Chuang (G' vs. G") plots and (b,d,f) van Gurp plots of copolymer bends with iPP, IPC1 and IPC2 respectively.

analysis of the rheological data based on van Gurp plots representing the relationship between complex modulus (G*) and phase angle delta (d) [44,45]. The van Gurp plots confirmed the findings from Han and Chuang analysis. In the case of EVA-MA containing blends, the time-temperature superposition principle was observed to have deviations (indicated by the deviation from the common curve), especially at lower frequency (or higher delta value), thus, indicating different degrees of phase immiscibility. Slight deviations at lower frequency were also observed for the iPP/EAO, IPC1/EBA and

IPC2/EBA blends, thus, also confirming the earlier findings from DSC. However, it has to be noted that overall extent of phase miscibility was still high as the individual transitions of the copolymers in the DSC thermograms of the blends were insignificant and did not signify any phase separated morphology.

3.3.2 Thermal Properties of PP/Copolymer Blends

The thermal degradation behavior of pure polymers and blends is presented in Figure 3.7. Table 3.2 also reports the initial/onset and maximum degradation temperatures ($T_{10\%}$ and $T_{50\%}$) of the materials.

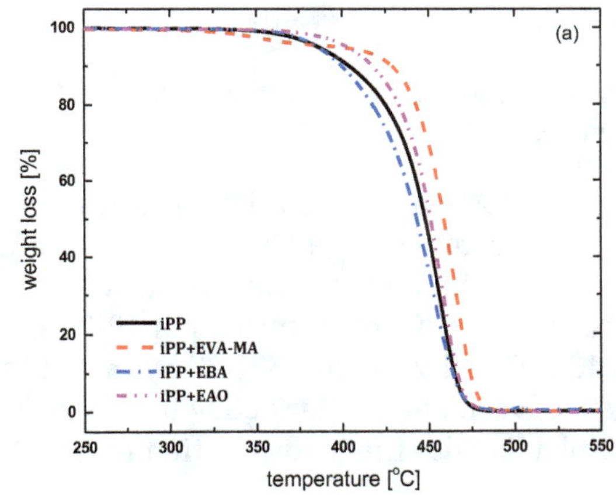

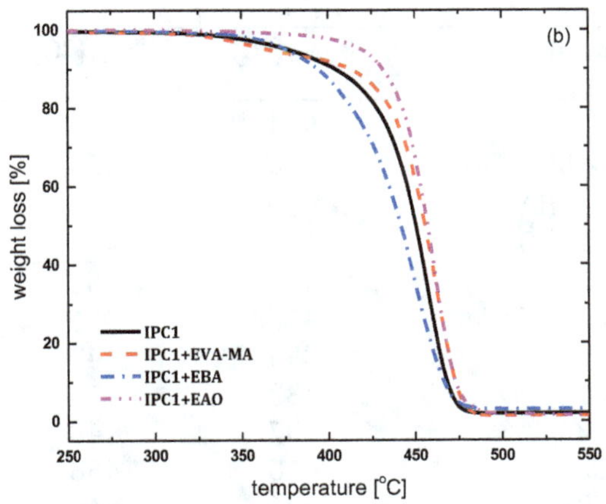

Figure 3.7 continued on next page

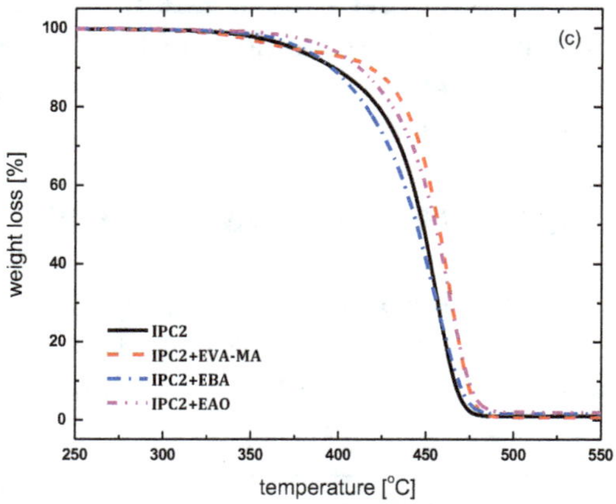

Figure 3.7 TGA thermograms of the PP blends with EVA-MA, EBA and EAO: (a) iPP, (b) IPC1 and (c) IPC2.

The addition of both EVA-MA and EAO significantly increased the onset degradation temperature in all PP grades. For example, the addition of EVA-MA and EAO into iPP resulted in an increase of $T_{10\%}$ from 404 °C to 428 °C and 418 °C respectively. For IPC1 blend, the $T_{10\%}$ increased from 404 °C to 415 °C and 433 °C by the addition of EVA-MA and EAO respectively. The similar behavior was also observed in the case of IPC2 blends with the incorporation of EVA-MA and EAO

Table 3.2 Thermal degradation performance of PP blends with EVA-MA, EBA and EAO

Sample	**$T_{10\%}$ (°C)**	**$T_{50\%}$ (°C)**
*i*PP	404	448
*i*PP+EVA-MA	428	459
*i*PP+EBA	400	443
*i*PP+EAO	418	451
IPC1	404	451
IPC1+EVA-MA	415	456
IPC1+EBA	395	442
IPC1+EAO	433	458
IPC2	397	448
IPC2+EVA-MA	418	458
IPC2+EBA	396	446
IPC2+EAO	412	455

(Table 3.2). Similar behavior was also observed for $T_{50\%}$ of all PP/copolymer blends with EVA-MA and EAO. The enhancement of thermal stability in the blends with EVA-MA and EAO exhibited that the earlier observed small extent of immiscibility did not hinder the blend performance. It is also interesting to note the enhancement of thermal stability on the addition of copolymers with much lower peak melting points. In addition, such an observation also indicates that efficient heat transfer took place from the PP chains to the copolymer phase, which would not have resulted in the absence of good degree of interfacial mixing between the PP and copolymer phases, even though molecular mixing of the phases did not take place. On the other hand, the addition of EBA into the iPP, and IPC1 exhibited a decrease in $T_{10\%}$ and $T_{50\%}$ of the PP blends by varying amounts. Probably, the faster degradation of the EBA copolymer resulted in the generation of reactive species (e.g. free radicals), which initiated the early degradation of the PP phase.

3.3.3 Diffraction Analysis of PP/Copolymer Blends

Wide angle diffraction patterns of PP and PP/copolymer blends are demonstrated in Figure 3.8. As evident in Figure 3.8(a), pure iPP exhibited characteristic α crystalline peaks (110, 040, 130, 111, 041, 160 and 020 planes) with the corresponding 2θ values of 14.58°, 17.30°, 19.02°, 21.50°, 22.23°, 25.82° and 29.03° respectively. In addition, iPP/copolymer blends also exhibited α crystalline peaks of iPP at the same positions. This indicated that the addition of copolymers to iPP did not alter the α crystalline nature of the PP. Moreover, the incorporation of copolymers created an additional β crystalline peak (300) at 16.5°. This confirmed the β nucleating effect of copolymers on the iPP matrix. Using Bragg's relationship, the inter-planar distances corresponding to α diffraction peaks were calculated as 6.16, 5.15, 4.69, 4.13, 4.00, 3.45 and 3.07 Å. The inter-planar distance corresponding to additional β crystal in the blends was calculated to be 5.36 Å. In the diffraction patterns of IPC1 and IPC1/copolymer blends in Figure 3.8(b), pure IPC1 exhibited characteristic peaks of α crystallites of PP (110, 040, 130, 160 and 020 planes), along with the diffraction peak of PE α crystalline phase (001) at corresponding 2θ values of 14.67°, 17.60°, 19.19°, 26.11°, 29.69°, and 22.33° respectively. In the IPC1/copolymer blends, same diffraction peaks of IPC1 were also observed, however, these diffraction peaks were observed to be shifted towards lower angles of 14.46°, 17.30°, 19.08°, 25.89°, 29.15°

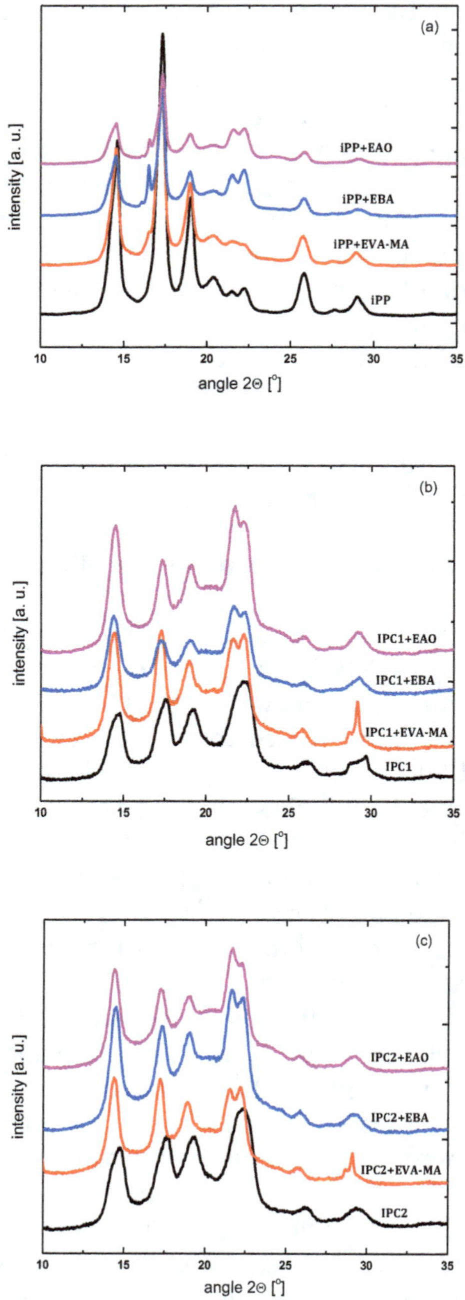

Figure 3.8 X-ray diffractograms of the PP blends with EVA-MA, EBA and EAO: (a) iPP, (b) IPC1 and (c) IPC2.

and 22.24° respectively. In addition, the diffraction peak of PP β crystallite (301) was observed at 2θ value corresponding to 21.70°. The inter-planar distances of the lattice planes d(110), d(040), d(130), d(001), d(160), d(020) and d(001), thus, shifted from 6.03, 5.04, 4.62, 3.41, 3.01 and 3.98 Å to 6.12, 5.12, 4.65, 3.44, 3.06 and 3.99 Å respectively. Similarly, as evident in Figure 3.8(c), pure IPC2 also exhibited characteristic peaks at 2θ values 14.77°, 17.72°, 19.29°, 26.22°, 29.36° and 22.33° corresponding to lattice planes of α crystallites of PP (110, 040, 130, 001, 160 and 020) and α crystallites of PE (001). Addition of copolymers into IPC2 also generated the same diffraction peaks of IPC2 but were shifted to lower angles of 14.46°, 17.30°, 18.98°, 25.89°, 29.15° and 22.24° respectively. This indicated the shift of inter-planar distances of lattice planes d(110), d(040), d(130), d(001), d(160), d(020) and d(001) from 5.99, 5.00, 4.60, 3.39, 3.04 and 3.98 Å to 6.12, 5.12, 4.67, 3.44, 3.06 and 3.99 Å respectively. In addition, similar to IPC1, the diffraction peak of PP β crystallite (301) was also observed in the IPC2 blends. These findings confirmed that the addition of copolymers to PP did not alter the α crystalline phase of PP, but also added additional β crystalline phase in the polymer. However, as also observed in DSC, the degree of crystallinity was still affected due to the varying degree of hindrance to the crystallization of PP chains in the presence of different copolymers.

3.3.4 Rheological Properties of PP/Copolymer Blends

Processability of the polymer blends is an important parameter that determines their end use applications. In order to study the processing behavior of PP/copolymer blends, melt rheological studies were carried out on shear rheometer at 180 °C. The complex viscosity (η*) of the polymer blends in comparison with pure polymers as a function of angular frequency is demonstrated in Figure 3.9. Among the three PPs, iPP exhibited a pseudo-plastic shear thinning behavior (decrease of η* with increasing ω) over the whole frequency range, whereas both IPC1 and IPC2 had a frequency independent Newtonian flow like behavior at lower angular frequency range (< 1 rad/s), followed by a shear thinning behavior. Blends of iPP with copolymers also exhibited a shear thinning behavior. The iPP/EVA-MA blend was also observed to have higher η* value when compared with pure iPP. The addition of EBA reduced the η* of iPP, whereas EAO copolymer generated almost similar η* as iPP. Thus, different copolymers had different extents of effect on the complex viscosity of the polymer.

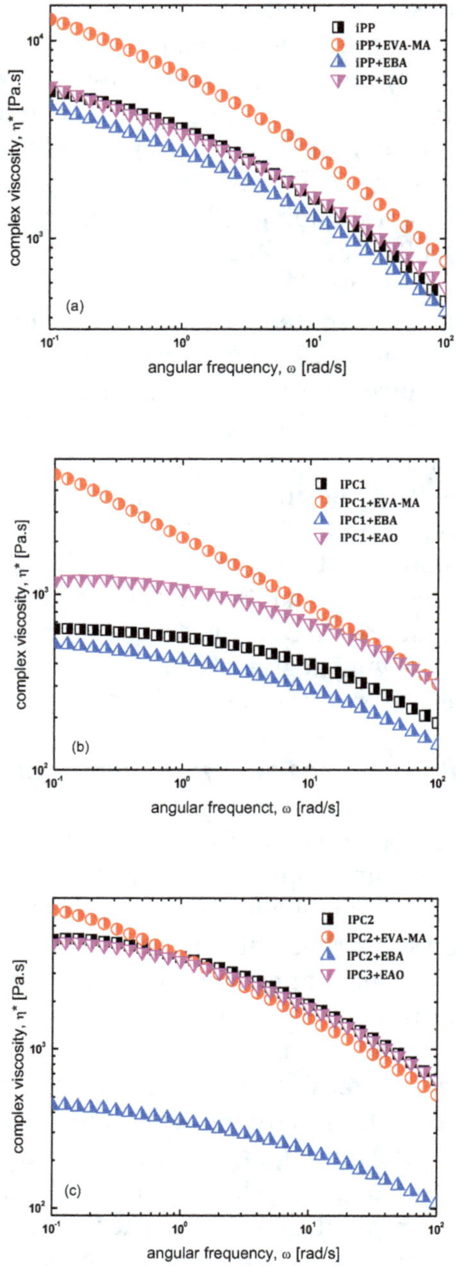

Figure 3.9 Log-log plots of the complex viscosity of the PP blends with EVA-MA, EBA and EAO as a function of angular frequency: (a) iPP, (b) IPC1 and (c) IPC2.

Similarly, in the case of IPC1/copolymer blends, the addition of both EVA-MA and EAO enhanced the η* of IPC1, whereas the IPC1/EBA blend exhibited lower η* values as compared with IPC1. Furthermore, both IPC1/EAO and IPC1/EBA blends exhibited a frequency independent Newtonian flow like behavior at angular frequencies lower than 1 rad/s, followed by a shear thinning behavior. On the other hand, the IPC1/EVA-MA blend had a shear thinning behavior for the entire frequency range. For the IPC2/copolymer blends, IPC2/EVA-MA blend exhibited higher value of η* at lower frequencies (up to 1 rad/s) as compared to IPC2. Afterwards, it exhibited a decrease in η* as compared to IPC2. Also, the blend followed a shear thinning behavior over the whole frequency range. On the other hand, the addition of EBA reduced the η* of IPC2, whereas the EAO had no effect on the polymer viscosity. Furthermore, both IPC2/EAO and IPC2/EBA blends exhibited a frequency independent behavior at lower frequencies (less than 1 rad/s), followed by a shear thinning behavior. The existence of higher η* resulted from the restricted chain relaxation and relative motion of molecular chains of PPs in the blends at the molten state due to the copolymer addition, which also indicated good degree of phase intermixing [46]. Probably, the larger domains of EVA-MA in the polymer matrices due to the earlier observed phase immiscibility would have acted as dispersed phase, thus, resulting in enhanced viscosity.

Figure 3.10 presents storage modulus of the polymers and blends as a function of angular frequency. G' of the pure polymers and their blends increased with angular frequency. Among the three PP grades, iPP exhibited the highest storage modulus. For instance, it had the storage modulus of 117 Pa at the angular frequency of 0.1 rad/s. The incorporation of EVA-MA into the iPP resulted in an increased value of G' over the entire angular frequency range. The G' of iPP/EAO blend was also observed to be greater than pure iPP at lower frequencies, followed by almost similar behavior at higher frequencies. On the other hand, the iPP/EBA blend exhibited slightly lower G' as compared to pure iPP. For the IPC1/copolymer blends, the addition of both EVA-MA and EAO exhibited greater storage modulus over the whole ω range. For instance, the IPC1/EVA-MA blend had an optimum G' value of 331 Pa at ω of 0.1 rad/s. At lower ω, G' of IPC1/EBA blend was also greater than that of pure IPC1, it had lower G' at higher ω (> than 1 rad/s) values. Similarly, in the IPC2/copolymer blends, EVA-MA resulted in greater G' at lower ω (up to 10 rad/s), followed by an almost similar G' as IPC2. The addition of EBA strongly reduced

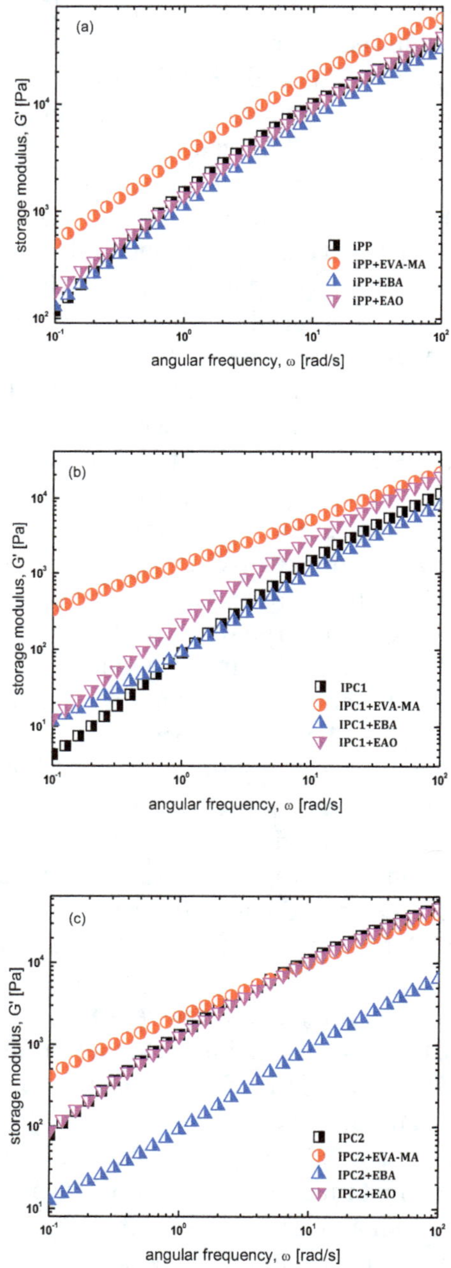

Figure 3.10 Log-log plots of the storage modulus of the PP blends with EVA-MA, EBA and EAO as a function of angular frequency: (a) iPP, (b) IPC1 and (c) IPC2.

the modulus of IPC2 throughout the ω range. However, the EAO copolymer did not affect the G′ of IPC2 significantly as the copolymer exhibited almost similar behavior as IPC2. The higher storage modulus of PP/copolymer blends indicated the delayed relaxation of the chains during melt shearing. The copolymers enhanced the solid elastic nature of the blends, especially EVA-MA, by forming dispersed phase in the polymer, which prolonged the relaxation of the polymer, thereby resulting in higher storage modulus which also supported the enhanced impact strength of IPC/EVA-MA blends [47]. Moreover, the blends exhibiting lower value of storage modulus as compared to the pure polymers would have undergone backbone chain entanglements due to higher degree of phase intermixing (as also observed in the phase miscibility studies), thus, resulting in some degree of matrix plasticization due to lower molecular weight of the copolymers [48].

The loss modulus of the PP grades and PP/copolymer blends as a function of angular frequency is presented in Figure 3.11. Similar to the storage modulus, the loss modulus was also observed to increase with angular frequency for all materials. Among the three PP grades, IPC1 was observed to exhibit the lowest loss modulus. In the case of iPP/copolymer blends, iPP/EVA-MA blend had greater loss modulus throughout the ω range, when compared with pure iPP. The addition of EBA lowered the loss modulus of iPP, whereas the PP/EAO blend showed almost similar behavior up to 10 rad/s, followed by enhanced loss modulus, when compared with iPP. In the case of IPC1/copolymer blends, both EVA-MA and EAO were exhibited higher storage modulus than the pure polymer over the whole ω range. On the other hand, the blend of IPC1 with EBA exhibited lower values of loss modulus for the entire ω range. Analyzing IPC2/copolymer blends, the addition of EVA-MA was observed to generate higher loss modulus value only up to 1 rad/s, followed by a lower modulus value over the entire ω range. The incorporation of EAO did not show significant effect on the loss modulus of IPC2, whereas the incorporation of EBA considerably reduced the loss modulus value of IPC2 over the whole ω range, similar to the storage modulus. Comparing with the storage modulus of the blends, the loss modulus was observed to be higher at lower frequencies indicating dominant viscous nature, whereas the viscous dominance was observed to decrease at higher angular frequency values.

The damping properties of the materials were studied from the variation of tan δ with ω, as presented in Figure 3.12. The tan δ values

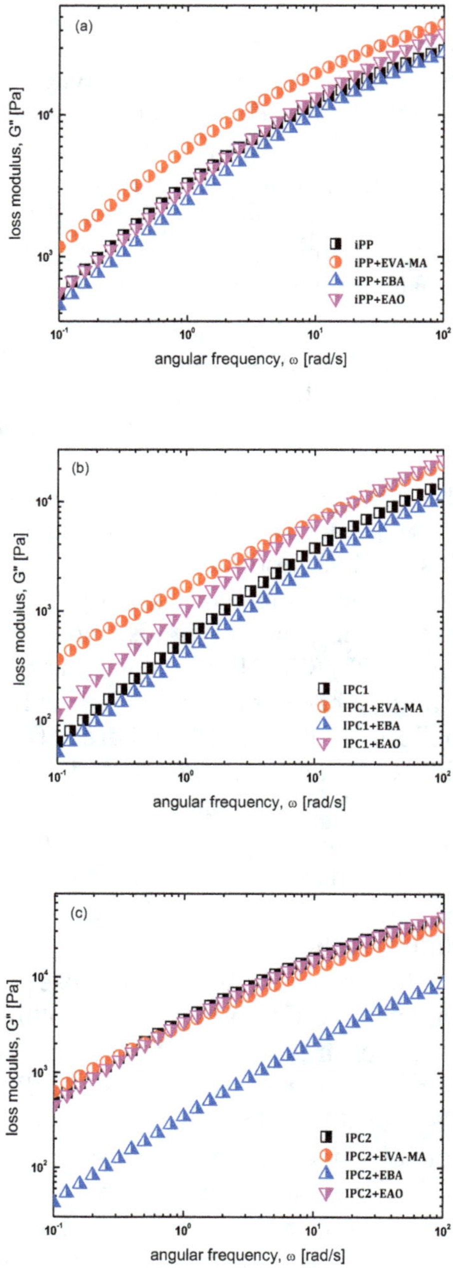

Figure 3.11 Log-log plots of the loss modulus of the PP blends with EVA-MA, EBA and EAO as a function of angular frequency: (a) iPP, (b) IPC1 and (c) IPC2.

of the PP polymers decreased with increasing ω indicating increasing dominance of elastic nature in the blends. In iPP/copolymer blends, the blends exhibited lower tan δ as compared to iPP, thus, indicating enhanced elastic effect in the blends due to the addition of copolymers. The iPP/EVA-MA blend exhibited lowest tan δ due to enhanced elastic nature due to the same reasons mentioned earlier for the enhanced viscosity and storage modulus. Furthermore, similar to the pure polymer, the blends exhibited decrease in tan δ over the whole ω range. Similar trend was also observed for the IPC1/copolymer blends as large reduction in the viscous nature was observed on the addition of the copolymers. Interestingly, IPC1/EVA-MA blend exhibited the lowest value of tan δ of 1 which was also observed to remain constant throughout the entire ω range. The tan δ of IPC2/copolymer blends was also observed to exhibit decreasing tendency over the whole angular frequency range, except the blend with EBA. Similar to other blend systems, the blend with EVA-MA also exhibited significant reduction in the viscous nature of the polymer. Thus, in addition to the thermal behavior, the addition of EVA-MA and EAO specifically enhanced the rheological performance of the blends. EBA, on the other hand, exhibited less optimal response due to the reduction in the overall thermal and rheological response of the materials. Furthermore, it can also be suggested that the overall blend performance was affected not only by the degree of phase intermixing, but also by possible matrix plasticization by the copolymer as well as effect on the polymer crystallinity and crystal structure, all of which originated form the molecular architecture of the copolymers.

3.3.5 Mechanical Properties of PP/Copolymer Blends

The tensile properties of the polymers and copolymer blends are presented in Table 3.3. The iPP/copolymer blends exhibited lower values of tensile strength as compared to pure iPP (39 MPa). In the case of IPC1/copolymer blends, the addition of copolymers did not significantly influence the tensile strength of IPC1 (22 MPa). On the other hand, in the case of IPC2/copolymer blends, the incorporation of EAO improved the tensile strength of IPC2 from 21 MPa to 25 MPa, whereas the addition of EVA-MA and EBA reduced the tensile strength value to 15 MPa and 16 MPa respectively. Thus, different copolymers were observed to impact the tensile strength of the polypropylene grades to different extents owing to the reasons mentioned earlier.

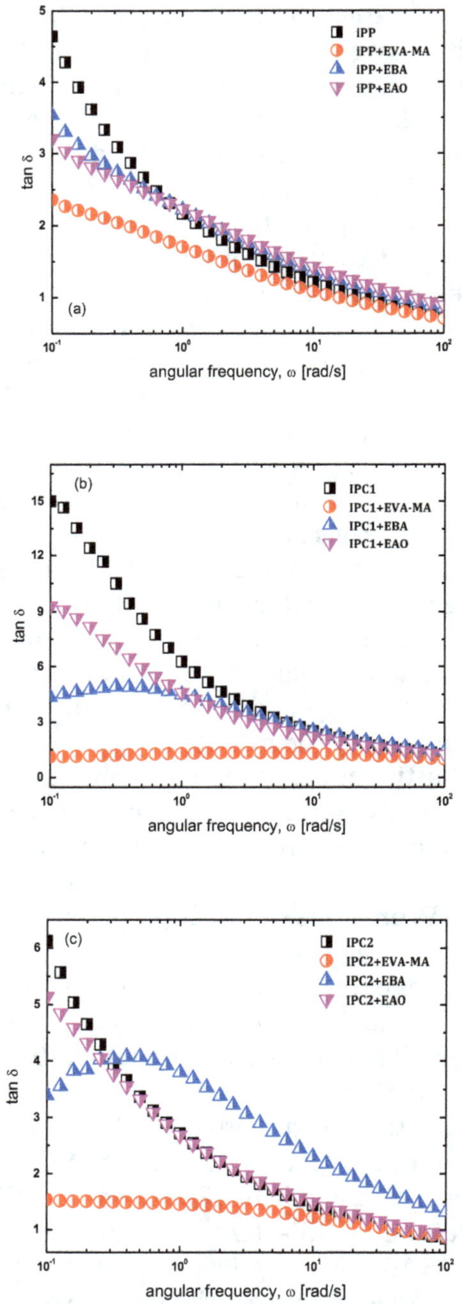

Figure 3.12 Log-log plots of the tan δ of the PP blends with EVA-MA, EBA and EAO as a function of angular frequency: (a) iPP, (b) IPC1 and (c) IPC2.

Table 3.3 Mechanical properties of PP blends with EVA-MA, EBA and EAO

Sample	Impact strength (KJ/m^2)	Tensile strength (MPa)	Tensile modulus (MPa)	Extension at break (%)
*i*PP	7	39	1011	24
*i*PP+EVA-MA	42	30	1143	43
*i*PP+EBA	74	26	1036	299
*i*PP+EAO	96	30	1084	263
IPC1	53	22	1018	35
IPC1+EVA-MA	32	20	910	14
IPC1+EBA	52	20	900	10
IPC1+EAO	102	22	881	26
IPC2	103	21	864	299
IPC2+EVA-MA	87	15	692	50
IPC2+EBA	80	16	584	27
IPC2+EAO	87	25	835	508

Among the three PP grades, both iPP and IPC1 exhibited higher tensile modulus value of about 1000 MPa. In iPP/copolymer blends, the addition of copolymers improved the tensile modulus of iPP. Specifically, the blend of iPP/EVA-MA exhibited an optimum value of 1143 MPa. As mentioned earlier, probably the EVA-MA phase acted as dispersed phase in PP which led to good degree of stress transfer from the PP chains to the EVA-MA, without extensive matrix plasticization. However, in the case of impact PP copolymers (IPC1 and IPC2), the addition of copolymers resulted in lower modulus value as compared to pure IPCs. It was also noted that the presence of copolymers in iPP resulted in a significant improvement in elongation at break, especially with EBA and EAO. The notion of EVA-MA acting as dispersed filler in iPP was also confirmed as the elongation of the blend with EVA-MA was much less enhanced as compared to blends with EBA and EAO. Overall, all the copolymers resulted in different degrees of simultaneous enhancement of stiffness and toughness of iPP. In IPC1/copolymer blends, a lower elongation at break was observed as compared with pristine IPC1. It indicated that the addition of copolymers resulted in the loss of both toughness and stiffness of IPC1. Probably, a lower weight fraction of copolymers is required to be studied in order to gain further insights into the effect of copolymers on the performance of IPC1. For IPC2/copolymer blends, the addition of EAO increased its elongation at break from 299 to 508% (~70% improvement), whereas the EVA-MA and EBA copolymers led to the reduction of the elongation at break of IPC2. As the modulus of the blend with EAO was only marginally lower as compared to pure IPC2, the observed increment in the stiffness indicated good balance of stiffness and toughness. The effect of the copolymers on the toughness of iPP and IPCs was also studied using Izod analysis and the impact strength of the materials is presented in Table 3.3. Similar to elongation at break, the incorporation of copolymers in iPP significantly increased the impact strength. For instance, incorporation of EAO increased the impact strength of iPP from 7 to 96 KJ/m^2. Furthermore, EBA and EVA-MA improved the impact strength of iPP from 7 to 74 and 42 KJ/m^2 respectively. The presence of EVA-MA as dispersed phase, thus, resulted in the least extent of impact toughness enhancement. For IPC1/copolymer blends, the addition of EAO further enhanced the impact strength of IPC1 (53 KJ/m^2) to 102 KJ/m^2, which is about 2 folds increment, even though the elongation of the break was lower than the pure polymer. On the other hand, the addition of EBA had no adverse effect on the impact strength. Similar to iPP blends, the incorporation of EVA-MA into IPC1 resulted in the least impact strength among blends, which was even lower than pure IPC1. Thus, the larger domain size of EVA-MA resulted in lowering of the impact strength. IPC2 had a higher impact strength as compared to IPC1. The incorporation of copolymers in IPC2 resulted in a decrease of the impact strength, however, the impact strength of the blends was still high. Overall, the iPP blends exhibited more optimal performance as compared to IPC blends, though most of the IPC bends had still higher impact strength than the iPP blends. As mentioned earlier, the observed behavior of the blends resulted from the combination of many factors including phase intermixing and domain sizes, plasticization, crystallinity, stress absorption by the copolymer phase, inhibition of crack growth, etc.

3.4 Conclusions

In the present work, the effect of 20 wt% copolymers such as maleic anhydride grafted ethylene vinyl acetate, ethylene butyl acrylate and ethylene alpha olefin on the thermal, mechanical and rheological properties of two impact polypropylene copolymers and pristine isotactic polypropylene was studied. The blends were obtained by industrially accepted melt blending method using twin screw extruder. The tensile properties, impact strength, thermal performance, melt rheological and diffraction behaviors of the PP/copolymer blends were observed to be strongly dependent on the nature of PP and copolymers. Combination of many factors including phase intermixing and domain sizes, plasticization, crystallinity, stress absorption by the copolymer phase, inhibition of crack growth, etc., collectively impacted the blend performance. The addition of both EVA-MA and EAO significantly improved the initial thermal degradation temperature ($T_{10\%}$) of all PP grades. Degree of crystallinity of iPP was also enhanced on the addition of the copolymers. Due to its presence as larger dispersed domains in the polymer matrix, the blend with EVA-MA had highest storage modulus and complex viscosity. On the other hand, EBA containing blends exhibited the lowest rheological properties. XRD analysis confirmed the formation of β crystallites in the PP/copolymer blends. Simultaneous improvements in toughness and stiffness of iPP/copolymer blends was also observed. Specifically, the tensile modulus of iPP with EVA-MA was improved around 13%, whereas the increment was 7% and 3% for corresponding EAO and EBA based iPP blends. The addition of EAO improved the impact strength of iPP by about 14 times. Similarly, the impact strength of the other blends as well as elongation at break were observed to be higher than the pure polymer. In the case of IPC/copolymer blends, the mechanical properties of the blends varied with nature of both IPC and copolymers. IPC1/EAO blend exhibited about two folds enhancement in the impact strength without affecting the tensile strength. IPC2 blends though had reduced impact strength as compared to the pure polymer, however, the impact strength of the blends was still high. IPC2/EAO blend also retained the modulus of the polymer (835 MPa for the blend as compared to 864 MPS for IPC2) and enhanced the elongation by more than 70%. Thus, many of the developed blends, especially with iPP, exhibited optimum enhancement in the performance as well alleviation of the drawbacks of PP. These materials are potential candidates for use in the tailored engineering applications requiring durability and stability. The blends with other compositions should also be analyzed to gain more insights about the impact of copolymers as well as to identify optimal blend compositions to achieve efficient material design.

References

1. Jancar, J. (1999) *Mineral Fillers in Thermoplastics I: Raw Materials and Processing*, Springer-Verlag, Germany.

2. Mittal, V. (2012) *Functional Polymer Blends: Synthesis, Properties and Performance*, CRC Press, USA.
3. Hippi, U. (2008) *Polyolefin Composites*, John Wiley & Sons, USA.
4. Jyotishkumar, P., Thomas, S., and Grohens, Y. (2015) Polymer blends: State of the art, new challenges, and opportunities. In: *Characterization of Polymer Blends: Miscibility, Morphology and Interfaces*, Thomas, S., Grohens, Y., and Jyotishkumar, P. (eds.), 1st edition, pp. 1-6.
5. Ibrahim, B. A., and Kadum, K. M. (2010) Influence of polymer blending on mechanical and thermal properties. *Modern Applied Science*, **4**, 157-161.
6. Wang, D., Li, Y., Xie, X.-M., and Guo, B.-H. (2011) Compatibilization and morphology development of immiscible ternary polymer blends. *Polymer*, **52**, 191-200.
7. Zhou, L., Wang, X., Lin, Y., Yang, J., and Wu, Q. (2003) Comparison of the toughening mechanisms of poly(vinyl chloride)/chlorinated polyethylene and poly(vinyl chloride)/acrylonitrile-butadiene-styrene copolymer blends. *Journal of Applied Polymer Science*, **90**, 916-924.
8. Mittal, V., Luckachan, G. E., and Matsko, N. B. (2014) PE/chlorinated-PE blends and PE/chlorinated-PE/graphene oxide nanocomposites: Morphology, phase miscibility, and interfacial interactions. *Macromolecular Chemistry and Physics*, **215**, 255-268.
9. Chaudhry, A. U., and Mittal, V. (2014) Blends of high-density polyethylene with chlorinated polyethylene: Morphology, thermal, rheological, and mechanical properties. *Polymer Engineering and Science*, **54**(1), 85-95.
10. Mittal, V. Suria, M., and Matsko, N. (2018) Polyethylene-copolymer blends: Morphology and properties. In: *Functional Polymer Blends and Nanocomposites*, Mittal, V. (ed.), Central West Publishing, Australia, pp. 35-69.
11. Keskin, R., and Adanur, S. (2011) Improving toughness of polypropylene with thermoplastic elastomers in injection molding. *Polymer - Plastics Technology and Engineering*, **50**, 20-28.
12. Dikobe, D., and Luyt, A. (2009) Morphology and properties of polypropylene/ethylene vinyl acetate copolymer/wood powder blend composites. *Express Polymer Letters*, **3**(3), 190-199.
13. Parameswaranpillai, J., Joseph, G., Shinu, K., Sreejesh, P., Jose, S., Salim, N. V., and Hameed, N. (2015) The role of SEBS in tailoring the interface between the polymer matrix and exfoliated graphene nanoplatelets in hybrid composites. *Materials Chemistry and Physics*, **163**, 182-189.
14. Tan, H., Li, L., Chen, Z., Song, Y., and Zheng, Q. (2005) Phase morphology and impact toughness of impact polypropylene copolymer. *Polymer*, **46**, 3522-3527.

15. Mittal, V., Pannirselvam, M., and Griffin, G. (2018) Polypropylene/polystyrene blend nanocomposites with graphene. In: *Functional Polymer Blends and Nanocomposites*, Mittal, V. (ed.), Central West Publishing, Australia, pp. 137-164.
16. Liu, G., and Qiu, G. (2013) Study on the mechanical and morphological properties of toughened polypropylene blends for automobile bumpers. *Polymer Bulletin*, **70**, 849-857.
17. Chen, F., Qiu, B., Wang, B., Shangguan, Y., and Zheng, Q. (2015) Balanced toughening and strengthening of ethylene-propylene rubber toughened isotactic polypropylene using a poly (styrene-b-ethylene-propylene) diblock copolymer. *RSC Advances*, **5**, 20831-20837.
18. Zebarjad, S., Lazzeri, A., Bagheri, R., Reihani, S. S., and Frounchi, M. (2003) Fracture mechanism under dynamic loading of elastomer-modified polypropylene. *Materials Letters*, **57**, 2733-2741.
19. Zebarjad, S., Bagheri, R., Reihani, S., and Lazzeri, A. (2003) Deformation, yield and fracture of elastomer-modified polypropylene. *Journal of Applied Polymer Science*, **90**, 3767-3779.
20. Choudhary, V., Varma, H., and Varma, I. (1991) Polyolefin blends: Effect of EPDM rubber on crystallization, morphology and mechanical properties of polypropylene/EPDM blends. 1. *Polymer*, **32**, 2534-2540.
21. Da Costa, H. M., Ramos, V. D., and Rocha, M. C. (2006) Analysis of thermal properties and impact strength of PP/SRT, PP/EPDM and PP/SRT/EPDM mixtures in single screw extruder. *Polymer Testing*, **25**, 498-503.
22. Hu, J., Zhang, Y., and Zhou, X. (2011) Study on modified recycling PP by thermoplastic elastomer SBS. *China Plastics Industry*, **39**, 37-40.
23. Das, V., Gautam, S. S., and Pandey, A. K. (2011) Effect of SBS content on low temperature impact strength, morphology and rheology of PP-cp/SBS blends. *Polymer - Plastics Technology and Engineering*, **50**, 825-832.
24. Bassani, A., and Pessan, L. (2002) Toughening of polypropylene with styrene/ethylene-butylene/styrene tri-block copolymer: Effects of reactive and nonreactive compatibilization. *Journal of Applied Polymer Science*, **86**, 3466-3479.
25. Balkan, O., Demirer, H., and Kayalı, E. S. (2011) Effects of deformation rates on mechanical properties of PP/SEBS blends. *Journal of Achievements in Materials and Manufacturing Engineering*, **47**, 26-33.
26. Zhu, L., Fan, H.-N., Yang, Z.-Q., and Xu, X.-H. (2010) Evaluation of phase morphology, rheological, and mechanical properties based on polypropylene toughened with poly (ethylene-co-octene). *Polymer - Plastics Technology and Engineering*, **49**, 208-217.
27. Yang, J., Zhang, Y., and Zhang, Y. (2003) Brittle-ductile transition of

PP/POE blends in both impact and high speed tensile tests. *Polymer*, **44**, 5047-5052.

28. Zuiderduin, W., Westzaan, C., Huetink, J., and Gaymans, R. (2003) Toughening of polypropylene with calcium carbonate particles. *Polymer*, **44**, 261-275.
29. Yordem, O. S., Simanke, A. G., and Lesser, A. J. (2011) Reinforcing and toughening of polypropylene with self-assembled low molar mass additives. *Polymer Engineering & Science*, **51**, 550-558.
30. Wu, C. L., Zhang, M. Q., Rong, M. Z., and Friedrich, K. (2005) Silica nanoparticles filled polypropylene: effects of particle surface treatment, matrix ductility and particle species on mechanical performance of the composites. *Composites Science and Technology*, **65**, 635-645.
31. Liang, J. Z. (2002) Toughening and reinforcing in rigid inorganic particulate filled poly (propylene): a review. *Journal of Applied Polymer Science*, **83**, 1547-1555.
32. Li, R., Zhang, X., Zhao, Y., Hu, X., Zhao, X., and Wang, D. (2009) New polypropylene blends toughened by polypropylene/poly (ethylene-co-propylene) in-reactor alloy: compositional and morphological influence on mechanical properties. *Polymer*, **50**, 5124-5133.
33. Mirabella, F. M. (1993) Impact polypropylene copolymers: fractionation and structural characterization. *Polymer*, **34**, 1729-1735.
34. Gahleitner, M., Tranninger, C., and Doshev, P. (2013) Heterophasic copolymers of polypropylene: Development, design principles, and future challenges. *Journal of Applied Polymer Science*, **130**, 3028-3037.
35. van Reenen, A. J., and Basson, N. C. (2012) Molecular composition and properties of impact propylene copolymers. *Express Polymer Letters*, **6**(5), 427-436.
36. Li, Y., Xu, J.-T., Dong, Q., Fu, Z.-S., and Fan, Z.-Q. (2009) Morphology of polypropylene/poly (ethylene-co-propylene) in-reactor alloys prepared by multi-stage sequential polymerization and two-stage polymerization. *Polymer*, **50**, 5134-5141.
37. Zhang, C., Shangguan, Y., Chen, R., Wu, Y., Chen, F., Zheng, Q., and Hu, G. (2010) Morphology, microstructure and compatibility of impact polypropylene copolymer. *Polymer*, **51**, 4969-4977.
38. Doshev, P., Lach, R., Lohse, G., Heuvelsland, A., Grellmann, W., and Radusch, H.-J. (2005) Fracture characteristics and deformation behavior of heterophasic ethylene–propylene copolymers as a function of the dispersed phase composition. *Polymer*, **46**, 9411-9422.
39. Starke, J., Michler, G., Grellmann, W., Seidler, S., Gahleitner, M., Fiebig, J., and Nezbedova, E. (1998) Fracture toughness of polypropylene copolymers: Influence of interparticle distance and temperature. *Polymer*, **39**, 75-82.
40. Bu, H., Qiu, W., Tan, Z., Li, Q., and Rong, J. (2014) Study on toughen-

ing of poly (4-methyl-1-pentene) with various thermoplastic elastomers. *Journal of Thermoplastic Composite Materials*, **28**(9), 1334-1342.

41. Liao, C. Z., and Tjong, S. C. (2011) Effects of carbon nanofibers on the fracture, mechanical, and thermal properties of PP/SEBS-g-MA blends. *Polymer Engineering & Science*, **51**, 948-958.
42. Chuang, H. K., and Han, C. D. (1984) Rheological behavior of polymer blends. *Journal of Applied Polymer Science*, **29**, 2205-2229.
43. Han, C. D., and Chuang, H. K. (1985) Criteria for rheological compatibility of polymer blends. *Journal of Applied Polymer Science*, **30**(11), 4431-4454.
44. Joshi, M., Butola, B. S., Simon, G., and Kukaleva, N. (2006) Rheological and viscoelastic behavior of HDPE/octamethyl-POSS nanocomposites. *Macromolecules*, **39**(5), 1839-1849.
45. van Gurp, M., and Palmen, J. (1998) Time-temperature superposition for polymeric blends. *Rheology Bulletin*, **67**(1), 5-8.
46. da Silva, A. L. N., Rocha, M. C., Coutinho, F. M., Bretas, R. E., and Farah, M. (2002) Evaluation of rheological and mechanical behavior of blends based on polypropylene and metallocene elastomers. *Polymer Testing*, **21**, 647-652.
47. Yan, X., Xu, X., Zhu, T., Zhang, C., Song, N., and Zhu, L. (2008) Phase morphological evolution and rheological properties of polypropylene/ethylene-octene copolymer blends. *Materials Science and Engineering A*, **476**, 120-125.
48. da Silva, A. L. N., Rocha, M. C., Coutinho, F. M., Bretas, R., and Scuracchio, C. (2000) Rheological and morphological properties of blends based on ethylene-octene copolymer and polypropylene. *Polymer Testing*, **19**, 363-371.

4

PP/EVA-MA Blend Nanocomposites with Graphene Nano-sheets

4.1 Introduction

Polypropylene (PP) is a stereo-regular polymer, with properties depending on crystalline structure and various tacticities. Moreover, PP possesses various beneficial properties like low density, high thermal resistance, resistance to chemical attack (insoluble in almost all solvents at room temperature), moisture resistance, high modulus, ease of processing and recyclability. Interestingly, PP manifests mechanical profile closely to the engineering thermoplastics. Owing to this extraordinary set of properties, PP is commonly used thermoplastic for structural applications [1]. Unfortunately, PP also suffers from drawbacks like low temperature brittleness and low impact strength, which limit its widespread application in the engineering field. In order to improve the impact strength, addition of elastomeric compounds has generally been employed [2-14], which leads to improved elongation, however, may negatively affect the modulus and crystallinity. On the other hand, incorporation of polymeric compatibilizers along with reinforcing fillers (like calcium carbonate, nano-clay, carbon nanotubes and graphene platelets, etc.) has been reported to achieve an optimal balance of modulus and toughness enhancement, though a negative impact on the elongation has been occasionally observed in the literature studies due to the compatibility issues [15-27]. Thus, the use of blend composites is a functional strategy to achieve a simultaneous enhancement in the stiffness and toughness of the materials. Also, in order to improve compatibility of the polymer and filler phases, approaches of reactive compatibilization and chemical modification of fillers also need to be effectively implemented.

Owing to its exceptional mechanical properties (Young's modulus 1 TPa) and high thermal conductivity (3000 $W.m^{-1}.K^{-1}$), use of graph-

Shital Pawar[a,], Vikas Mittal[a] and Nadejda Matsko[b]*
[a]Khalifa University of Science and Technology, Abu Dhabi, UAE; [b]Graz Centre of Electron Microscopy Graz, Austria
**Central address: University of Calgary, Calgary, Canada*

ene presents opportunities for significant reinforcement of polymers [28-32]. However, due to its nano-scale dimensions and polarity mismatch with the polymer, its dispersion in the polymer matrix is challenging. In order to improve dispersion and compatibility with matrix, various functional groups can be generated/attached on graphene sheets by oxidization of graphite platelets. These groups can further be used for reactive compatibilization with either polymer matrix directly or one of the components of the polymer blend (or compatibilizer), thus, leading to advanced nanocomposites with uniform filler dispersion and selective localization. For instance, Mittal *et al.* [33] reported nanocomposites of bio-polyamide (PA) with varying amounts of thermally reduced graphene, generated by melt mixing. As shown in Figure 4.1, free amide bonds in the amorphous phase of PA were observed to undergo chemical reaction with the surface groups of graphene, thus, resulting in better interfacial interactions. Due to this reactive compatibilization, polyamide nanocomposites

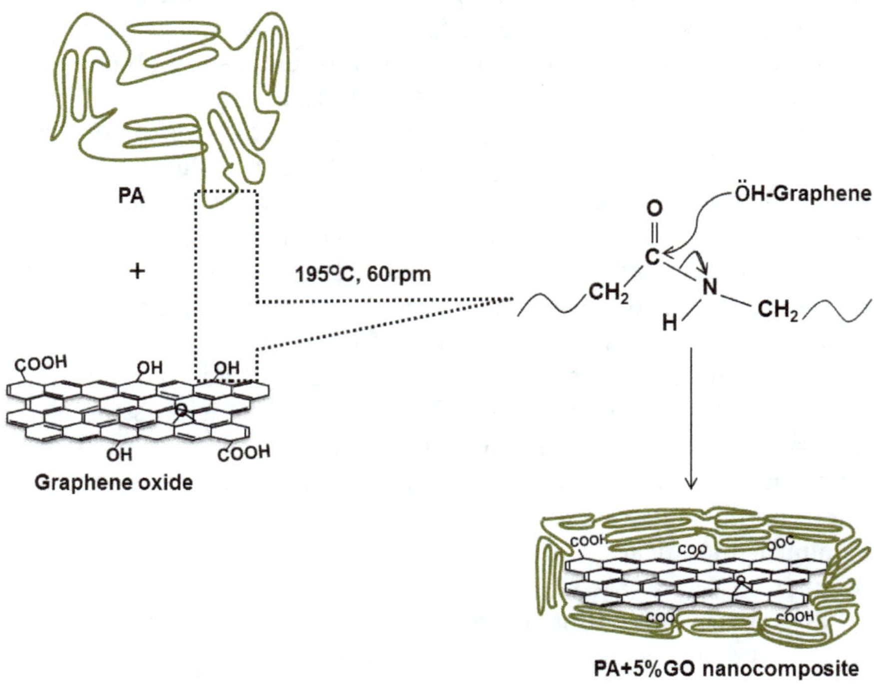

Figure 4.1 Representation of interaction of the polyamide chains with graphene surface. Reproduced from Reference 33 with permission from Elsevier.

had exfoliated morphology, and the tensile modulus of the nanocomposite with 5 wt% filler content was 1.8 times higher than the pure polymer. The polyamide nanocomposites were also observed to exhibit enhanced thermal stability owing to polymer immobilization on filler surface as well uniform filler dispersion in the polymer, as the peak degradation temperature in PA composite with 5 wt% graphene content was enhanced by 10 °C, and was also a function of the filler content. The findings, thus, confirmed the role of filler-matrix interactions and compatibilization in achieving filler dispersion and superior nanocomposite properties.

In another study reporting the compatibilization of polymer and filler through the chemical interaction of the functional groups present on the filler surface with the compatibilizer, Vengatesan *et al.* [34] developed nanocomposites using polypropylene-grafted-maleic anhydride (PP-g-MA) as compatibilizer with different weight ratios (0.5, 1.0, 1.5, 2.5, 5.0, and 10.0 wt%) of amine functionalized mullite fibers (AMUF) via solution blending method. Chemical grafting of AMUF with PP-g-MA resulted in enhanced filler dispersion in the polymer as well as effective polymer-filler interactions. Figure 4.2 shows the SEM images of the fracture surface of the PP matrix and PP-AMUF nanocomposites. In Figure 4.2(a) for pure PP, a smooth fracture surface was observed. The fracture surface in Figure 4.2(b) and (c) exhibited a plastic deformation during the Izod impact test. The voids and cracks were minimal in these 1 wt% and 5 wt% AMUF-filled PP nanocomposites. These results suggested that covalently bound AMUF along with its uniform dispersion provided dissipation of large amount of energy, thus, leading to high toughness of the PP nanocomposites. On the other hand, the fractography of 10 wt% AMUF filled PP nanocomposite exhibited a large number of voids, agglomeration, and cracking (Figure 4.2(d)). This behavior resulted from the aggregation of AMUF within the PP matrix at higher filler fractions, which was also evidenced by the SEM elemental mapping.

Vengatesan *et al.* [35] also reported the effect of amine functionalized reduced graphene oxide (r-AGO) on the properties of high density polyethylene (HDPE) matrix, using chlorinated polyethylene (CPE) as a compatibilizer (Figure 4.3). r-AGO chemically interacted with CPE, which was confirmed by Fourier-transform infra-red spectroscopy (FT-IR) analysis. For instance, in the CPE spectrum, the peak at 668 cm^{-1} corresponded to the C-Cl stretching frequency. In addition, the two peaks at 2926 cm^{-1} and 2853 cm^{-1} corresponded to the asymmetric and symmetric characteristic –CH_2 stretching vibrations

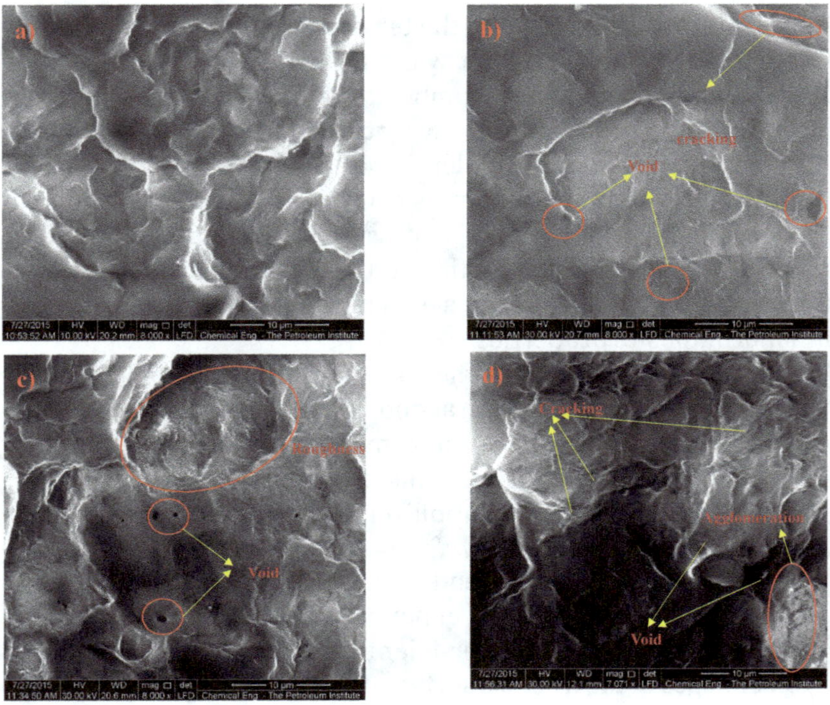

Figure 4.2 SEM images of fracture surface of (a) pure PP, (b) PP-AMUF2 (1 wt % AMUF), (c) PP-AMUF5 (5 wt % AMUF) and (d) PP-AMUF6 (10 wt % AMUF). Reproduced from Reference 34 with permission from Wiley.

of CPE. For the filler/CPE hybrid, a reduction in the intensity of peak at 668 cm^{-1} for C-Cl stretching frequency was observed, which indicated the reaction between r-AGO and CPE. Furthermore, the peak at 1029 cm^{-1} corresponding to –CNO- stretching frequency also represented the nucleophilic reaction between the amine groups in r-AGO and Cl group in CPE. The improved interfacial interactions between the polymer and filler resulted in high degree of filler dispersion in the nanocomposites. Young's modulus of the composites increased with filler content and a maximum increment was observed at 3 wt% of r-AGO content. The tensile strength of the polymer was also retained in the composites. 3 wt% of r-AGO filled HDPE nanocomposites also exhibited lowest surface resistivity indicating enhancement of conducting pathways in the composites on filler addition. Due to the similar architecture and molecular nature of HDPE and polypropylene, similar functional approaches for the filler compatibilization through surface immobilization of the polymer chains can, thus, also

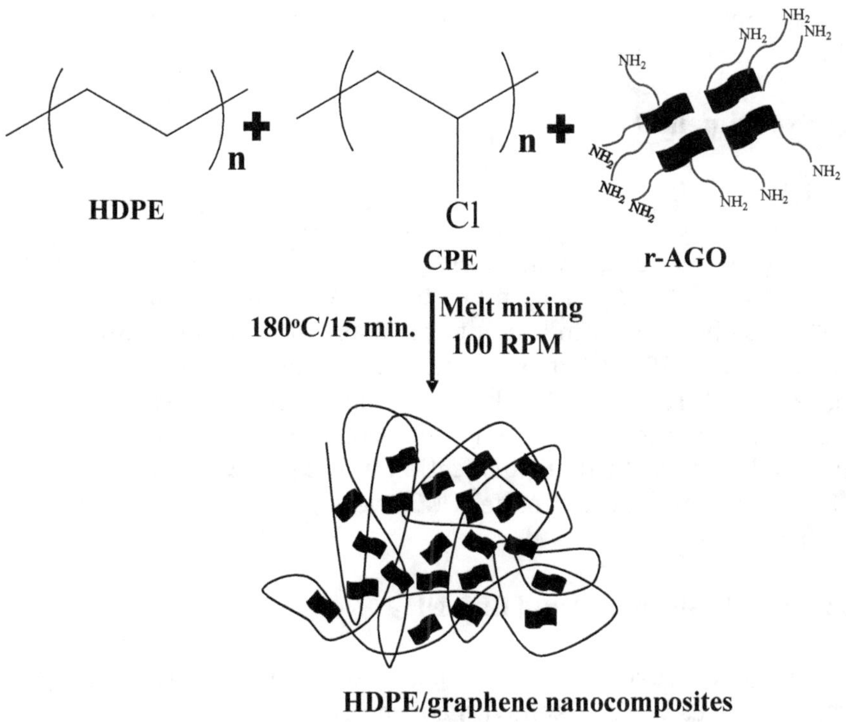

Figure 4.3 Schematic representation of the generation of polyethylene nanocomposites employing chemical interaction of compatibilizer with the filler [35].

be applied for generating high performance polypropylene composites.

In this study, the approach of generating blend nanocomposites using reactive compatibilization method has been further explored to achieve significant enhancement in the composite properties, especially at a low filler content. For this purpose, a 80:20 blend of PP with ethylene vinyl acetate-g-maleic anhydride (EVA-MA) was used. The surface of the graphite oxide platelets was functionalized with amine groups so as to achieve chemical interaction with EVA-MA chains. The chemical interaction was characterized using different structural analyses, and the resulting property enhancements were compared with composite generated without reactive compatibilization as well as pure polymer blend. The blend and nanocomposites were prepared using industrially accepted melt mixing method using optimized processing conditions so as to achieve good degree of phase mixing, but avoiding thermal degradation of the polymers at high processing temperature.

4.2 Experimental

4.2.1 Materials

Isotactic PP (HD951CF) was received from Abu Dhabi Polymers (Borouge), UAE. EVA-MA polymer with a trade name of Dupont Fusabond C190 was obtained from Dupont. It had a melt flow index (MFI, g/10 min at 2.16 kg, 190 °C) of 16, specific gravity of 0.950 and peak melting temperature of 71 °C. Solvents required for graphite oxide synthesis and surface modification were obtained from local suppliers and used without further purification. Ethylene diamine (EDA) and N,N'-dicyclohexylcarbodiimide (DCC) were procured from Sigma Aldrich.

4.2.2 Synthesis of GrO and GrO-NH_2

Graphite oxide (GrO) was synthesized through acid treatment, where graphite flakes were mixed with concentrated mixture of H_2SO_4 and H_3PO_4 (9:1 ratio). After homogenization of the solution, $KMnO_4$ was added slowly. Subsequently, the reaction mixture was transferred to the closed reactor and reaction was carried out for 24 h at room temperature. After completion of the reaction, mixture was diluted with deionised (DI) water and stirred vigorously for 2 h. Hydrogen peroxide (35%) solution was added dropwise till the reaction mixture turned bright yellow in color. The reaction product was filtered to obtain solid mass, which was washed several times with DI water, HCl and ethanol. The obtained graphite oxide was finally vacuum dried at room temperature.

Graphite oxide was modified with amine functional groups using ethylene diamine. Briefly, 100 mg GrO was dispersed in 50 mL THF using bath sonication for 1 h. *N,N'*-Dicyclohexylcarbodiimide was used as catalyst for the functionalization reaction. DCC and EDA were added in GrO dispersion and stirred for 1 h at room temperature. Concentration of reaction components (GrO:DCC:EDA) employed for the reaction was 1:20:20. Finally, the reaction mixture was refluxed at 80 °C for 24 h. The obtained surface modified GrO (designated as GrO-NH_2) was washed with DI water and vacuum dried at room temperature.

4.2.3 Generation of PP/EVA-MA Blend and Blend Nanocomposites

PP/EVA-MA (80/20) blend was generated by melt mixing using mini twin conical screw extruder (MiniLab HAAKE Rheomex CTW5, Germany). A mixing temperature of 190 °C for 10 min at 60 rpm under inert atmosphere with batch size of 5 g was used. The screw length and screw diameter were 109.5 mm and 5/14 mm conical, respectively. Both polymers were vacuum dried overnight at 60 °C before melt processing. Similarly, the blend nanocomposites with GrO and GrO-NH_2 were generated using the same processing conditions with 0.1, 0.5 and 1 wt% filler content. All three components were physically mixed prior to feeding to the compounder. The disc and dumbbell-shaped test samples were prepared by mini injection molding machine (HAAKE MiniJet, Germany) at a processing temperature of 190 °C. The injection pressure was 1000 bar for 6 s, whereas holding pressure was 400 bar for 3 s. The temperature of the mold was kept at 55 °C.

4.2.4 Material Characterization

IR spectra of the fillers, blend and nanocomposites were collected on Nicolet iS10 spectrometer equipped with SmartiTR diamond ATR accessory (angle of incidence of 45O), DTGS KBr detector and KBr beam splitter. It had a diamond ATR crystal (index of refraction 2.4 at 1000 cm^{-1}) and a depth penetration of 2 μm at 1000 cm^{-1} for sample with refractive index of 1.5. Spectra were recorded by the OMNIC software in the 4000-525 cm^{-1} region with a resolution of 4 cm^{-1} from 16 scans. Raman spectra were recorded on a Jobin Yvon Horiba LabRAM spectrometer with backscattered confocal configuration using a HeNe laser (633 nm). A long working distance objective with magnification 50x was used both to collect the scattered light and to focus the laser beam on the sample surface. CCD (charge-coupled device) detector exposure time was 10 s and an average of 2 cycles was used to increase S/N ratio. Raman shifts were calibrated using Si wafer signal at 520.7 cm^{-1}. Perkin Elmer elemental analyzer was used to quantify the elemental composition of graphite oxide before and after surface modification with amine groups.

Wide angle X-ray diffraction analysis of the fillers, blend and composites was performed on Panalytical Powder Diffractometer (X'Pert PRO) using CuKα radiation (λ = 1.5406 A°) in reflection mode. Zero-background holder was used to minimize the noise. The samples were step-scanned from 10-70° 2Θ at room temperature using a step size of 0.02° 2Θ and a step time of 10 s.

DHR-3 rheometer from TA Instruments was used to characterize the rheological properties such as storage modulus (G′) and complex viscosity elasticity (η*). Disc shaped samples of 25 mm diameter and 2 mm thickness were measured at 190 °C using a gap opening of 1.2 mm. Flow sweep experiment was carried out apriori and all experiments were performed in linear viscoelastic region (LVR) at 0.5% strain from ω = 0.1 to 100 rad/s.

Thermal properties of the blend and nanocomposites were recorded using Discovery thermogravimetric analyzer (TGA) from TA Instruments. Nitrogen was used as a carrier gas and the scans were obtained from 50 to 700 °C at a heating rate of 10 °C/min. Calorimetric properties of pure polymers, blend and nanocomposites were recorded on a Discovery differential scanning calorimeter (DSC) from TA Instruments under nitrogen atmosphere. The scans were obtained from 30 to 180 °C (first heating), 180 to -50 °C (first cooling) and -50 to 180 °C (second hating) using heating and cooling rates of 10 °C/min and 5 °C/min, respectively. Second heating runs were used for the comparison purpose.

Tensile analysis of the blend and composites was carried out using universal testing machine (Instron, USA) (ASTM D 638). The dumbbell-shaped samples were used. A loading rate of 10 mm/min was used and the tests were carried out at room temperature. The tensile strength and modulus of the composites were calculated using Win Test Analysis software. An average of five sample values is reported.

For the microscopy analysis, ultrathin sections (30-90 nm) as well as block faces of the composite samples were obtained using a Leica Ultracut E microtome (Leica, Austria) equipped with a diamond knife (Diatome, Switzerland) at -120 °C. The so-obtained sections for transmission electron microscopy (TEM) analysis were collected on 400 mesh formvar electron microscopy grids (coated with holey carbon film), and were examined in a Philips CM 20 (Philips/FEI, Germany) electron microscope at 200 kV at room temperature without additional staining. TEM image processing was performed using DigitalMicrograph software (Gatan, USA). Electron energy loss spectroscopy (EELS) of the blend and composites was performed in TEM mode and the spectra were visualized using DigitalMicrograph software (Gatan, USA). The surface of the block faces after cryo-ultramicrotomy was observed in Hitachi field emission in-lens S-900 high resolution scanning electron microscope at accelerating voltages of 10-20 kV. In addition, the block faces of specimens were also investigated at ambient conditions using ZEISS Axioplan light microscope (Zeiss, Germany) equipped with Zeiss Axio Cam ICc 1 CCD camera in reflected polarized light.

4.3 Results and Discussion

Chemically binding the filler phase with the polymer can lead to uniform filler dispersion as well as enhancement of composite properties at lower filler fraction. It, thus, helps to retain the polymer density, which is an important parameter for many structural applications of polymer nanocomposites. In addition, the requirement of lower filler fraction also helps to retain the polymer flexibility, which has been reported in the literature to significantly reduce due to large extent of filler in the conventional composites. In the case of blend based composites, such chemical binding of the filler with one of the polymer phases can also lead to better phase intermixing, and reduction in domain sizes through interfacial compatibilization, thereby leading to synergistic combination of component properties. Thus, in the current study, the EVA-MA interaction with functionalized graphite oxide was explored in order to generate polypropylene based functional nanocomposites with enhanced properties in comparison with the composite systems where such interactions are absent.

Presence of oxygen functional groups on graphite surface after oxidation was confirmed using FT-IR and elemental analysis. Figure 4.4a depicts FT-IR spectra for oxidized graphite sheets (GrO) where newly emerged peak at 1736 cm^{-1} was characteristic stretching of carbonyl functional group. Similarly, two distinct peaks evident at 850 and 1213 cm^{-1} were attributed to epoxy groups. These observations clearly supported the generation of oxygen functional groups on graphite surface for carrying out the amine functionalization reaction. Furthermore, functionalization was confirmed using elemental analysis (Table 4.1), where graphite oxide powder exhibited significant extent of oxygen. Amine functionalization of graphene sheets was also confirmed using FT-IR and elemental analysis. As evident in Figure 4.4a, the FT-IR spectrum of GrO-NH_2 exhibited the emergence of two new peaks evident at 1346 and 1166 cm^{-1} which were signature stretching characteristic of C-N bonds. The peak for amide bond at 1608 cm^{-1} also confirmed the successful tethering of amine molecules on the graphite oxide surface by chemical reaction. Further, the

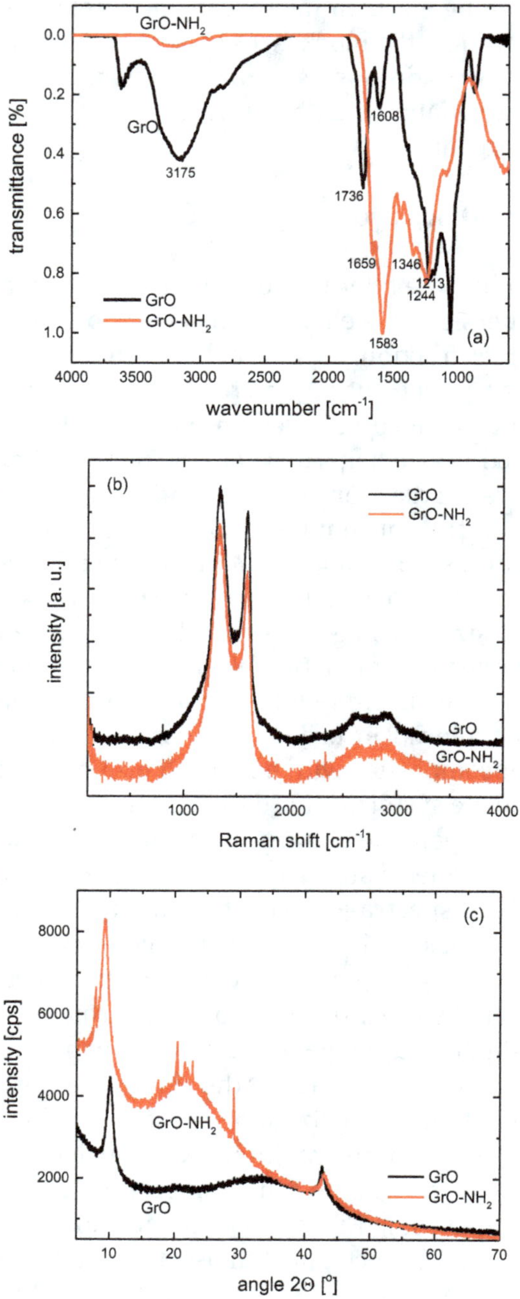

Figure 4.4 (a) FTIR, (b) Raman spectroscopy and (c) X-ray diffraction analysis of GrO and GrO-NH_2.

Table 4.1 Elemental analysis for graphite oxide and amine functionalized graphite oxide

Material	N%	C%	O%
GrO	0.56	40.05	53.89
GrO-NH_2	9.36	47.84	37.77

elemental analysis confirmed the presence of nitrogen in the GrO-NH_2 to be 9.4%, thus, confirming the presence of nitrogen containing moieties on graphite oxide surface. Figure 4.4b demonstrates the Raman spectra of GrO and GrO-NH_2. GrO exhibited the D band at 1335 cm^{-1}, whereas a weaker G band was observed at 1592 cm^{-1}. In the case of GrO-NH_2, the characteristic D and G bands were also observed to appear at the same Raman shift. The intensity ratio of the D/G bands (I_D/I_G), regarded as degree of measurement of the disorder, was slightly increased in GO-NH_2 probably due to the restoration of the sp^2 network in GrO after amine functionalization as well as due to the presence of edge oriented structural defects. The successful attachment of amine groups on GrO surface was also confirmed through X-ray diffraction. As shown in Figure 4.4c, the characteristic 2Θ diffraction peak for GrO was observed at 10.9°, which was shifted to a lower diffraction angle at 9.9°, thus, indicting the intercalation of amine groups inside the interlayers of GrO. It should be noted that the interlayer spacing of GrO did not increase to the large extent, however, it was expected as the molecular size of the amine attached to the GrO surface was also small. Enhancement of even a small amount of interlayer distance can be expected to provide better opportunities for the functional groups to interact with the polymer during composite generation. Overall, the structural characterization along with elemental analysis confirmed the modification of GrO surface.

In the IR spectra of PP+EVA-MA and its composites, the peaks between 2837 and 2948 cm^{-1} were attributed to the C–H asymmetric and symmetric stretching frequencies for both PP and EVA-MA. The peak at 1457 cm^{-1} represented the asymmetric C-H bending vibration of $-CH_2$ and $-CH_3$ groups of PP and EVA-MA. The peak at 1374 cm^{-1} indicated the symmetric C-H bending vibration of $-CH_2$ and $-CH_3$ groups and the peak at 1166 cm^{-1} represented the C-C stretching of C-CH_3 groups of PP and EVA-MA. The band at 1735 cm^{-1} represented the –C=O stretching of MA in PP/EVA-MA. The bands at 1019 and 1236 cm^{-1} indicated the ester C-O stretching of EVA-MA. The amide linkage in the region 1600-1650 cm^{-1} was observed in Figure 4.5(a)

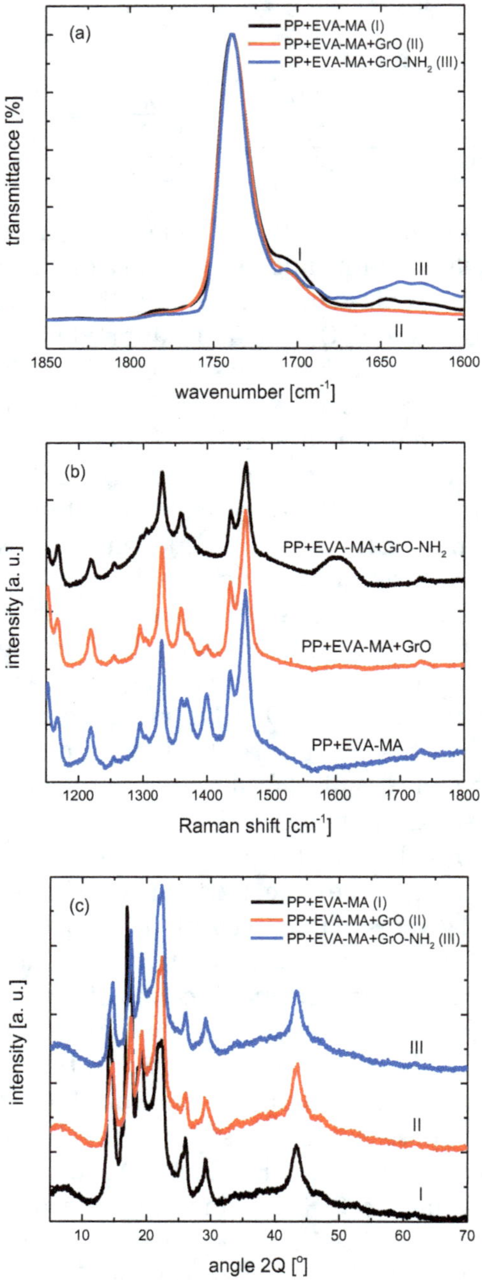

Figure 4.5 (a) FTIR, (b) Raman spectroscopy and (c) X-ray diffraction analysis of PP+EVA-MA, PP+EVA-MA+1%GrO and PP+EVA-MA+1%GrO-NH_2.

for the PE+EVA-MA+1%GrO composite, which was attributed to the filler, as observed earlier. This band was observed to increase in intensity in the case of PE+EVA-MA+1%GrO-NH_2 composite owing to the reaction of NH_2 groups on the graphene surface with the MA groups in EVA-MA. Thus, it indicated that some degree of chemical linking of filler platelets took place with the EVA-MA phase. Even though the observed signal was not sharp, however, a small extent of reactive compatibilization can lead to high degree of interfacial compatibilization of filler and EVA-MA phases, thus, leading to enhanced dispersion levels, which would subsequently affect the composite properties. Since the contribution of graphite oxide in nanocomposites was difficult to detect in IR, Raman spectra of PE+EVA-MA, PP+EVA-MA+1%GrO and PP+EVA-MA+1%GrO-NH_2 were also recorded and are provided in Figure 4.5(b). Characteristic G band of graphite oxide was observed at ~1590 cm^{-1} for PP+EVA-MA+GrO-NH_2, however, it was completely missing in the case of PP+EVA-MA+GrO composite. The D band was observed to be merged with PP signals in the 1325 cm^{-1} to 1350 cm^{-1} region. PP+EVA-MA+1%GrO-NH_2 also exhibited the absence of Raman shifts at 1300 cm^{-1} and 1400 cm^{-1}, in comparison with PE+EVA-MA and PP+EVA-MA+1%GrO, probably owing to the chemical interactions between GO-NH_2 and EVA-MA. As observed in Figure 4.5(c), the X-ray diffraction of the PP+EVA-MA blend and its composites with 1% filler content did not exhibit the presence of characteristic diffraction peak of GO or GO-NH_2, due to the low filler content. Good degree of filler intermixing with the polymers may have also resulted in the loss of diffraction peak of the filler platelets.

Further insights into the morphology of the PP+EVA-MA blend and composites were gained through microscopy. Figure 4.6 demonstrates the TEM micrographs of the PP+EVA-MA blend. Though EVA-MA was not molecularly miscible with PP under the processing conditions used in the study, however, its dispersion in PP was observed to be uniform (observed as darker phase in the TEM micrographs). Specifically, EVA-MA was dispersed as smaller domains inside the PP matrix due to its 20% fraction in the blend. Thus, it also indicated that due to its uniform dispersion in PP, EVA-MA can automatically lead to an enhanced degree of dispersion of GrO-NH_2 platelets owing to the chemical interaction with them. The optical analysis of the blend also represented the similar morphology, as observed in TEM. Figure 4.7 represents the microscopy analysis of the PP+EVA-MA+GrO and PP+EVA-MA+GrO-NH_2 with 1% filler content. Composites with other

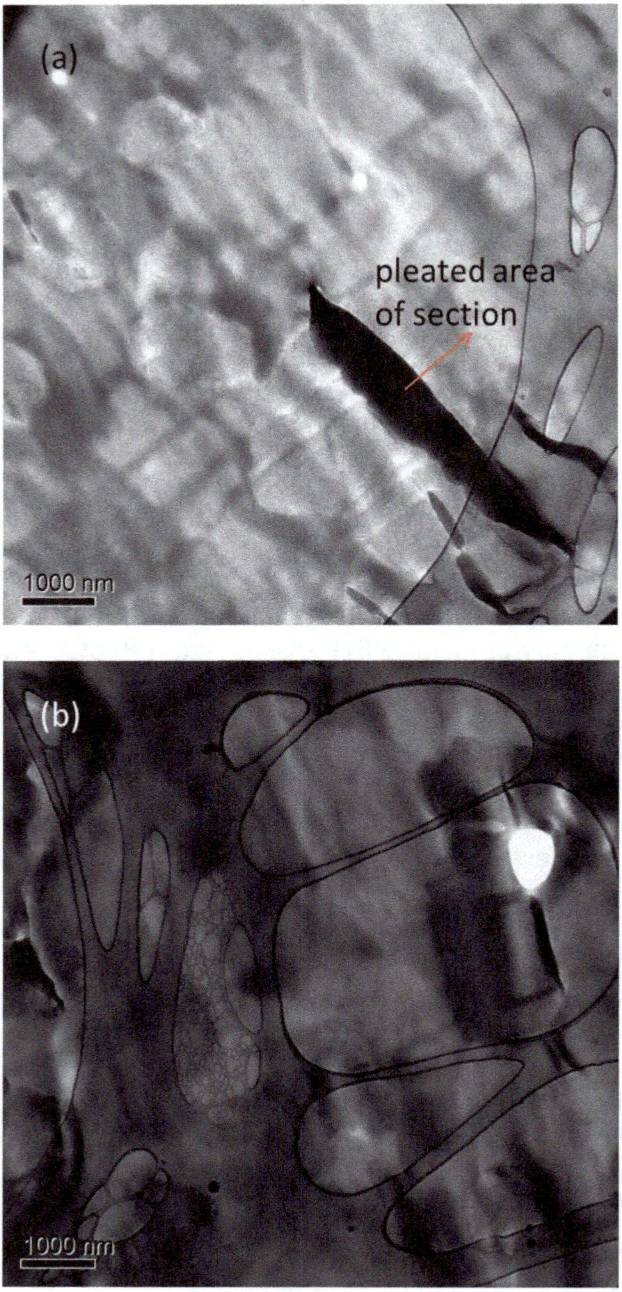

Figure 4.6 TEM micrographs of PP+EVA-MA blend.

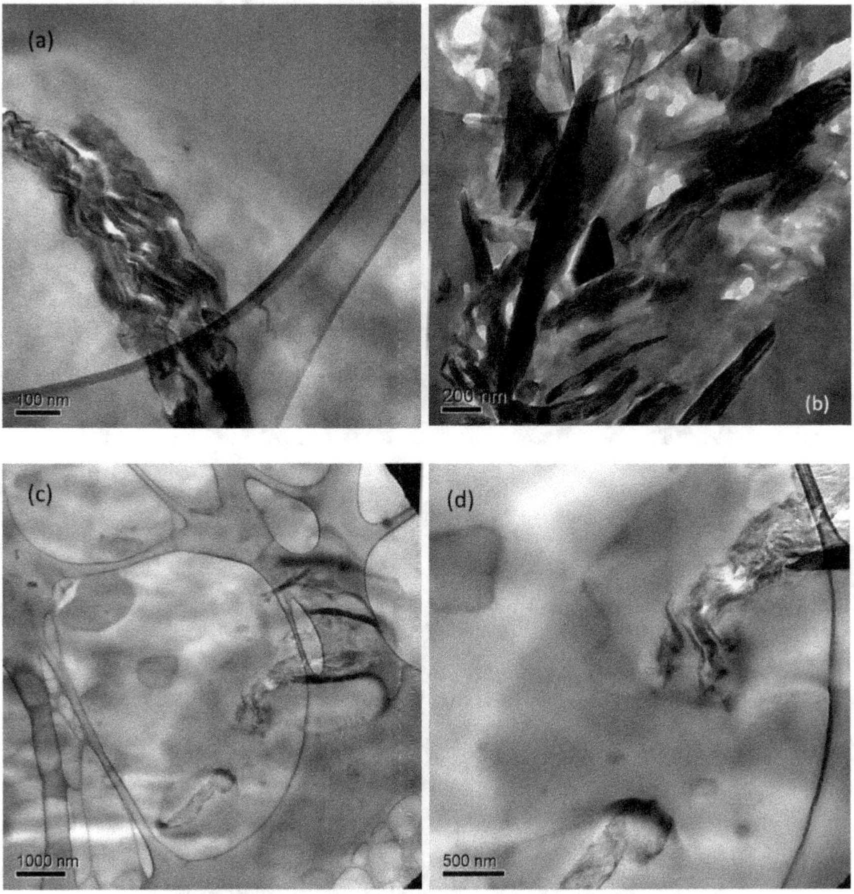

Figure 4.7 TEM micrographs of (a,b) PP+EVA-MA+1%GrO and (c,d) PP+EVA-MA+1%GrO-NH_2.

filler fractions also exhibited similar morphology. In the case of PP+EVA-MA+GrO composite, the filler was largely observed in the form of thick stacks mostly localized at the interface of PP and EVA-MA phases, though localization of these platelets in the PP phase was also observed. On the other hand, the filler in the case of PP+EVA-MA+GrO-NH_2 nanocomposite was observed to be largely embedded in the EVA-MA (darker dispersed phase in the TEM micrographs). In addition, the stack thickness was observed to be lesser than observed in the earlier case of PP+EVA-MA+GrO composite. Similar finding was also observed from the SEM analysis presented in Figure 4.8. The filler was largely in the PP phase as well as at PP/EVA-MA interface in the case of PP+EVA-MA+GrO composite, whereas the filler platelets

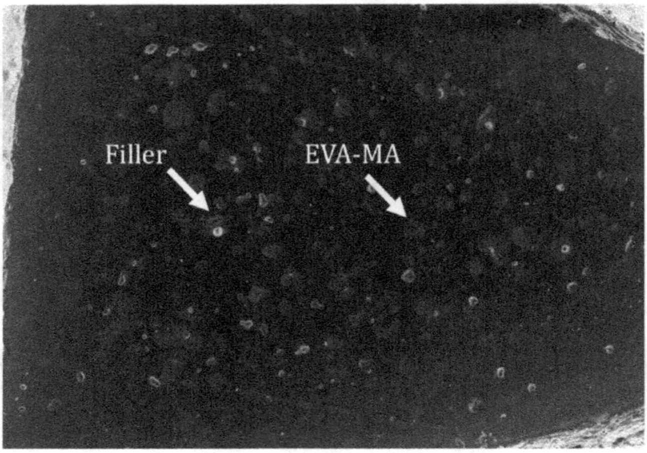

(a)

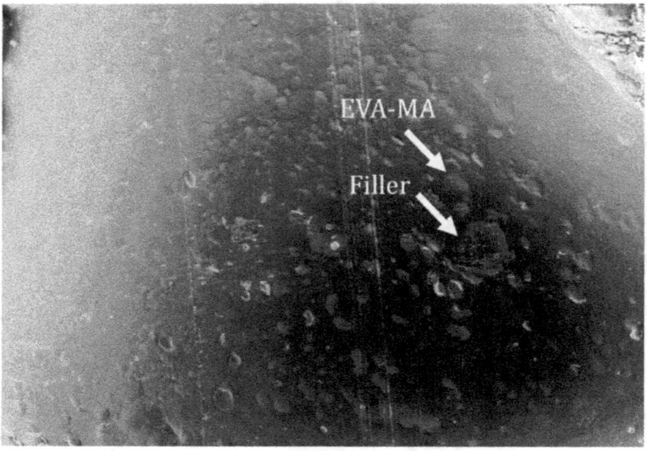

(b)

Figure 4.8 SEM images of (a,b) PP+EVA-MA+1%GrO and (c,d) PP+EVA-MA+1%GrO-NH_2.

were largely localized in the EVA-MA phase in the case of PP+EVA-MA+GrO-NH_2 nanocomposite. Such a morphology would only result due to the strong interaction of EVA-MA phase with the filler platelets. As also observed in the TEM micrographs, the SEM micrographs also revealed uniform distribution of EVA-MA phase as spherical domains inside PP, which due to chemical interaction with GrO-NH_2 filler resulted in better filler distribution as well as delamination inside the polymer. Thus, the component distribution was altered depending on the filler-polymer interactions. Figure 4.9 also compares

PP+EVA-MA+GrO and PP+EVA-MA+GrO-NH_2 nanocomposites at different magnifications. As clearly observed, the PP+EVA-MA+GrO composite did not exhibit any exfoliation of the GrO platelets. On the other hand, PP+EVA-MA+GrO-NH_2 nanocomposite exhibited a good degree of filler delamination in the matrix, as observed earlier. Thus, the microscopy findings confirmed the earlier findings from FTIR about the interaction of EVA-MA with GrO-NH_2 in the composite. The observed enhancement in composite morphology in the case of PP+EVA-MA+GrO-NH_2 nanocomposite is, thus, also expected to lead to superior composite performance.

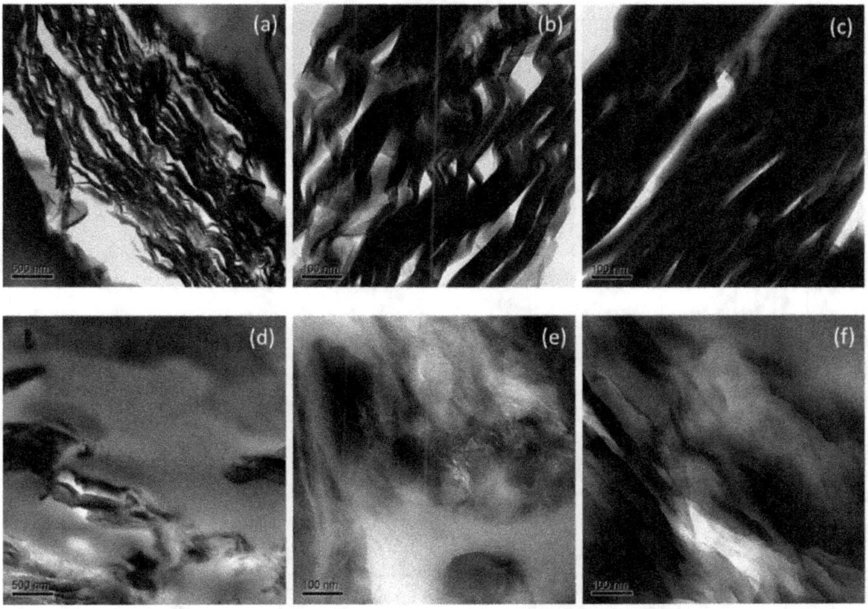

Figure 4.9 Comparison of the morphology of PP+EVA-MA+1%GrO (a-c) with PP+EVA-MA+1%GrO-NH_2 (d-f) at different magnifications.

Electron energy loss spectroscopy in the TEM mode was also carried out on PP, PP+EVA-MA blend and nanocomposites with 1 wt% filler content to gain further insights about the polymer filler interactions, as demonstrated in Figure 4.10. PP and EVA-MA exhibited signals for both nitrogen and oxygen in very low concentration, though the presence of N in these materials was unexpected. The presence of some additive added occasionally to commercial polymers may have also resulted in such weak signal. The PP+EVA-MA+GO composite exhibited increased concentration of O due to the contribution from

GrO platelets along with EVA-MA. On the other hand, no specific increase in the N signal was observed. The concentration of N in the PP+EVA-MA+GrO-NH_2 composite was much higher than the corresponding GO composite due to the functionalization of GrO with amine groups. In addition, the shape of the electron energy-loss near-edge structure (ELNES) N K edge indicated that N was most probably present as mixture of NH_2 and some other forms. This also possibly confirmed the change of NH_2 form to amide form, as observed earlier in IR analysis. However, it was not possible to identify if this signal indicated change of NH_2 form during GrO functionalization with diamine or due to the reaction of NH_2 groups on graphene surface with EVA-MA, as both reactions formed an amide linkage. Similar to the PP+EVA-MA+GO composite, the PP+EVA-MA+GrO-NH_2 composite also exhibited O signal due to the EVA-MA and filler contributions. Thus, though it was not possible to absolutely confirm the reaction between amine functionalized graphite oxide and EVA-MA from the EELS studies, however, these findings largely corroborated the earlier findings from structural analysis and microscopy.

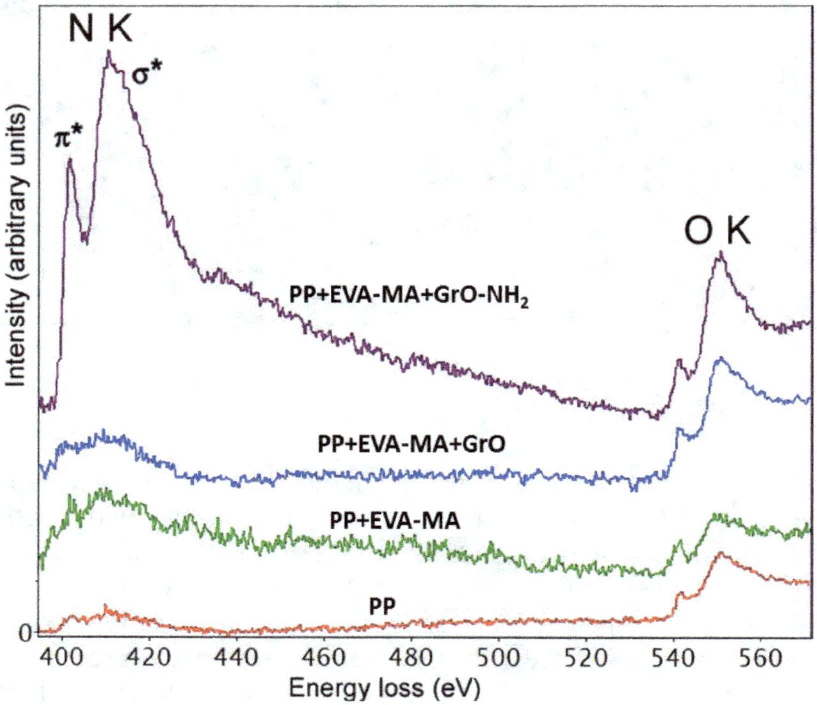

Figure 4.10 Energy loss spectra of the materials.

Rheological, thermal and mechanical properties of the composites were studied further to analyze the effect of polymer-filler interactions in the case of PP+EVA-MA+GrO-NH_2 composites, in comparison with PP+EVA-MA+GrO composites as well as PP+EVA-MA blend. In the case of polymer nanocomposites, the viscoelastic properties in the melt state are an effective tool to analyze the dispersion of nanoparticles in polymer matrices. Thus, in the current study too, the dispersion and network formation of nano-filler in PP+EVA-MA matrix was studied using frequency sweep, where all rheological parameters were recorded as a function of oscillation frequency in the range of 0.1 to 100 rad/s. Figure 4.11 shows the storage modulus of the blend and nanocomposite materials as a function of frequency. The storage modulus of the materials increased with angular frequency, however, the rate of increase slowed down at higher frequencies due to shear thinning effect. The PP+EVA-MA blend exhibited lower storage modulus than the nanocomposites over the whole frequency range. The composites with GrO enhanced the storage modulus gradually as a function of filler content, thus, indicating matrix reinforcement on the

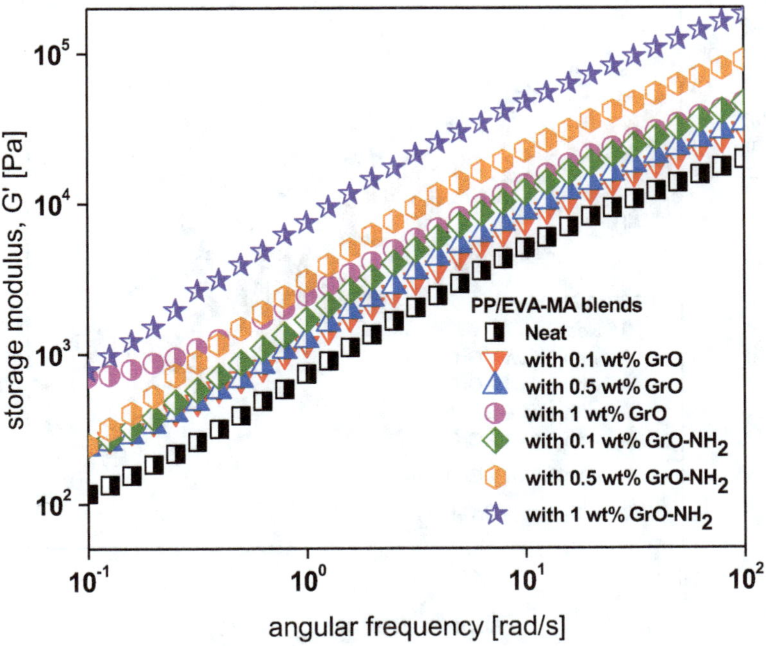

Figure 4.11 Storage modulus of the blend and composites as a function of angular frequency.

addition of GrO. In comparison, the incorporation of GrO-NH_2 filler in PP+EVA-MA led to even higher extent of storage modulus increment. For instance, the storage modulus of PP+EVA-MA at angular frequencies of 1 and 10 rad/s was measured to be 733 and 5035 Pa, which was enhanced to 7427 and 46403 Pa respectively for the composite with 1 wt% GrO-NH_2. Comparing the storage modulus of PP+EVA-MA+GrO and PP+EVA-MA+GrO-NH_2 composites, the composite with 0.5 wt% GO-NH_2 filler was observed to have higher storage modulus than the composite with even 1 wt% GrO. It indicated that better polymer-filler interfacial interactions achieved due to chemical reaction between EVA-MA and GO-NH_2 filler resulted in effective reinforcement of the polymer matrix due to efficient stress transfer. The loss modulus of the composites was also lower than storage modulus over the complete frequency range due to the dominance of elastic nature over the viscous nature.

Figure 4.12 shows the complex viscosity of the blend and nanocomposites as a function of angular frequency. As expected, the PP+EVA-MA blend and blend nanocomposites exhibited shear

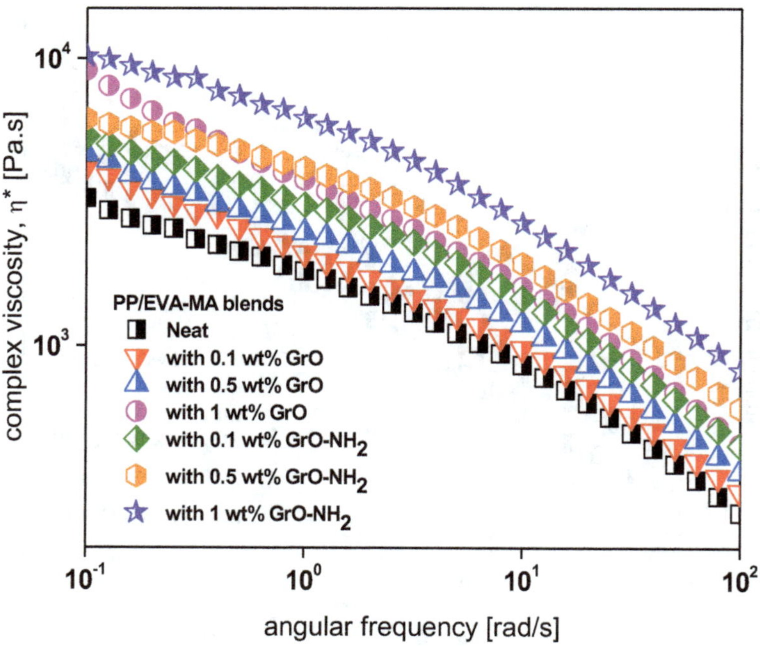

Figure 4.12 Complex viscosity of the blend and composites as a function of angular frequency.

thinning behavior with increasing angular frequency. As observed in the case of storage modulus, the complex viscosity of the nanocomposites increased with the addition of filler in the polymer matrix. The increment in viscosity for both GO and GO-NH_2 filler systems was a function of filler content in the composites. Comparing the GrO and GrO-NH_2 filled nanocomposites systems, the GrO-NH_2 composites were observed to exhibit larger extent of increase in viscosity owing to effective delamination of filler platelets and immobilization of the polymer chains on the surface of filler platelets. Though the viscosity of the composites increased significantly in comparison with PP+EVA-MA blend, however, the composites were still processable using the same processing conditions as the blend.

Figure 4.13(a) depicts the DSC melting thermograms of the pure polymers, polymer blend as well as composites. Crystalline melting of PP was observed at 167 °C, whereas EVA-MA exhibited the endothermic crystalline melting peak at 67.5 °C. The thermograms of PP+EVA-MA blend as well as nanocomposites with GrO and GrO-NH_2 exhibited a small hump at 67.5 °C corresponding to the melting transition of EVA-MA, along with the transition corresponding to the melting of PP crystals. These findings supported the earlier findings of absence of molecular mixing between PP and EVA-MA. Also, the melt transition corresponding to PP did not shift in the blend as well as blend nanocomposites, thus, indicating that the EVA-MA phase as well as nanoparticles had no effect on the melting behavior of PP. The melt enthalpy of polypropylene was observed to reduce slightly in the blend as compared to pure polypropylene due to the hindrance to the crystallization of PP chains. The melt enthalpy of PP remained constant on the addition of the filler platelets to the blend, irrespective of the type and amount of filler. Figure 4.13(b) presents the crystallization exotherms of the materials. Neat PP and EVA-MA exhibited peak crystallization temperatures of 126.5 °C and 51 °C respectively. Similar to the melting behavior, the peak crystallization temperature of PP remained unchanged in the blend as well as nanocomposites, thus, indicating that the filler particles did not exhibit any nucleating effect on PP crystallization. In addition, the thermograms of blend and nanocomposites exhibited a small hump at 51 °C corresponding to the crystallization of EVA-MA. Overall, the DSC findings confirmed that the filler particles did not change the melting and crystallization behavior of the polymer blend, even though the nanocomposites with GrO and GrO-NH_2 fillers exhibited different phase morphologies in microscopy analysis.

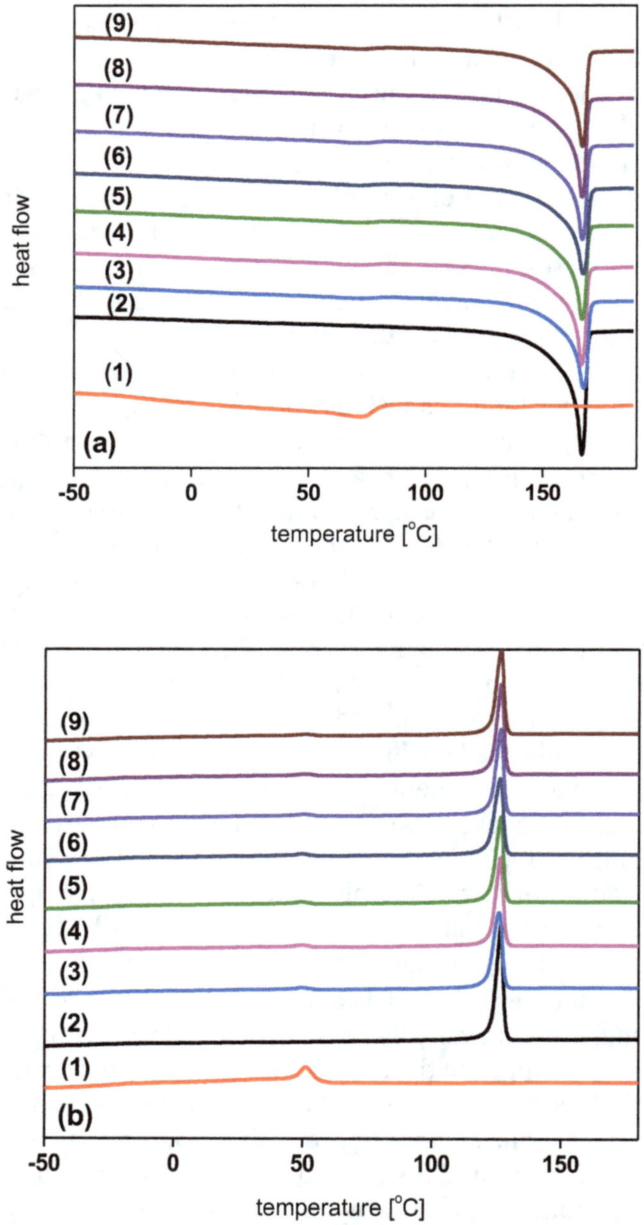

Figure 4.13 (a) Melting and (b) crystallization thermograms of the materials. Samples denoted as (1) neat EVA-MA, (2) neat PP, (3) PP+EVA-MA 80/20 blend, and blends with (4, 5 and 6) 0.1, 0.5, 1 wt% GrO respectively and blends with (7, 8 and 9) 0.1, 0.5, 1 wt% GrO-NH_2 respectively.

The thermal degradation behavior of the PP+EVA-MA+GrO and PP+EVA-MA+GrO-NH_2 nanocomposites with 1 wt% filler content is also compared with PP+EVA-MA in Figure 4.14. The thermogram of the PP+EVA-MA blend exhibited two step degradation pattern with a low temperature degradation step having onset of degradation at 350 °C attributed to EVA-MA, whereas the higher temperature degradation step with an onset of degradation at 425 °C was attributed to pure polypropylene. The incorporation of GrO and GrO-NH_2 fillers in the blend led to an improvement in the thermal stability as the lower temperature degradation step due to EVA-MA was absent in the thermograms, thus, indicating better heat transfer between the polymer chains and filler platelets. There was no difference in the behavior of the composites with GrO and GrO-NH_2 fillers. In addition, similar improvement in the thermal performance was exhibited by all the composites, indicating that even 0.1 wt% filler was able to impart significant thermal resistance. Dynamic TGA of the nanocomposites was also carried out by analyzing the degradation as a function of time at 190 °C (compounding temperature) and the composites were observed to have insignificant weight loss for the test duration equivalent to the compounding time, thus, confirming the durability of the

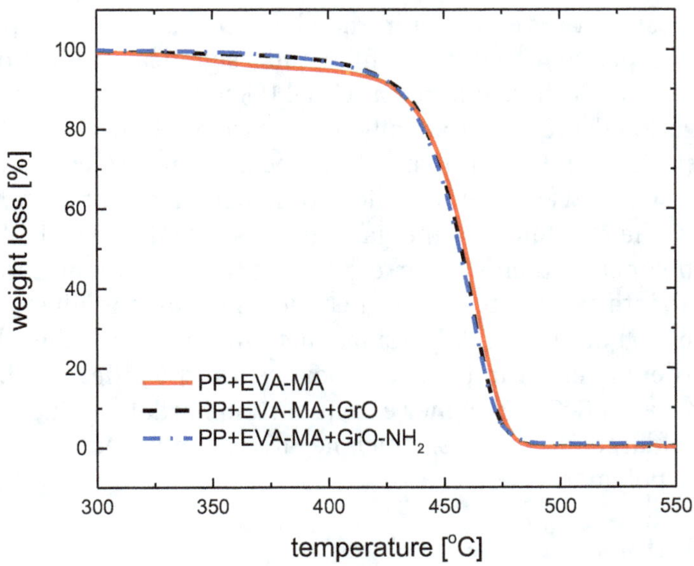

Figure 4.14 TGA thermograms of PP+EVA-MA, PP+EVA-MA+1%GrO and PP+EVA-MA+1%GrO-NH_2.

composites for high temperature processing and applications.

The tensile properties of the nanocomposites are compared with the PP+EVA-MA blend as a function of filler content in Figure 4.15. The tensile modulus of the blend was measured as 916 MPa, which was enhanced to 1202 MPa for the PP+EVA-MA-GrO composite with 1 wt% filler content, thus exhibiting an increase of 30% in the modulus (Figure 4.15(a)). On the other hand, the corresponding composite with 1 wt% GrO-NH_2 filler had the tensile modulus of 1343 MPa, which was an increment of >45%. It, thus, confirmed the earlier findings from rheological analysis that the chemical interaction between the EVA-MA and GrO-NH_2 resulted in better stress transfer between the polymer chains and filler platelets. As no change in the extent of polymer crystallinity was observed in the blend and composites, thus, the observed increment in tensile modulus was solely attributed to the polymer-filler interactions and filler dispersion. Similar behavior was also observed in the case of tensile strength, as shown in Figure 4.15(b). In comparison to 29 MPa for the PP+EVA-MA blend, the tensile strength of the nanocomposites with 1% GrO and GrO-NH_2 fillers was observed to be 30.9 and 33.4 MPa respectively. It again underlined the impact of reactive compatibilization on the composite properties. As observed in Figure 4.15(c), the elongation of the composites decreased with the addition of filler. In the case of PP+EVA-MA+GrO nanocomposites, the reduction in the elongation was considerable. Even the composite with 0.1 wt% filler had a significant reduction in elongation to 7.7%, in comparison with 41% for the polymer blend. The elongation decreased gradually in the case of PP+EVA-MA+GrO-NH_2 composites, and still remained appreciable at 21% even in the composite with 1 wt% filler. It indicated that similar to other tensile properties, the toughness of the nanocomposites with GO-NH_2 filler was much superior than the corresponding GrO containing composites due to interfacial interactions between EVA-MA and filler platelets leading to better filler dispersion. This subsequently reduced the strain hardening effect in these composites. On the other hand, the composites with GrO had higher extent of stain hardening due to the presence of large filler stacks, which negatively impacted the toughness of the polymer.

4.4 Conclusions

In the current study, blend nanocomposites and reactive compatibilization approaches have been successfully demonstrated to generate

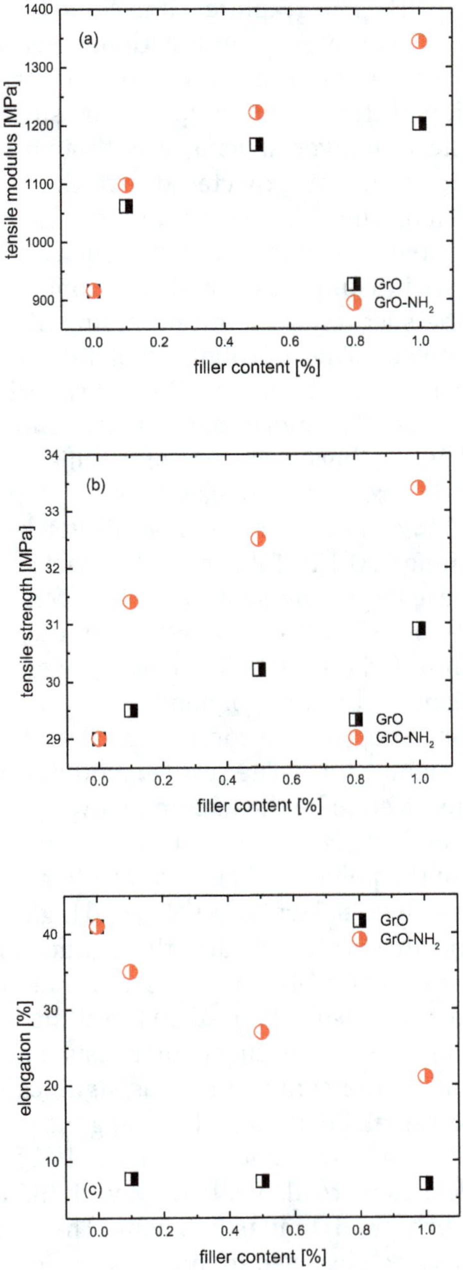

Figure 4.15 Tensile properties of the composites as a function of filler content. The relative probable error for tensile modulus and tensile strength was ±2%, whereas it was ±5% for elongation.

functional PP/EVA-MA nanocomposites with amine functionalized graphite oxide, GrO-NH_2. The graphite oxide could be modified using ethylene diamine to generate amine functionalities on the surface. The structural characterization as well as elemental analysis confirmed the tethering of the amine groups on the surface of graphite oxide platelets. The interlayer spacing was though increased by a small degree, however, it was expected due to the small size of the ethylene diamine molecules. The nanocomposites generated with GrO-NH_2 also exhibited an increase in the intensity of amide linkage in the FTIR analysis in comparison with the composites with GrO, thus, confirming the successful chemical binding of EVA-MA chains with the filler platelets. The morphology of PP+EVA-MA was observed to be uniformly mixed with small sized spherical domains of EVA-MA in the PP phase. The morphology of the composites was significantly affected depending on the type of the filler. The composites with GrO filler had the presence of thick stacks of GrO platelets largely localized at the PP/EVA-MA interface or in PP itself. In the case of composites containing GrO-NH_2 filler, the filler stacks were only a few layers thick and were largely present in the EVA-MA phase owing to the chemical interactions between the two. Thus, due to uniform dispersion of EVA-MA in PP, the GrO-NH_2 platelets were also observed to be uniformly dispersed in the polymer matrix. The observed differences in the morphology of the composites also affected the composites properties significantly. The storage modulus of the GrO-NH_2 composites was observed to be the maximum owing to the polymer-filler interactions and effective filler dispersion leading to higher stress transfer from the polymer chains to the filler platelets. For instance, the composite with 0.5 wt% GrO-NH_2 had higher storage modulus than the composite with 1 wt% GrO, thus, indicating effective reinforcement in the case of GrO-NH_2 composites. Similarly, the viscosity of the composites was also enhanced on the addition of filler, with GrO-NH_2 based composites exhibiting higher resistance to flow. Even though the viscosity of the composites was significantly enhanced, the composites still remained processable using the same processing conditions as the PP+EVA-MA blend. The thermal degradation properties of the composites were also enhanced with filler, though both GrO and GrO-NH_2 exhibited similar behavior. The composites with even 0.1 wt% filler were observed to impart significant thermal stability to the polymer blend. The DSC thermograms also confirmed the absence of molecular mixing between PP and EVA-MA. In addition, the peak melting and crystallization temperatures of the polymer were not affected in the blend and composites indicating that EVA-MA and fillers did not impact the melting and crystallization behavior of PP. However, a small reduction in the melt enthalpy of PP was observed in the blend and corresponding blend composites due to the hindrance to crystallization of PP chains in the presence of EVA-MA. The specific polymer-filler interactions and subsequent enhanced degree of filler dispersion were also observed to result in considerable enhancement in the mechanical properties of GrO-NH_2 based composites. For instance, the composite with 1 wt% GrO-NH_2 filler led to an increment of >45% in the tensile modulus, in comparison to 30% for the composite with 1 wt% GrO. The GrO-NH_2 based composites also exhibited larger tensile strength than the corresponding GrO based composites. The elongation of the composites was significantly reduced in GrO based composites to around 7% even with 0.1 wt% filler content. On the other hand, the GrO-NH_2 based composites had much higher elongation with 1 wt% composite exhibiting the elongation of 21%. The GrO-NH_2 based composites, thus, represent advanced composites with optimal properties achieved at low weight fractions of filler and have the potential to compete with many engineering plastics.

Acknowledgement

The authors are indebted to Dr. Gisha E. Luckachan for support in structural characterization (X-ray, Raman, and IR) of the materials.

References

1. Mittal, V. (2007) Polypropylene-layered silicate nanocomposites: Filler matrix interactions and mechanical properties. *Journal of Thermoplastic Composite Materials*, **20**, 575-599.
2. Zhang, M., Liu, Y., Zhang, X., Gao, J., Huang, F., Song, Z., Wei, G., and Qiao, J. (2002) The effect of elastomeric nano-particles on the mechanical properties and crystallization behavior of polypropylene. *Polymer*, **43**, 5133-5138.
3. van der Wal, A., Verheul, A., and Gaymans, R. (1999) Polypropylene–rubber blends: 4. The effect of the rubber particle size on the fracture behaviour at low and high test speed. *Polymer*, **40**, 6057-6065.
4. Wahit, M., Hassan, A., Mohd Ishak, Z., and Czigany, T. (2009) Ethylene-octene copolymer (POE) toughened polyamide 6/polypropylene nanocomposites: Effect of POE maleation. *EXPRESS Polymer*

Letters, **3**, 309-319.

5. Martins, C., Larocca, N., Paul, D., and Pessan, L. (2009) Nanocomposites formed from polypropylene/EVA blends. *Polymer*, **50**, 1743-1754.
6. Bassani, A., and Pessan, L. (2002) Toughening of polypropylene with styrene/ethylene-butylene/styrene tri-block copolymer: Effects of reactive and nonreactive compatibilization. *Journal of Applied Polymer Science*, **86**, 3466-3479.
7. Jafari, S., and Gupta, A. (2000) Impact strength and dynamic mechanical properties correlation in elastomer-modified polypropylene. *Journal of Applied Polymer Science*, **78**, 962-971.
8. Zebarjad, S., Lazzeri, A., Bagheri, R., Seyed Reihani, S., and Frounchi, M. (2003) Fracture mechanism under dynamic loading of elastomer-modified polypropylene. *Materials Letters*, **57**, 2733-2741.
9. Wang, Z. (1996) Toughening and reinforcing of polypropylene. *Journal of Applied Polymer Science*, **60**, 2239-2243.
10. Wei, G. X., and Sue, H. J. (2000) Fracture behavior of styrene-ethylene-propylene rubber-toughened polypropylene. *Polymer Engineering & Science*, **40**, 1979-1988.
11. Bai, H., Wang, Y., Song, B., and Han, L. (2008) Synergistic toughening effects of nucleating agent and ethylene–octene copolymer on polypropylene. *Journal of Applied Polymer Science*, **108**, 3270-3280.
12. Lim, J., Hassan, A., Rahmat, A., and Wahit, M. (2006) Rubber-toughened polypropylene nanocomposite: Effect of polyethylene octene copolymer on mechanical properties and phase morphology. *Journal of Applied Polymer Science*, **99**, 3441-3450.
13. Abreu, F., Forte, M., and Liberman, S. (2005) SBS and SEBS block copolymers as impact modifiers for polypropylene compounds. *Journal of Applied Polymer Science*, **95**, 254-263.
14. Bassani, A., Pessan, L., and Hage, E. (2001) Toughening of polypropylene with styrene/ethylene-butylene/styrene tri-block copolymer: Effects of mixing condition and elastomer content. *Journal of Applied Polymer Science*, **82**, 2185-2193.
15. Ma, Y., Wu, D., Liu, Y., Li, X., Qiao, H., and Yu, Z.-Z. (2014) Electrically conductive and super-tough polypropylene/carbon nanotube nanocomposites prepared by melt compounding. *Composites Part B: Engineering*, **56**, 384-391.
16. Song, P., Cao, Z., Cai, Y., Zhao, L., Fang, Z., and Fu, S. (2011) Fabrication of exfoliated graphene-based polypropylene nanocomposites with enhanced mechanical and thermal properties. *Polymer*, **52**, 4001-4010.
17. Yuan, B., Bao, C., Song, L., Hong, N., Liew, K. M., and Hu, Y. (2014) Preparation of functionalized graphene oxide/polypropylene nanocomposite with significantly improved thermal stability and studies

on the crystallization behavior and mechanical properties. *Chemical Engineering Journal*, **237**, 411-420.

18. Song P., Liu, L., Huang, G., Fu, S., Yu, Y., and Guo, Q. (2013) Facile Fabrication of Polyolefin/Carbon Nanotube Composites via in Situ Friedel–Crafts Polyalkylation: Structure and Properties. *Industrial & Engineering Chemistry Research*, **52**, 14384-14395.
19. Kuila, T., Bose, S., Hong, C. E., Uddin, M. E., Khanra, P., Kim, N. H., and Lee, J. H. (2011) Preparation of functionalized graphene/linear low density polyethylene composites by a solution mixing method. *Carbon*, **49**, 1033-1037.
20. Shin, K.-Y., Hong, J.-Y., Lee, S., and Jang, J. (2012) Evaluation of anti-scratch properties of graphene oxide/polypropylene nanocomposites. *Journal of Materials Chemistry*, **22**, 7871-7879.
21. Hsiao, M.-C., Liao, S.-H., Lin, Y.-F., Wang, C.-A., Pu, N.-W., Tsai, H.-M., and Ma, C.-C. M. (2011) Preparation and characterization of polypropylene-graft-thermally reduced graphite oxide with an improved compatibility with polypropylene-based nanocomposite. *Nanoscale*, **3**, 1516-1522.
22. Page, D. J., and Gopakumar, T. G. (2006) Properties and crystallization of maleated polypropylene/graphite flake nanocomposites. *Polymer Journal*, **38**, 920-929.
23. Clemons, C. (2010) Elastomer modified polypropylene–polyethylene blends as matrices for wood flour–plastic composites. *Composites Part A: Applied Science and Manufacturing*, **41**, 1559-1569.
24. Souza, A., and Demarquette, N. (2002) Influence of coalescence and interfacial tension on the morphology of PP/HDPE compatibilized blends. *Polymer*, **43**, 3959-3967.
25. Modesti, M., Lorenzetti, A., Bon, D., and Besco, S. (2005) Effect of processing conditions on morphology and mechanical properties of compatibilized polypropylene nanocomposites. *Polymer*, **46**, 10237-10245.
26. Liu, S.-P., Ying, J.-R., Zhou, X.-P., Xie, X.-L., and Mai, Y.-W. (2009) Dispersion, thermal and mechanical properties of polypropylene/magnesium hydroxide nanocomposites compatibilized by SEBS-g-MA. *Composites Science and Technology*, **69**, 1873-1879.
27. Lee, H.-s., Fasulo, P. D., Rodgers, W. R., and Paul, D. (2005) TPO based nanocomposites. Part 1. Morphology and mechanical properties. *Polymer*, **46**, 11673-11689.
28. Mukhopadhyay, P., and Gupta, R. K. (2011) Trends and frontiers in graphene-based polymer nanocomposites. *Plastics Engineering*, **67**, 32-42.
29. Mittal, V. (2014) Functional polymer nanocomposites with graphene: A review. *Macromolecular Materials and Engineering*, **299**, 906-931.

30. Kuilla, T., Bhadra, S., Yao, D., Kim, N. H., Bose, S., and Lee, J. H. (2010) Recent advances in graphene based polymer composites. *Progress in Polymer Science*, **35**, 1350-1375.
31. Cai, D., and Song, M. (2010) Recent advance in functionalized graphene/polymer nanocomposites. *Journal of Materials Chemistry*, **20**, 7906-7915.
32. Rafiee, M. A., Rafiee, J., Wang, Z., Song, H., Yu, Z. Z., and Koratkar, N. (2009) Enhanced me- chanical properties of nanocomposites at low graphene content. *ACS Nano*, **3**, 3884-3890.
33. Mittal, V., Chaudhry, A. U., and Luckachan, G. E. (2014) Biopolymer - thermally reduced graphene nanocomposites: Structural characterization and properties. *Materials Chemistry and Physics*, **147**, 319-332.
34. Vengatesan, M. R., Singh, S., Pillai, V. V., and Mittal, V. (2016) Crystallization, mechanical, and fracture behavior of mullite fiber-reinforced polypropylene nanocomposites. *Journal of Applied Polymer Science*, doi:10.1002/APP.43725
35. Vengatesan, M. R., Al Asafen, H. Y., Varghese, A. M., and Mittal, V. (2017) Effect of amine functionalized reduced graphene oxide on the thermal, mechanical, rheological and electrical properties of CPE compatibilizer HDPE. *In preparation*.

5

Polypropylene/Polystyrene Blend Nanocomposites with Graphene

5.1 Introduction

The physical polymer-reinforcement means such as blending or filler incorporation have gained favorable position for acquiring new polymeric materials with altered properties [1-3]. The properties of the resultant materials depend on features like interaction between the phases, compatibilization, processing conditions, composition, etc. The generation of polymer blends specially offers advantages such as property profile combinations, aid in processing, use of existing processing technologies, low capital costs, etc., as compared with the development of new monomers and polymers of required properties [4-7]. The advances in nanofillers and nanofibers have also made it possible to produce light weight nanocomposites with superior physical and mechanical properties at a low filler concentration than conventional composites. This is owing to the much higher number of interfacial contacts of inorganic particles with the polymer chains in such nanocomposites, which generate a completely different interfacial morphology as compared with the bulk polymer [8,9]. In the class of polymer nanocomposites, further development of blend composites, combining the useful features of polymer blending as well as filler incorporation has also resulted in many advanced materials recently.

For instance, Mittal and Matsko [10] recently reported poly(3-hydroxybutyrate-co-3-hydroxyvalerate) (PHBV) and poly(butylene adipate-co-terephthalate) (PBAT) based blend nanocomposites with vermiculite and graphene as reinforcing fillers. The authors observed co-continuous blend morphology in the developed blend composites. The filler phase (graphene and vermiculite platelets) was observed to be localized in the tougher PBAT phase due to its early melting, as shown in Figure 5.1, which helped to retain the toughness of the blend nanocomposites. Overall, owing to the generated composite

V. Mittal[a,], M. Pannirselvam[b] and Gregory Griffin[b]*
[a]Khalifa University of Science and Technology), Abu Dhabi, UAE; [b]Royal Melbourne Institute of Technology, Melbourne, Australia

morphology, simultaneous enhancement in both stiffness and toughness in the blend nanocomposites was achieved, which otherwise is quite challenging to attain by employing the conventional methodologies.

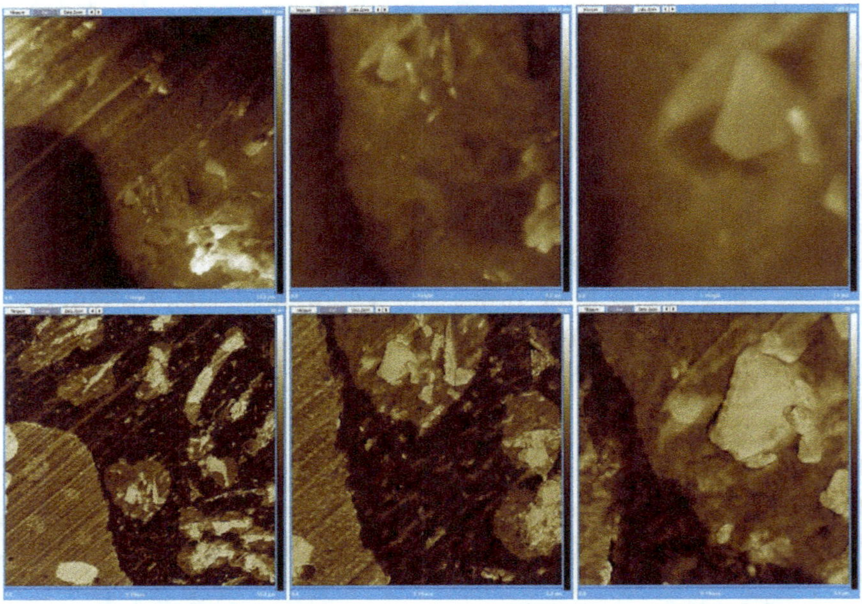

Figure 5.1 AFM height (top row) and phase (bottom row) images of the PHBV/PBAT/5% graphene composite [10]. In the images, the light phase corresponded to PHBV, whereas the dark phase contained PBAT. The filler platelets were observed as the bright platy inclusions.

Polyolefins are the largest group of industrial thermoplastics in terms of production and have superior rank among commodity plastics owing to their use in a large number of applications. The number of individual polyolefin members is limited and the number of applications based on such materials is constantly increasing. Various polyolefin-based blends and composites are, thus, required to meet the increasing demand of materials with required properties and processing suitable for specific applications [1,3]. A large variety of reinforcing fillers such as silica, clay, nanotubes and graphene have been explored as fillers in polyolefin matrices. Graphene, which consists of one atomic thick sheets of covalently sp^2-bonded carbon atoms in a hexagonal arrangement, has already received attention of researchers for the generation of polymer nanocomposites [11,12]. Its choice as a filler is due to its excellent electrical and mechanical properties, which are significantly better than other inorganic filler materials. A single defect-free graphene layer has Young's modulus of ~1.0 TPa, intrinsic strength ~42 N/m, thermal conductivity ~4840-5300 W/(m.K), electron mobility exceeding 25,000 $cm^2/V.s$, excellent gas impermeability and specific surface area of ~2630 m^2/g [11]. All these properties make this material superior to carbon nanotubes and layered silicates, which have been commonly used in polymer nanocomposites.

Among various polyolefins, polypropylene (PP) and polyethylene (PE) have outstanding properties including low cost, recyclability, good processability, high thermal and chemical resistance, non-toxicity, etc., and compete with many engineering plastics over a wide spectrum of application areas. However, these polymers suffer from limitations such as high oxygen permeation, shrinkage and relatively lower stiffness as compared to many engineering polymers [13-18]. To overcome these limitations, blends of these polyolefins with stiff polymers like polystyrene (PS) represent effective alternatives. For instance, Diaz *et al.* [19] and Xu *et al.* [20] reported the polyethylene/polystyrene blends and stabilized these using reactive compatibilization and incorporation of copolymer based compatibilizers. Other alternative is to combine these blends additionally with a functional filler, as mentioned earlier. For instance, Inberg *et al.* [21] reported the long glass fiber reinforcement of PP/PS blends. The authors observed that on the addition of 30 wt% of fibers, the modulus increased by a factor of 3.5, whereas the fracture stress was observed to increase by a factor of 2.5. In addition, the unnotched Izod impact strength in the blend composites increased by a factor of 10. In a recent study, Mittal *et al.* [22] also studied functional PE/PS blend nanocomposites with different fillers like silica, alumino-silicate and graphene. One step melt mixing process could be used for generating the composites, which resulted in the kinetic embedding of the filler particles in the PE phase due to its earlier melting during compounding. A fraction of the filler was also observed at the interface between PE and PS phases, as shown in Figure 5.2. The tensile modulus in the blend composites increased with filler content with maximum extent of increase provided by graphene platelets. High aspect ratio platelets of graphene were observed to be beneficial in generating good degree of stress transfer. In another study, Parameswaranpillai *et al.* [23] also reported polypropylene/polystyrene/carbon nanofiber nanocomposites and tailored the interface in the blend composite through

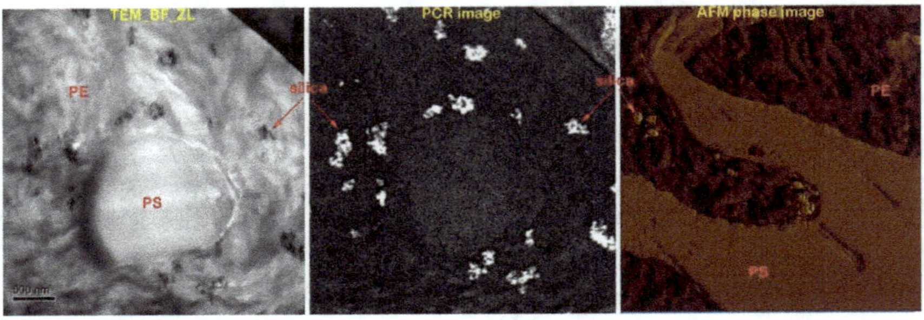

(a)

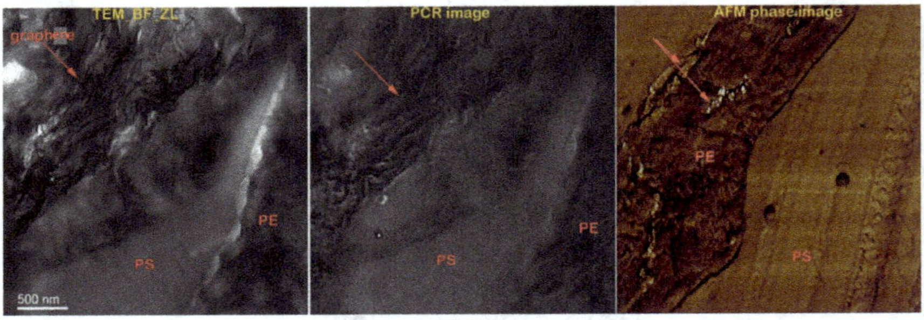

(b)

Figure 5.2 Morphological analysis of (a) $PE_{75}PS_{25}$/Silica and (b) $PE_{75}PS_{25}$/Graphene composites [22].

the addition of polystyrene-block-poly(ethylene-ran-butylene)-block-polystyrene.

In the current study, the blend nanocomposites of polypropylene and polystyrene have been further explored by using graphene as filler. Two different blend compositions were employed to study the effect on composite morphology and properties. In addition, in order to analyze the effect of additional compatibilizer on filler dispersion and subsequent composite performance, a small amount of ethylene-co-acrylic acid copolymer was also incorporated in the blend composites. As many earlier studies have reported the localization of the filler in the tougher polymer phase in the blend nanocomposites to be beneficial in retaining the overall toughness of the composites [10,22], therefore, processing conditions were also tuned in the current study so as to achieve similar morphology. In addition, due to the risk of thermal degradation of polymers above 200 °C, the compounding of the materials was carried out at 200 °C for optimal duration, so as to achieve good degree of phase intermixing, however, avoiding the thermal degradation of the polymers.

5.2 Experimental

5.2.1 Materials

Injection-molding PP grade RA130E-1498 was received from Abu Dhabi Polymers Company Limited. PS with average molecular weight (M_w) of 400,000 was procured from Sigma. Ethylene-co-acrylic acid (EAA) with a trade name of ESCOR 5050 was a product of Exxon Mobil. It has an acrylic acid content of 9 wt% and melt flow index (MFI, 190 °C and 2.16 kg) of 8.4 g/10 min. The EAA copolymer had a specific gravity of 0.936 and peak melting temperature of 97 °C. Graphene sample with trade name Grafen-iGP was kindly provided by Grafen Chemical Industries, Turkey. The polymers and filler were used as received, without further modification.

5.2.2 Generation of PP/PS Blends and Nanocomposites

Mini twin conical screw extruder (MiniLab HAAKE Rheomex CTW5, Germany) was used for the melt mixing of PP, PS and graphene to generate PP/PS (70/30 and 30/70) blends and their composites with 2 wt% graphene. In the composites incorporating EAA, 5% of the copolymer was added along with PP/PS (70/30 and 30/70) blends and 2% graphene. One step melt mixing process was used for the generation of blends and nanocomposites by pre-mixing the components physically followed by feeding into the compounder. The materials were dried overnight at 60 °C before performing the melt mixing. The screw length and screw diameter were 109.5 mm and 5/14 mm conical respectively. A mixing temperature of 200 °C and rotational speed of 70 rpm were used. Approximately 20 g of each sample was dry mixed prior to extrusion, and an initial mass of 6 g was added to the compounder. After 5 min, 3 g was flushed from the machine so that another 3 g could be added. This process was repeated until 20 g material was blended. This method ensured that the compounder would not be overloaded and also provided small batches to ensure more intense mixing without any significant thermal degradation of the polymer.

For the subsequent characterization, the disc, dumbbell and bar-shaped test samples were prepared by hot pressing the compounds at 200 °C for 5 min. The sample were cooled down to 50 °C by using flowing water through the plates by keeping a controlled cooling rate of 5 °C.min^{-1}.

5.2.3 Material Characterization

For the bright field transmission electron microscopy (TEM) analysis of the nanocomposite samples, Philips CM 20 (Philips/FEI, Eindhoven) electron microscope at 120 kV and 200 kV accelerating voltages was used. Thin sections of 70-90 nm thickness were microtomed from the sample blocks using a Leica Ultracut E microtome (Leica, Austria) equipped with a diamond knife (Diatome, Switzerland) at -120 °C. The so-obtained sections were supported on 100 mesh grids sputter coated with a 3 nm thick carbon layer. In addition, the block faces of the specimens were also investigated at ambient conditions using ZEISS Axioplan light microscope (Zeiss, Germany) equipped with Zeiss Axio Cam ICc 1 CCD camera in reflected polarized light.

X-ray diffraction analysis was conducted using a Bruker AXS D4 Endeavour system using Cu-Kα radiation operated at 40 kV and 40 mA and a Lynxeye linear strip detector. All polymer samples (hot pressed) were loaded onto the sample holder. Microscope slides were used to flatten the surface of all samples before loading. All samples were tested between 6 and 90° 2Θ with variable slit v20 with step size 0.02°/s. The instrument was calibrated using a NIST standard of Al_2O_3 (corundum ceramic block). Endeavour is equipped with a large sample magazine that allows a wide range of samples to be docked in a single stream (manual). In the current study, a search/match technique was used to compare the constituents present in the samples (ICDD Database, Bruker DIFFRAC.EVA software version 3.0). It should be noted that background was removed from all spectra before further processing. Diffraction peaks were not smoothened for any spectra as smoothing might vary area of the peaks of interest. Raman spectra were recorded on a Jobin Yvon Horiba LabRAM spectrometer (in the range 200-3200 cm^{-1}, data interval 2 cm^{-1}) with backscattered confocal configuration using a HeNe laser (633 nm). A long working distance objective with magnification 50x was used both to collect the scattered light and to focus the laser beam on the sample surface. CCD (charge-coupled device) detector exposure time was 200 s and an average of 2 cycles was used to increase S/N ratio. Raman shifts were calibrated using Si wafer signal at 520.7 cm^{-1}.

Thermal degradation properties of the samples were analyzed using TA Discovery thermogravimetric analyzer (TGA). Nitrogen was used as carrier gas and the scans were obtained from 50 to 750 °C at a heating rate of 10 °C.min^{-1}. Differential scanning calorimetric (DSC) measurements on the polymer blends and nanocomposites were performed using TA Discovery DSC under nitrogen atmosphere. The second heating scans were obtained by heating the samples from -20 °C to 200 °C at a heating rate of 10 °C.min^{-1}, followed by cooling from 200 °C to -20 °C at a cooling rate of 10 °C.min^{-1}.

Tensile analysis of the samples was carried out using universal testing machine (Instron, USA) (ASTM D638). The dumbbell-shaped samples were used. A loading rate of 10 mm.min^{-1} was used and the tests were carried out at room temperature. The tensile strength and modulus of the composites were calculated using Win Test Analysis software. An average of five sample values was reported. ARES rheometer was used to characterize the rheological properties such as storage modulus (G′), loss modulus (G″), as well as complex viscosity (η^*). Disc-shaped samples of 8 mm diameter and 2 mm thickness were measured using a gap opening of 1.2 mm. Strain sweep scans were recorded at 180 °C with ω = 1 rad.s^{-1} and 0.1% to 100% strain. For the comparison, temperature and frequency sweep scans (dynamic testing) of all the samples were recorded at 1% strain from ω = 0.1 to 100 rad.s^{-1} in the temperature range of 180-200 °C. DMA (dynamic mechanical analyzer) RSA3 of TA Instruments was used to study the change in modulus with temperature (-30 to 120 °C). Dynamic temperature ramp test was performed at a frequency of 5 rad.s^{-1}, ramp rate 3 °C.min^{-1} and 1% strain for all samples.

5.3 Results and Discussion

In the current study, blend nanocomposites of polypropylene and polystyrene with two different blend compositions were generated to study the effect of graphene filler on the composite morphology and properties. In addition, the blend composites were also generated using a small amount of polar copolymer as compatibilizer.

Figure 5.3 demonstrates the TEM micrographs of the PP/PS 70/30 and 30/70 blends. There was a difference in the phase morphology of the two blends. In the case of PP/PS 70/30 blend, though PP formed the continuous phase surrounding the large sized PS domains, however, it was also observed to be dispersed as spherical domains inside the PS phase. The size of these PP domains was 1-2 μm in diameter, indicating some phase intermixing between the PP and PS phases. As there is no thermodynamic interaction between the two polymers, therefore, the observed morphology would have solely resulted from the kinetic mixing in the compounder. In the case of PP/PS 30/70 blend, the polymer phases were more elongated in nature, though some degree of mixing of PP within the PS phase was still visible.

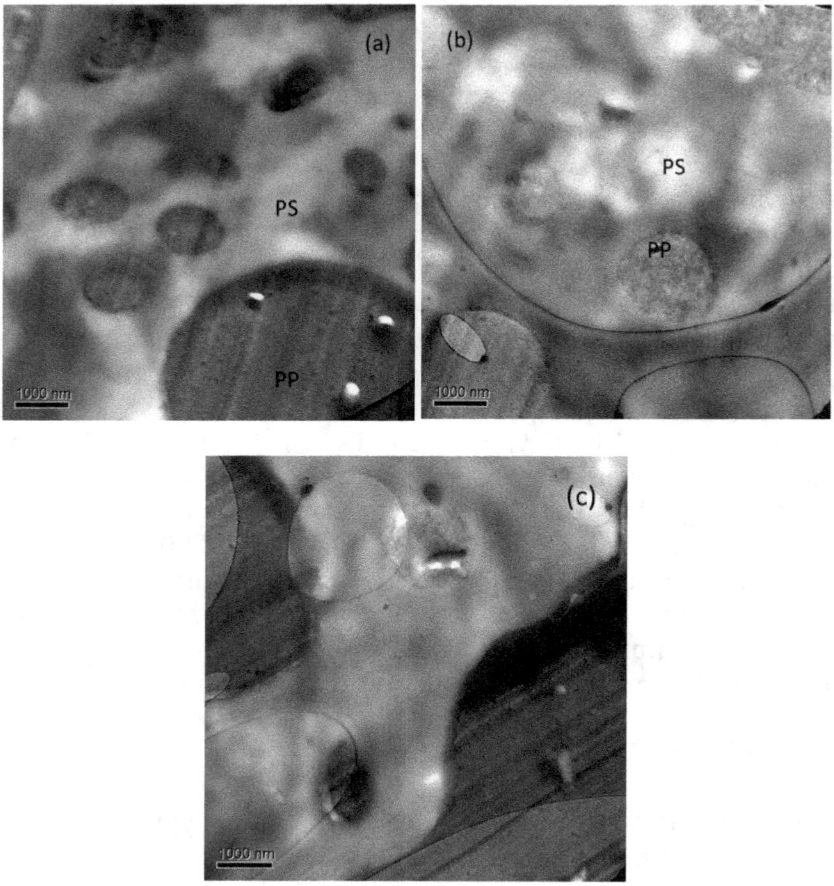

Figure 5.3 TEM images of the (a,b) PP/PS (70/30) and (c) PP/PS (30/70) blend.

Figures 5.4-5.5 present the optical images of the PP/PS (70/30)+2%G+5%EAA and PP/PS (30/70)+2%G+5%EAA blend composites. In these images, the dark phase was semi-crystalline PP, whereas amorphous PS was observed as the light phase. As observed in the TEM images of the PP/PS 70/30 blend, the composite exhibited the dispersion of the PS phase as large domains inside PP and a small amount of PP was also dispersed inside the PS domains. The filler particles (observed as white signal in the optical images) were observed to be largely localized inside the PP phase. Similarly, the PP/PS 30/70

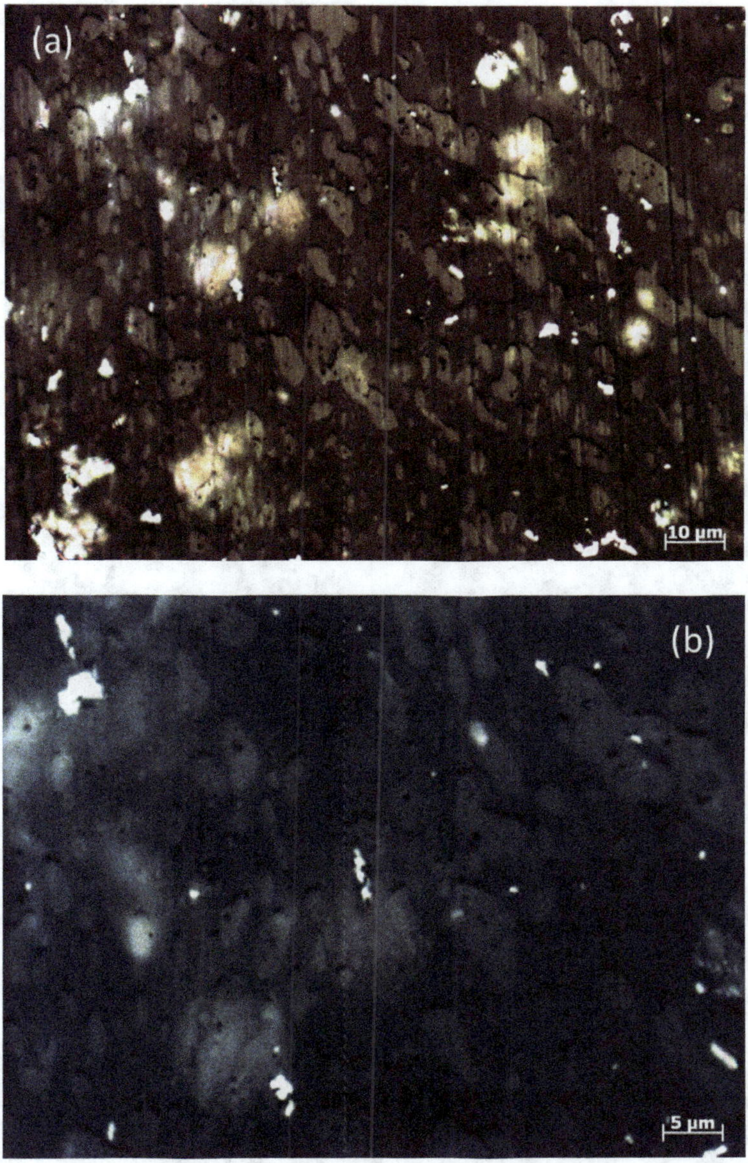

Figure 5.4 Optical images of the PP/PS (70/30)+2%G+5%EAA blend nanocomposite.

+2%G+5%EAA blend nanocomposite exhibited elongated domains with filler particles dispersed inside the PP phase. The observed composite morphology resulted from the one-step melt mixing approach used to generate the composites. As PP melted before PS inside the compounder, thus, the filler particles would have been mixed with the PP melt before the PS phase formed the melt. Therefore, the filler particles could be localized largely inside the PP phase. Occasionally, the filler platelets were also observed at the interface between the

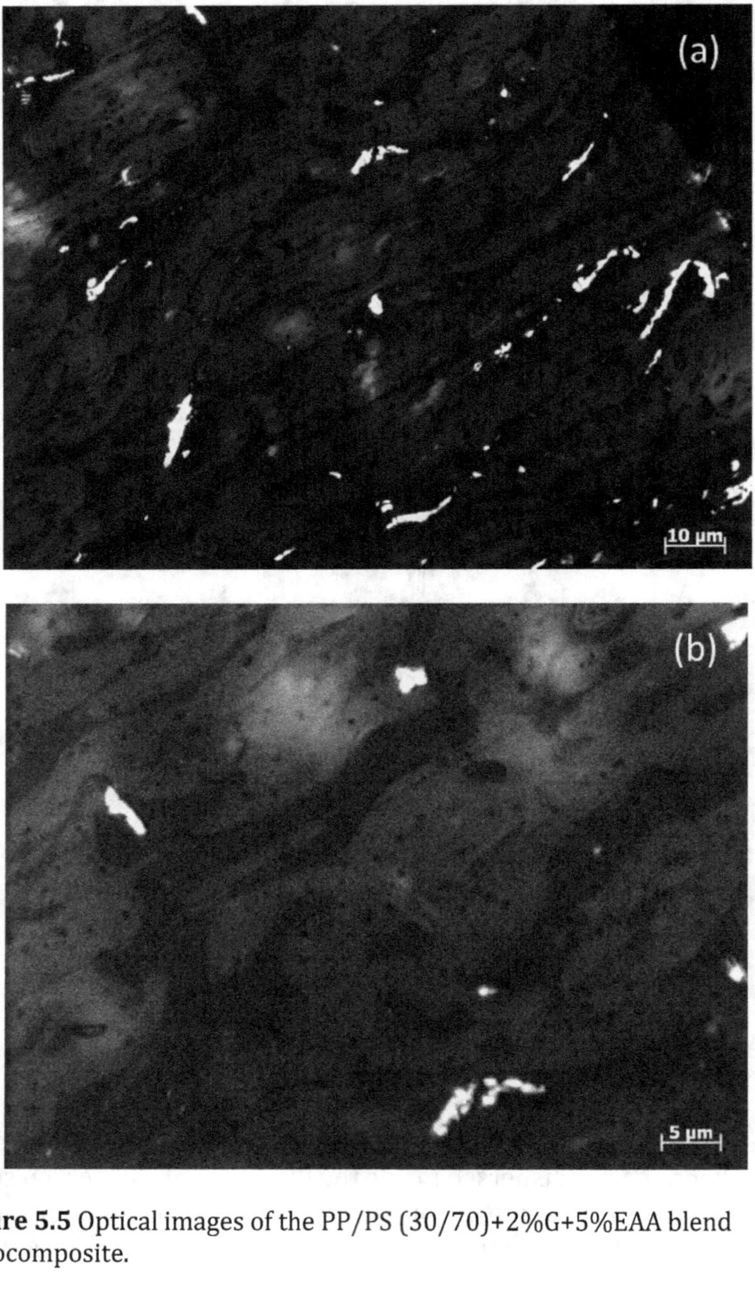

Figure 5.5 Optical images of the PP/PS (30/70)+2%G+5%EAA blend nanocomposite.

PP and PS phases, thus, providing some degree of interface compatibilization, which has been reported in the literature to be beneficial in enhancing phase intermixing by reducing the domain sizes. EAA could not be specifically observed in the composite morphology. As EAA had much lower melting point, thus, it is expected to have melt mixed with filler particles, followed by its mixing in the PP phase. As mentioned earlier, localization of the filler particles inside the PP phase is expected to be beneficial in retaining the overall flexibility of the composites. The morphology of the composites was also evaluated further using TEM, as shown in Figure 5.6. Similar to the findings from optical microscopy, the filler was localized in the PP phase and

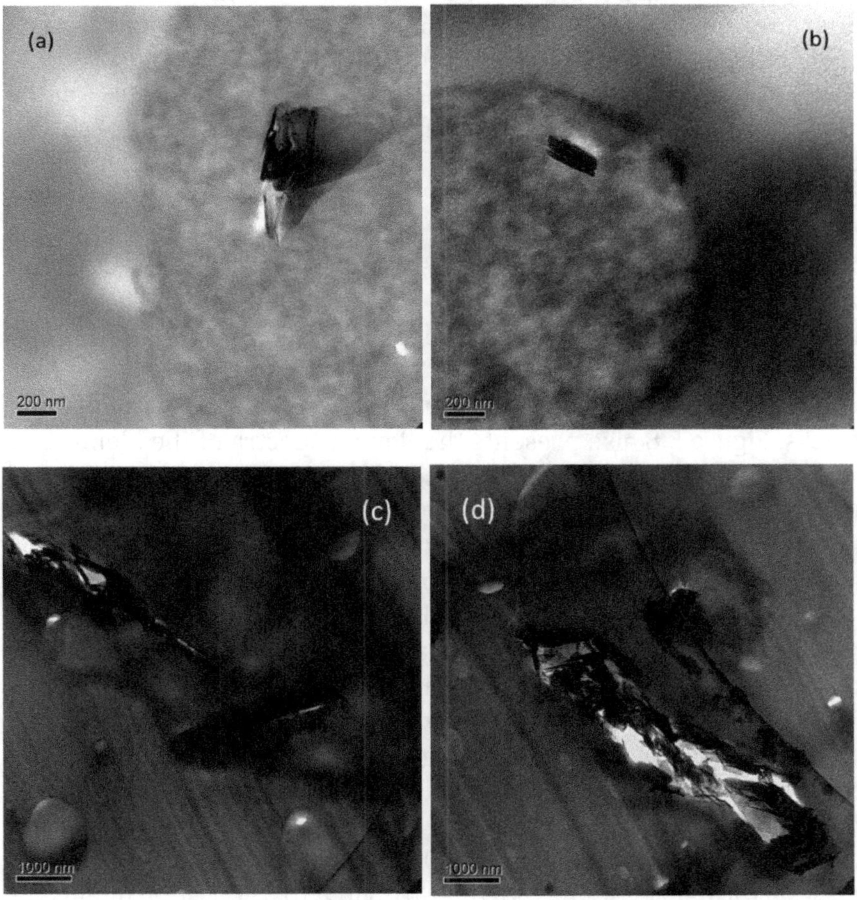

Figure 5.6 Transmission electron micrographs of the (a,b) PP/PS (70/30)+2%G+5%EAA and (c,d) PP/PS (30/70)+2%G+5%EAA blend nanocomposite.

the filler particles were also observed at the interface between PP and PS on many occasions (Figure 5.6 c and d). In addition, the filler was observed to be present mostly in the form of stacks rather than single layers, thus, indicating the need to further delaminate the filler platelets. Probably higher content of EAA is required to enhance the compatibilization of filler and the polymer phases. Another alternative would be to generate a masterbatch of EAA and graphene, which can then be added to PP and PS for effective mixing [24]. However, it should be noted that observed composite morphology can still be expected to result in enhanced composite properties owing to overall good degree of filer dispersion and large aspect ratio graphene platelets. The phase distribution in the composites was also retained even after multiple processing cycles, indicating stable morphology even though the components were mixed only kinetically.

Figure 5.7 also demonstrates the X-ray diffractograms of the blends and blend composites. No significant difference was observed between the diffraction patterns of blends and their composites indicating that the crystal structure of PP was not affected in the presence of PS, filler and EAA, even though the degree of crystallinity and crystal size may have been affected due to the hindrance to the crystallization of PP chains in the presence of other components. The diffraction peak corresponding to graphene was also observed in the composites as a sharp signal at $2\Theta \sim 27°$, indicating that the periodicity was retained in the filler due to the presence of multilayered platelet stacks. Figure 5.8 also presents the Raman spectra of the blends and composites. The characteristic D and G bands of graphene overlapped with the signals originating from the polymer phases. In addition, no new Raman shift was observed, which was expected due to the physical mixing of the composite components.

Insights into the thermal degradation of the blends and nanocomposites were gained through TGA, as demonstrated in Figure 5.9 for the PP/PS 70/30 and 30/70 blends and their composites. In PP/PS 70/30 blend, the onset of degradation was observed around 350 °C, which was observed to be unchanged in the composite of the 70/30 blend with 2% graphene. However, an increment in the onset of degradation temperature was observed in the case of composite with 5 wt% EAA. Generally, the compatibilizers tend to deteriorate the thermal performance of the polymers owing to their low molecular weight. However, the observed increment in the thermal degradation performance indicated that the filler dispersion must have been enhanced in comparison with the composite without EAA addition, thus

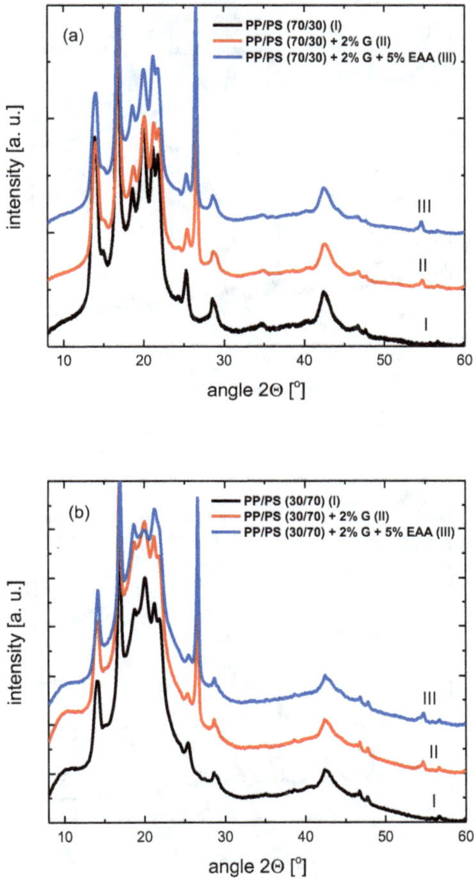

Figure 5.7 X-ray diffractograms of (a) PP/PS (70/30) blend and its composites; (b) PP/PS (30/70) blend and its composites.

enabling the polymer chains to have effective heat transfer with the high aspect ratio graphene platelets through increased extent of interfacial contacts. Though the enhancement in the onset degradation temperature in the composite was about 15 °C, however such an increase is also significant for generating effective thermal resistance. Similar enhancement was also observed in the case of PP/PS 30/70 blend composite system. In fact, the composite without EAA was also observed to enhance the thermal stability of the polymer in this case, though the composite with EAA still exhibited higher extent of thermal resistance enhancement. This confirmed that the addition of EAA was beneficial in achieving better thermal resistance owing to higher

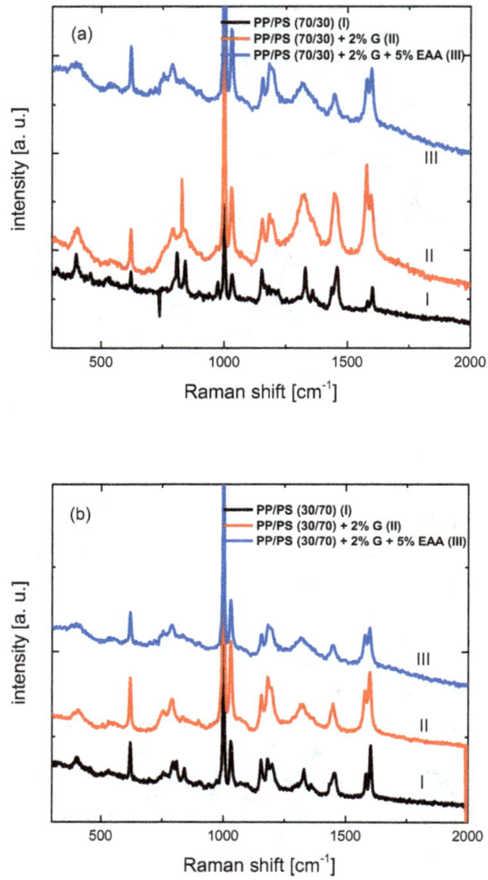

Figure 5.8 Raman spectra of (a) PP/PS (70/30) blend and its composites; (b) PP/PS (30/70) blend and its composites.

extent of filler dispersion. In addition, enhancement of thermal stability in the composite without EAA could be attributed to the blend composition. Lower extent of PP in the blend would have resulted in higher extent of its reinforcement due to filler dispersion solely in PP phase, in comparison with the blend with 70 wt% PP, thereby, leading to an enhancement in the overall thermal performance. Thus, the blend nanocomposites were suitable for high temperature processing and applications. Dynamic TGA carried out the compounding temperature of 200 °C as a function of time also revealed the stability of the composites, as no significant weight loss occurred for the test duration equivalent to the compounding time.

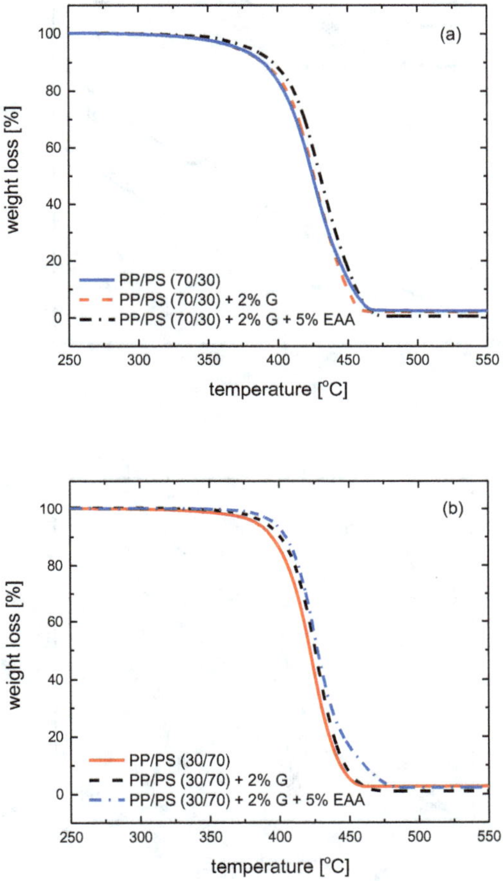

Figure 5.9 TGA thermograms of (a) PP/PS (70/30) blend and its composites; (b) PP/PS (30/70) blend and its composites.

Figures 5.10-5.11 also present the melting and crystallization thermograms of PP/PS blends and their composites. As observed in Figure 5.10(a), the melting behavior of the PP phase in the PP/PS 70/30 blend remained unaffected in the PP/PS (70/30)+2%G composite. However, the melting behavior was observed to be affected on the addition of EAA in the composites, as indicated by the decreased peak melting temperature. The melt enthalpy of PP (normalized to PP weight in the composite) was also observed to decrease in this case. It, thus, confirmed that even though the crystal structure of PP was not affected in the composite with EAA (as observed from X-ray), the degree of crystallinity was affected probably due to the hindrance to

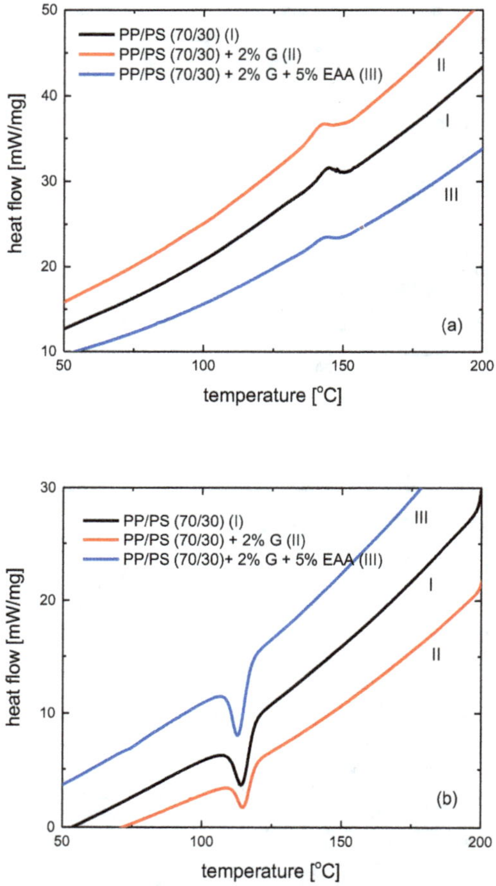

Figure 5.10 DSC (a) melting and (b) crystallization thermograms of PP/PS (70/30) blend and its composites.

crystallization in the presence of EAA. It, thus, also indicated the EAA was present in the PP phase, as opined earlier through findings from microscopy. Similar to the melting thermograms, the crystallization thermogram of the PP/PS (70/30)+2%G+%EAA composite also revealed slight reduction in the peak crystallization temperature (Figure 5.10(b)), thus, confirming again the hindrance to polymer crystallization. No change in the peak crystallization temperature was observed in the case of PP/PS (70/30)+2%G composite in comparison with the PP/PS (70/30) blend. In the case of PP/PS (30/70) blend, the endothermic transition corresponding to the crystalline melting of PP phase was significantly suppressed, indicating the higher extent

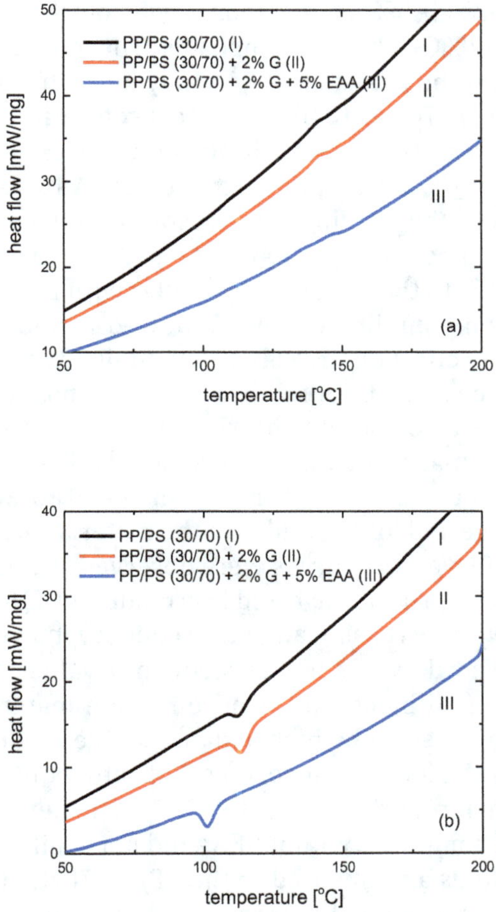

Figure 5.11 DSC (a) melting and (b) crystallization thermograms of PP/PS (30/70) blend and its composites.

of hindrance to PP crystallization in the blend. The composite without EAA also exhibited similar transition at the same peak melting temperature. However, the composite with EAA exhibited further suppression of the peak melting temperature as well as melt transition in the composite. Similarly, the crystallization behavior was also impacted in the case of PP/PS (30/70)+2%G+5%EAA composite as a large decrease in the peak crystallization temperature of PP was observed in this case. In comparison, the peak crystallization temperature in the composite without EAA was not affected and was observed to be the same as the PP/PS 30/70 blend. It should also be noted that the extent of reduction in the peak crystallization temperature of the PP/PS (30/70)+2%G+5%EAA composite was much higher than the PP/PS (70/30)+2%G+5%EAA composite. The effect of enhanced filler dispersion in the presence of EAA may have affected the PP/PS 30/70 composite more due to the higher concentration of filler in the PP phase in this case, thereby resulting in much higher hindrance to the PP crystallization. Overall, the presence of EAA was observed to significantly affect the melting and crystallization behavior of PP, therefore, confirming that the EAA was present in the PP phase, which resulted due to the melt mixing of filler with EAA during initial stages of melt compounding process, followed by melt mixing with PP and PS. Such an effect on the polymer crystallinity is also expected to impact the mechanical performance of the composites.

Figures 5.12-5.14 elaborate the rheological performance of the composites in comparison with the pure blends. The measurements were performed over the temperature range of 180-200 °C. As expected, the storage and loss modulus of the materials decreased with temperature, however, the curves were observed to merge at higher angular frequencies. The storage and loss modulus of the composites also increased with increasing angular frequency, however, the rate of increase slowed down at higher frequency values due to shear thinning effect. In all blend and nanocomposite materials, the loss modulus was observed to be higher than storage modulus over the complete angular frequency range, thus, indicating the dominance of the viscous behavior. In the case of PP/PS 70/30 blend nanocomposites system, the composite without EAA did not exhibit increment in the storage modulus as compared to the PP/PS 70/30 blend. On the other hand, an increment in the storage modulus was observed in the composite with EAA, thus, confirming the earlier findings of enhanced filler dispersion in the presence of EAA, thereby leading to effective stress transfer from the polymer chains to the platelets. In the case of PP/PS 30/70 blend nanocomposites system, a large increase in the modulus was observed in the composites with and without EAA, therefore, confirming the findings from TGA about the enhancement in the performance due to the localization of filler in the lower fraction PP phase, which together with higher modulus PS phase resulted in overall much higher degree of increment. As observed in Figure 5.14, the tan δ of the composites decreased in comparison with the pure blends, thus, indicating the increment in the elastic contribution. The decrease in tan δ was more significant in the PP/PS 70/30 composite with EAA as well as both PP/PS 30/70 composites due to

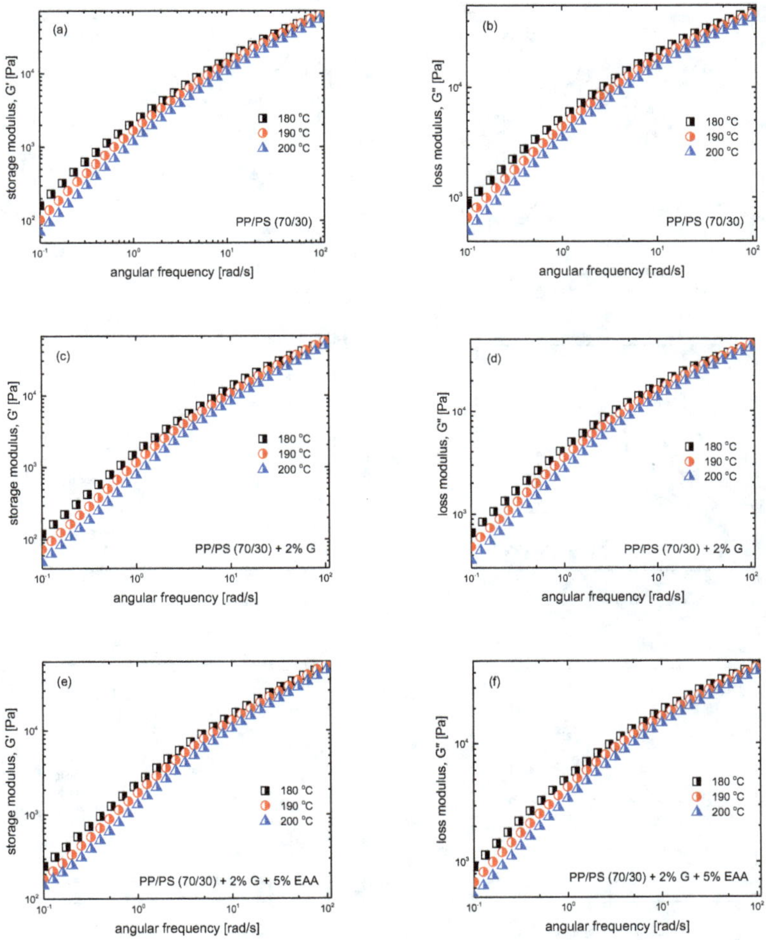

Figure 5.12 (a,c,e) Storage and (b,d,f) loss modulus of PP/PS (70/30), PP/PS (70/30)+2%G and PP/PS (70/30)+2%G+5%EAA respectively, as a function of angular frequency in the temperature range of 180-200 °C.

the same reasons observed earlier.

Dynamic mechanical analysis of the PP/PS blends and their nanocomposites is also presented in Figure 5.15 as a function of temperature (-30 to 120 °C). In the PP/PS 70/30 composite system, the composites with EAA exhibited an increment in the storage modulus at lower temperatures, whereas the composite without EAA did not exhibit much enhancement in the modulus. These findings corroborated with earlier observations from the TGA and rheological analysis

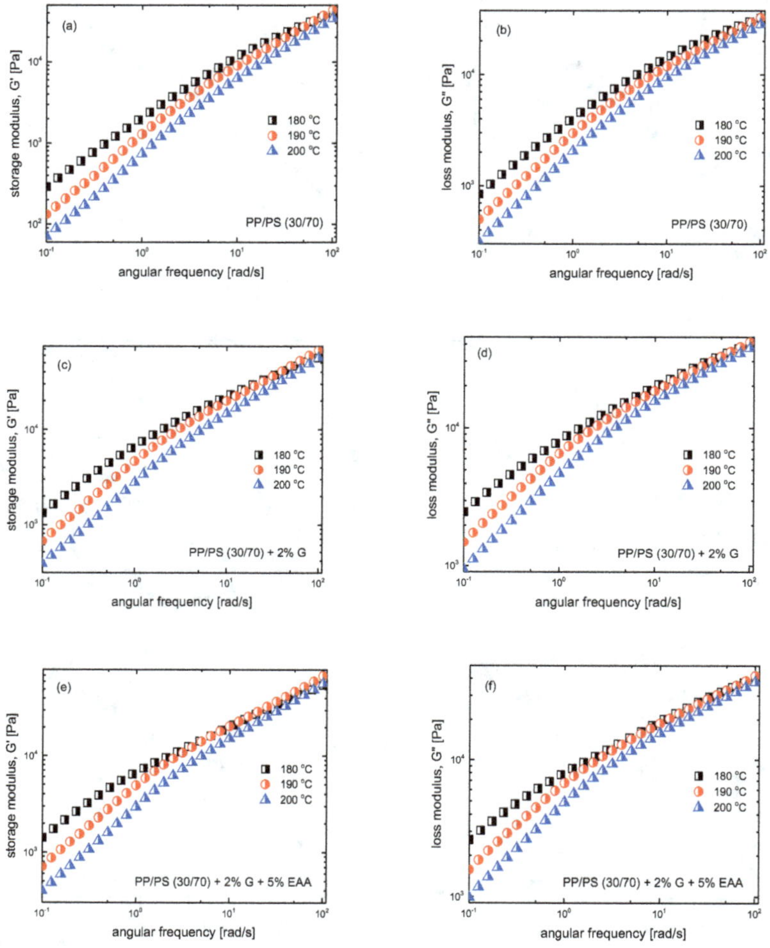

Figure 5.13 (a,c,e) Storage and (b,d,f) loss modulus of PP/PS (30/70), PP/PS (30/70)+2%G and PP/PS (30/70)+2%G+5%EAA respectively, as a function of angular frequency in the temperature range of 180-200 °C.

where higher extent of filler dispersion in the presence of EAA was observed to lead to effective heat and stress transfer. It should be noted that the modulus of the composite increased even though the crystallinity of the composite with EAA was observed to decrease due to the hindrance to crystallization caused by EAA. Thus, the observed increment in modulus in the EAA composite resulted from the combined action of decrease in properties due to reduced crystallinity and increased modulus due to enhanced extent of filler dispersion. In

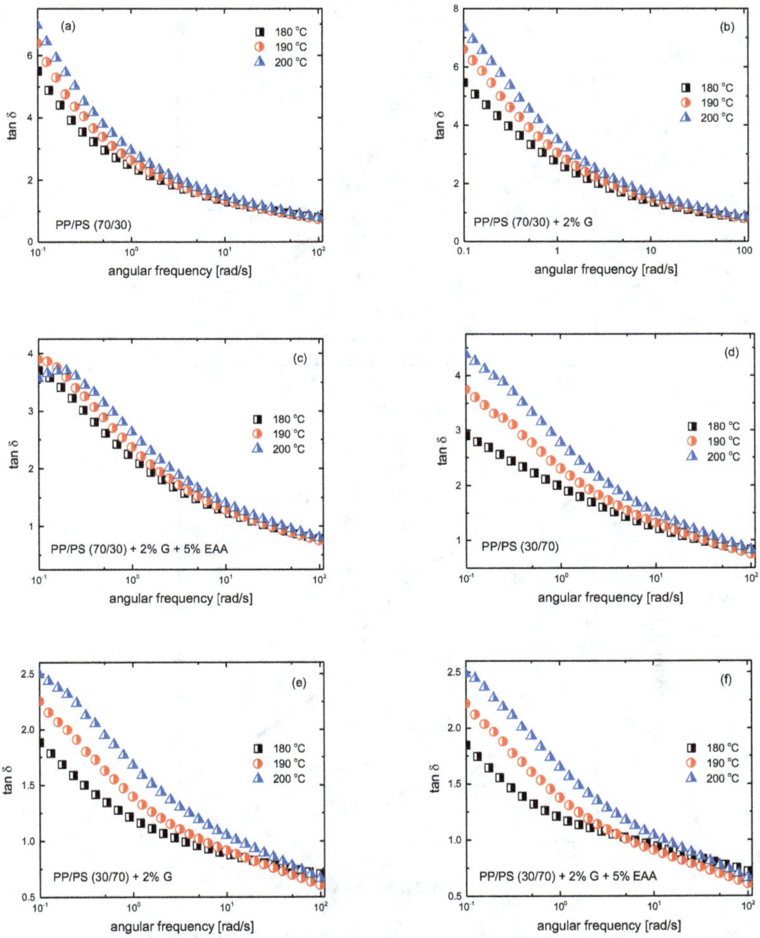

Figure 5.14 tan δ of the materials as a function of angular frequency in the temperature range of 180-200 °C: (a) PP/PS (70/30), (b) PP/PS (70/30)+2%G, (c) PP/PS (70/30)+2%G+5%EAA, (d) PP/PS (30/70), (e) PP/PS (30/70)+2%G and (f) PP/PS (30/70)+2%G+5%EAA.

the case of PP/PS 30/70 composites, both composites (with and without EAA) resulted in an enhancement of storage modulus at lower temperatures, which also correlated with the earlier findings. Reinforcement of the lower fraction PP phase with the filler led to an overall enhancement of the modulus of the composites. The composites also exhibited sharp reduction in the modulus around 30 °C, irrespective of the blend composition. Even though the modulus decreased on

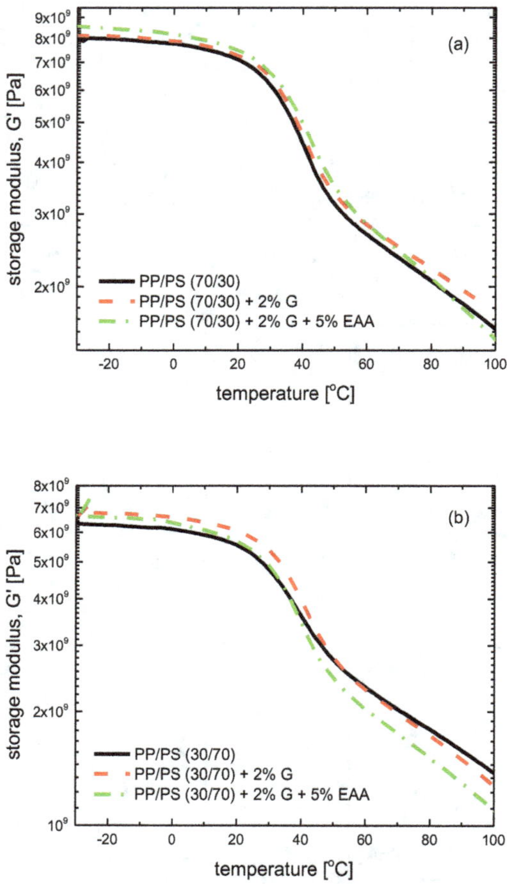

Figure 5.15 Dynamic mechanical analysis of (a) PP/PS (70/30) blend and its composites; (b) PP/PS (30/70) blend and its composites, as a function of temperature.

on increasing the temperature further, however, the modulus of the composites still remained appreciable even at 100 °C.

Figure 5.16 and Table 5.1 depict the tensile properties of the blends and nanocomposites. The PP/PS 70/30 blend exhibited a tensile modulus of 1517 MPa, which increased to 1589 MPa in the composite with 2 wt% graphene. The modulus was further enhanced marginally to 1618 MPa in the composite where EAA was also added, due to enhanced level of filler dispersion. The effect of enhanced filler dispersion on the addition of EAA copolymer was also observed on tensile strength, strain at break as well as stress at break. All of these

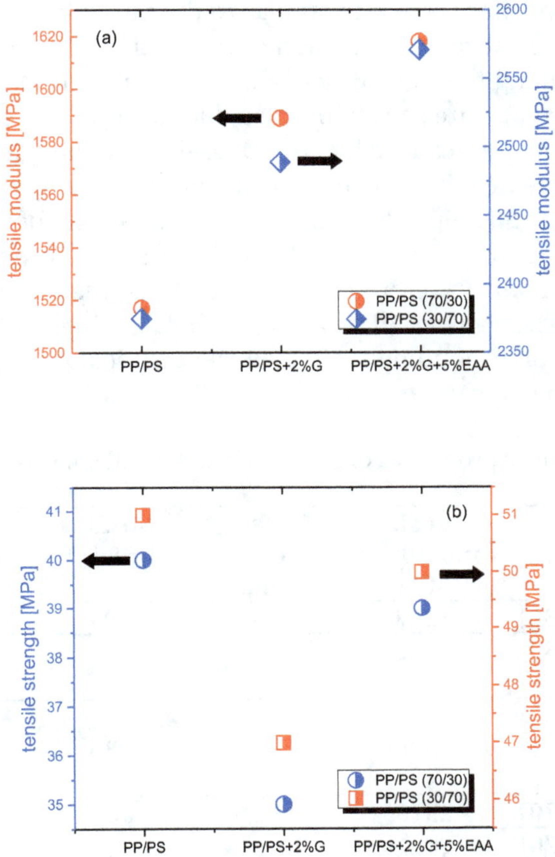

Figure 5.16 Tensile properties of (a) PP/PS (70/30) blend and its composites; (b) PP/PS (30/70) blend and its composites.

quantities were observed to be improved in the case of PP/PS (70/30)+2%G+5%EAA composite, in comparison with the composite without EAA. For instance, the strain at break of the composite with EAA was almost similar to the pure blend, which indicated that the modulus of the composites was enhanced without negatively impacting the flexibility. In the case of PP/PS 30/70 composites, the tensile modulus for the EAA containing composite exhibited higher extent of increment than the composite without EAA due to the combined effect of reinforcement of PP phase by the filler as well as higher extent of filler dispersion in the presence of EAA, which was achieved even though the crystallinity of PP phase was significantly reduced. The tensile strength, strain at break and stress at break for the EAA containing composites were also similar to the pure blend. It, thus, indicated that the specific localization of the filler in the PP phase led to largely retaining the flexibility of the composites in the case of EAA containing composites, probably also due to some degree of plasticization of the matrix caused by EAA. Overall, even though the filler did not exfoliate extensively as observed from microscopy analysis, the EAA addition still helped to achieve better composite performance by enhancing the overall extent of filler dispersion, thereby leading to the enhancement in interfacial contacts, thus, resulting in effective heat and stress transfer. In addition, the localization of filler in the PP phase also correspondingly affected the composite properties, as a function of blend composition.

Table 5.1 Tensile properties of the blends and blend nanocomposites

Material	Tensile modulus, MPa	Tensile strength, MPa	Strain at break, %	Stress at break, MPa
PP/PS (70/30)	1517	40	9.2	38
PP/PS (70/30) + 2% G	1589	35	6.5	33
PP/PS (70/30) + 2% G + 5% EAA	1618	39	8.1	37
PP/PS (30/70)	2375	51	6.3	46
PP/PS (30/70) + 2% G	2489	47	4.8	42
PP/PS (30/70) + 2% G + 5% EAA	2571	50	5.9	46

5.4 Conclusions

The current study demonstrated the successful generation of blend nanocomposites of polypropylene and polystyrene with graphene and ethylene-co-acrylic acid copolymer. PS/PP blends with two different compositions were employed. Single step melt mixing method was used for the generation of blend nanocomposites. The morphology of the blends was affected by the blend composition. The blend with PP/PS 70/30 ratio exhibited spherical domains of PS inside the

PP phase. In addition, smaller spherical domains of PP were also present inside the PP phase. The blend with PP/PS 30/70 exhibited the morphology with more elongated conformation of the components. The composites exhibited the localization of the filler particles in the PP phase due to the earlier melting of PP during the melt mixing process. Occasionally, the filler particles were also observed to be present at the interfaces of PP and PS phases. In addition, the presence of EAA was suggested to be in the PP phase owing to the much earlier melting of EAA during the compounding process. The filler did not delaminate extensively in the composites. The X-ray diffractograms of the composites also revealed the characteristic diffraction peak of graphene, indicating that the periodicity of the filler platelets was largely retained in the composites. The thermal degradation of the PP/PS (70/30)+2%G+5%EAA was observed to be enhanced in comparison with the PP/PS (70/30)+2%G composite, thereby, signaling the effect of EAA on the filler dispersion. The composites with PP/PS (30/70) blend had also enhanced thermal stability, though the addition of EAA led to higher extent of increment. The reinforcement of minority PP phase by the filler would have resulted in effective thermal resistance in the case of PP/PS (30/70)+2%G composite, whereas the composite with EAA additionally benefited from the positive interaction of EAA with filler surface thereby leading to enhanced extent of filler dispersion. The peak melting and crystallization temperatures were observed to decrease in the PP/PS (70/30)+2%G+5%EAA composite due to the additional hindrance caused by EAA chains to the PP crystallization. The melt enthalpy was also observed to decrease in this composite, thus, indicating reduced extent of crystallinity, whereas no such effect was observed for the PP/PS (70/30)+2%G composite. More significant effect on melting and crystallization of PP was observed in the case of PP/PS (30/70)+2%G+5%EAA. Probably, the effect of enhanced filler dispersion in the presence of EAA affected the PP/PS 30/70 composite more due to the higher concentration of filler in the PP phase. These findings also corroborated the earlier suggestion from microscopy analysis that EAA was present in the PP phase. The storage modulus of the PP/PS (30/70) based composites was observed to enhance significantly owing to the superior reinforcement of the minority PP phase, as mentioned earlier. In addition, the EAA containing composite of PP/PS (70/30) blend also exhibited higher storage modulus. The tan δ of the composites also decreased as compared to the pure blends due to the increment in the elastic contribution. Similar to the storage modulus observed from melt rheology, the dynamic mechanical analysis also exhibited an increment in the storage modulus of the composites due to the same reasons. It should also to be noted that the increment in the modulus of EAA containing composites was achieved even though the crystallinity was negatively impacted. Thus, the observed enhancement resulted from the competing factors of reduction in crystallinity leading to deterioration of composite performance and enhanced extent of filler dispersion resulting in increment of composite properties. Similar to rheology and dynamic mechanical analysis, the tensile analysis of the composites also revealed increment in the modulus due to the incorporation of EAA. Higher extent of increment in the modulus was observed in the case of PP/PS (30/70)+2%G+5%EAA. In addition, the composites with EAA also retained the tensile strength, elongation at break as well as stress at break due to filler localization in PP. Some degree of plasticization resulting from the addition of EAA may have also helped to enhance the flexibility of the EAA containing composites. Overall, the generated blend nanocomposites exhibited superior properties than the corresponding blends and are more useful alternatives than not only the parent materials but also other thermoplastics due to optimal properties. In addition, the morphology of the blend nanocomposites was also stable even after few high temperature processing cycles, though the components were only kinetically mixed. The composite performance can be enhanced further by employing strategies such as enhancing the content of polar copolymer, generating masterbatch of the filler and copolymer prior to melt mixing with PP and PS, and surface functionalization of graphene surface to generate chemical moieties which can chemically interact with the copolymer chains leading to reactive compatibilization.

Acknowledgement

The authors are indebted to Dr. Nadejda Matsko and Bradley Burren for support in material analysis.

References

1. Jancar, J. (1999) *Mineral Fillers in Thermoplastics I: Raw Materials and Processing*, Springer-Verlag, Germany.
2. Mittal, V. (2012) *Functional Polymer Blends: Synthesis, Properties &*

Performance, CRC Press, USA.
3. Hippi, U. (2008) *Polyolefin Composites*, John Wiley & Sons, USA.
4. Fekete, E., Foldes, E., and Pukanszky, B. (2005) Effect of molecular interactions on the miscibility and structure of polymer blends. *European Polymer Journal*, **41**, 727-736.
5. Lipatov, Y. S. (2002) Polymer blends and interpenetrating polymer networks at the interface with solids. *Progress in Polymer Science*, **27**, 1721-1801.
6. Litmanovich, A. D., Plate, N. A., and Kudryavtsev, Y. V. (2002) Reactions in polymer blends: interchain effects and theoretical problems. *Progress in Polymer Science*, **27**, 915-970.
7. Sundararaj, U., and Macosko, C. W. (1995) Drop breakup and coalescence in polymer blends: The effects of concentration and compatibilization. *Macromolecules*, **28**, 2647-2657.
8. Nwabunma, D., and Kyu, T. (2008) *Polyolefin Composites*, Wiley, USA.
9. Liu, H., and Brinson, L. C. (2008) Reinforcing efficiency of nanoparticles: A simple comparison for polymer nanocomposites. *Composites Science and Technology*, **68**(6), 1502-1512.
10. Mittal, V., Burren, B., and Matsko, N. (2017) Eco-friendly bio-nanocomposites of PHBV/PBAT blend with graphene and vermiculite. In: *Biopolymer Based Advanced Materials*, Mittal, V. (ed.), Central West Publishing, Australia, pp. 57-85.
11. Mukhopadhyay, P., and Gupta, R. K. (2011) Trends and frontiers in graphene-based polymer nanocomposites. *Plastics Engineering*, **67**, 32-42.
12. Mittal, V. (2014) Functional polymer nanocomposites with graphene: A review. *Macromolecular Materials and Engineering*, **299**, 906-931.
13. Katz, H. S., and Milewski, J. V. (1987) *Handbook of Fillers for Plastics*, Van Nostrand, USA.
14. Rothon, R. (1995) *Particulate-Filled Polymer Composites*, Longman, UK.
15. Brydson, J. A. (1975) *Plastic Materials*, Newnes-Butterworths, UK.
16. Karger-Kocsis, J. (1995) *Polypropylene Structure, Blends and Composites*, Chapman & Hall, UK.
17. Elias, H.-G. (2003) *Anwendungen von Polymeren*, Wiley-VCH, Germany.
18. Robertson, G. L. (1993) *Food Packaging - Principles and Practice*, volume 6 of Packaging and Converting Technology, Marcel Dekker, Inc., USA.
19. Diaz, M. F., Barbosa, S. E., and Capiati, N. J. (2007) Reactive compat-

ibilization of PE/PS blends. Effect of copolymer chain length on interfacial adhesion and mechanical behavior. *Polymer*, **48**, 1058-1065.

20. Xu, S.-A., and Cham, C.-M. (1998) Polystyrene/high density polyethylene blends compatibilized by a tri-block copolymer I. Properties and morphology. *Polymer Journal*, **30**, 552-558.
21. Inberg, J. P. F., Hunse, P. H., and Gaymans, R. J. (1999) Long fiber reinforcement of polypropylene/polystyrene blends. *Polymer Engineering and Science*, **39**(2), 340-346.
22. Mittal, V., Chaudhry, A. U., Patwary, F., and Matsko, N. (2017) Filler distribution and properties of polyethylene/polystyrene blend composites with 0-D and 2-D nanomaterials. In: *Functional Polymer Blends and Nanocomposites*, Mittal, V. (ed.), Central West Publishing, Australia, pp. 1-34.
23. Parameswaranpillai, J., Dubey, V. K., Sisanth, K. S., Jose, S., Zachariah, A. K., Siengchin, S., Salim, N. V., and Hameed, N. (2016) Tailoring of interface of polypropylene/polystyrene/carbon nanofibre composites by polystyrene-block-poly(ethylene-ran-butylene)-block-polystyrene. *Polymer Testing*, **51**, 131-141.
24. Chaudhry, A. U., and Mittal, V. (2013) High-density polyethylene nanocomposites using masterbatches of chlorinated polyethylene/graphene oxide. *Polymer Engineering and Science*, **53**, 78-88.

6

Effect of Hygrothermal Ageing on Mechanical, Thermal, Structural and Morphological Properties of Polypropylene and its Blends with Different Copolymers

6.1 Introduction

Polymeric materials are extensively useful for a wide range of industrial, structural, engineering, energy and environmental applications owing to their benefits over other conventional materials. These benefits include satisfactory performance, adjustable properties, light weight structures, low cost, superior corrosion resistance, user-friendly processing, manufacturing and handling, etc. [1-6]. In particular, polypropylene (PP) is a versatile commodity thermoplastic material comprising of many of the above mentioned benefits [7,8]. In order to successfully meet the increasing demand of the polymeric materials with superior property profiles, a large number of physical and chemical approaches for the modification of PP have been developed. Among these approaches, blending PP with other polymeric materials including elastomers, copolymers, thermoplastics, etc., as well as reinforcing PP with functional fillers having nano- or micro-dimensions, have displayed wide commercial acceptance [9-16].

Polymer blending is one of the most common approaches for the development of high performance polymeric materials in order to meet the continuously increasing demands for various functional applications [17-19]. In the case of PP, the blending technique presents importance to acquire enhanced impact performance, which opens opportunities for its use in engineering applications. Accordingly, various copolymers such as styrene-butadiene rubber (SBR) [20], styrene-butadiene-styrene (SBS) [21,22], styrene-ethylene-butylene-styrene (SEBS) [23,24], ethylene-propylene rubber (EPR) [25,26], ethylene-propylene-diene monomer (EPDM) [27,28], ethylene-octene copolymer (EOC) [29-31], ethylene-1-hexene copolymer (EHC) [32] and ethylene-vinyl acetate (EVA) [33] have been blended with polypropylene for the generation of the toughened

Anish Varghese and Vikas Mittal
Khalifa University of Science and Technology, Abu Dhabi, UAE

grades. Many copolymers also enhance the other physical properties of PP.

The service life of PP based materials in various applications, particularly in outdoor, greatly depends on their performance in hostile environments like humidity, varying temperature and pressure, sunlight, ultraviolet (UV) irradiation, ice or marine conditions, pollutants, chemicals, mechanical loading, biological surroundings, etc., along with their combination [34-36]. Accelerated ageing under simulated hostile environmental conditions in laboratory can be used to predict the material behavior instead of analyzing the long-term natural ageing. Using the accelerated ageing, information related to the life time, maintenance and replacement of the materials can be extracted [37,38]. According to the previous studies, ageing induced changes in the durability of the polymeric materials are related to diminishing mechanical and physical properties owing to the significant structural changes, especially under hygrothermal conditions [39-41]. In addition, the chemical nature of the material as well as time, temperature and medium of immersion present significant influence on the structural changes during hygrothermal ageing [42,43]. Usually, the influence of moisture leads to the detrimental effects such as swelling and degradation of polymers even under normal conditions [6,44-46]. In this respect, accelerated hygrothermal ageing of water immersed samples imitates very aggressive condition which causes moisture diffusion. The introduction of temperature also speeds up such moisture diffusion by following Fickian and/or non-Fickian path [35,44,47-49].

In this study, three sets of PP/copolymer blends in 90/10 and 80/20 weight ratios have been developed. The copolymers chosen to toughen PP by using direct melt blending in a mini twin screw extruder included ethylene vinyl acetate with maleic anhydride grafting (EVA-MA), ethylene butyl acrylate (EBA) and ethylene alpha olefin (EAO). The effect of hygrothermally induced ageing on the properties of neat PP and blend samples was subsequently studied. Simulated hygrothermal ageing conditions were created by immersing the samples in distilled water (DW) and synthetic saline water (SW) at 70 and 90 °C for a period of 32 days. Accordingly, water uptake was measured as a function of hygrothermal ageing time periodically with an interval of 4 days. Specifically, the effect of hygrothermal conditions on impact strength, tensile and thermal properties, structural behavior and morphology of neat PP and PP/copolymer blends was characterized.

6.2 Experimental

6.2.1 Materials

PP homopolymer HD915CF was procured from Abu Dhabi Polymers Company Limited, UAE. The melt flow rate of PP was ca. 8 g/10 min at 230 °C and 2.16 kg load. Also, the reported density value for the PP grade was in the range of 900-910 kg.m^{-3}. Three different commercially procured copolymers included EVA-MA (Fusabond C250, DuPont), EBA (Elvaloy AC 34035, DuPont) and EAO (AFFINITY EG 8200G, Dow Chemical Company). More details of these copolymers are provided in Table 6.1. The materials were used as received from the suppliers.

Table 6.1 Material data for PP and copolymers used in the study

Material	**Repeat unit**	**Appearance**	**Details**
PP homopolymer	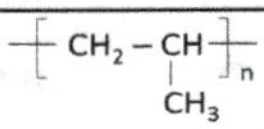		Borouge HD915CF **Density:** 900-910 kg.m^{-3} **Melt flow rate:** 8 g/10 min (230 °C, 2.16 kg) **Melting point (MP):** 168.5 °C
Ethylene-vinyl acetate copolymer with maleic anhydride grafting (EVA-MA)	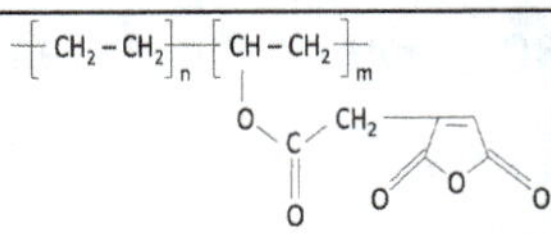		DuPont Fusabond C250 **Density:** 962 kg. m^{-3} **Melt flow rate:** 1.4 g/10 min (190 °C, 2.16 kg) **MP:** 71 °C
Ethylene-butyl acrylate copolymer (EBA)	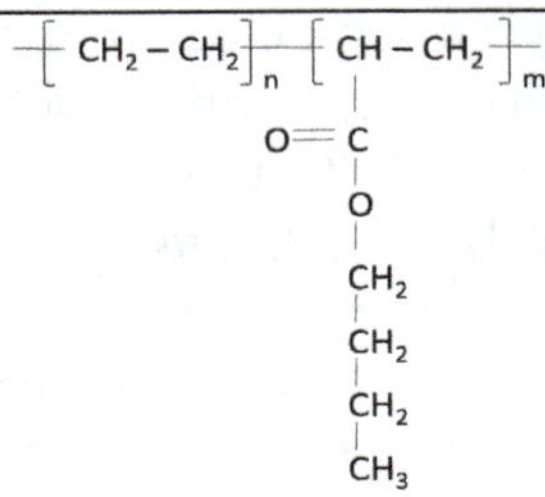		DuPont Elvaloy AC 34035 **Density:** 930 kg.m^{-3} **Melt flow rate:** 40 g/10 min (190 °C, 2.16 kg) **MP:** 90 °C

Ethylene-alpha olefin copolymer (EAO)	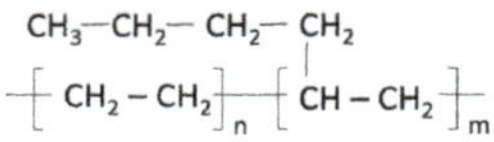		**Brand name:** Dow AFFINITY EG 8200G **Density:** 868 kg.m^{-3} **Melt flow rate:** 0.50 g/10 min (190 °C, 2.16 kg) **MP:** 56 °C

6.2.2 Fabrication of PP/Copolymer Blends

Melt mixing in a co-rotating conical twin screw extruder (HAAKE MiniLab, Thermo Scientific, Germany) was performed to fabricate the PP/copolymer blends with weight ratios of 90/10 and 80/20. 5 g batch of the blend samples was generated using 180 °C temperature, 100 rpm screw speed and 15 min blending time. For accurate comparison, neat PP was also subjected to similar processing conditions. Followed by melt extrusion, the samples were injection molded to form specific test specimens, i.e., dumbbells for tensile test and rectangular bars for Izod impact testing, in accordance with the ISO standards with the aid of a lab-scale mini injection molding machine (HAAKE MiniJet PRO, Thermo Scientific, Germany). For injection molding, processing temperature of 180 °C, mold temperature of 125 °C, injection pressure of 430 bar for 10 s and holding pressure of 500 bar for 6 s were used.

6.2.3 Hygrothermal Ageing of Samples

The blend samples were subjected to accelerated hygrothermal ageing conditions to imitate the real life undesirable environmental conditions. Here, the samples were subjected to artificial ageing in distilled water as well as synthetic saline water containing baths at two different temperatures of 70 and 90 °C for a period of 32 days according to the standard ASTM D570-90. Ageing of the samples in DW was useful to imitate the substantial humid environment, which was employed as a reference. SW with 3% sodium chloride concentration was used to examine the ageing behavior under the influence of sea water and to make comparison of the performance of the samples with the DW aged samples.

6.2.4 Characterization

Water Uptake

In order to evaluate the water uptake, the weight of the samples was assessed before ageing (W_0) and periodically after 4 days during ageing (W_t) with the aid of an electronic weighing balance. The periodic weight measurement during ageing was carried out by taking out the samples and wiping the surface water using a dirt-free dry cotton cloth, followed by weighing and re-immersion for the continuation of ageing. An average of five specimens was generally taken for each case. Thus, the weight gain ratio (W_g) of the samples during accelerated ageing was calculated as

$$W_g = \frac{W_t - W_0}{W_0} \times 100$$

where *t* is the immersion time.

Fourier Transform Infrared (FTIR) Spectroscopy

Structural changes after ageing were analyzed with the aid of FTIR spectrometer BRUKER TENSOR II in transmission mode between 4000 and 400 cm^{-1} wavenumber span with a resolution of 4 cm^{-1}. FTIR spectra of the samples were generated after 32 scans using a diamond attenuated total reflectance (ATR) crystal.

Tensile Properties

In accordance with ISO 527, the tensile properties such as strength, modulus and breaking extension of the samples were assessed using Instron 3345 universal testing machine (UTM). An average of five test specimens (with 75 mm x 5 mm x 2 mm dimensions and 35 mm span length) was recorded. The mechanical performance of the samples was examined at ambient temperature using a crosshead movement rate of 10 $mm.min^{-1}$.

Izod Impact Strength

In accordance with ISO 180, the Izod impact testing of the unaged and aged samples was performed on a Resil impactor. Un-notched

Izod impact strength values of five rectangular bar shaped test specimens with 80 mm x 10 mm x 4 mm dimensions were averaged for each case. The test was performed at ambient temperature with a hammer speed of 3.64 m.s^{-1}.

Differential Scanning Calorimetry (DSC)

DSC analysis of the unaged and aged samples was performed on a Discovery Series differential scanning calorimeter from TA instruments to evaluate the thermal parameters such as peak melting temperature (T_m), melting enthalpy (ΔH_f), peak crystallization temperature (T_c), crystallization enthalpy (ΔH_c) and percentage crystallinity (X_c). DSC thermograms were generated by scanning ca. 3-8 mg of the samples in dry nitrogen flow (50 mL.min^{-1}). At a rate of 10 °C.min^{-1}, the samples placed in the aluminum pans were scanned by following two sets of heating and cooling cycles within the temperature range of -50 °C to +200 °C, and the second set of analysis was used to generate the test results.

Scanning Electron and Optical Microscopy

The effect of hygrothermal ageing on the fracture micro-morphology of the samples was studied with the aid of a scanning electron microscope (SEM) (FEI Quanta, FEG250, USA) operated using an accelerating voltage of 10 kV. Prior to the microscopy analysis, the impact fractured surface of ca. 3 mm thick test specimens placed on aluminum stubs using conductive carbon tape was sputter coated with gold in order to render them conductive. Ageing induced surface changes in the samples were also examined using an optical microscope (OLYMPUS BX51M, Japan) with an eyepiece magnification of 10x.

Wide Angle X-ray Diffraction (WAXD)

Ageing induced structural changes in the samples were analyzed using an X'Pert PRO Panalytical powder diffractometer by applying CuKα irradiation with a wavelength of 1.5406 Å. The analysis was carried out by employing 40 mA current and 45 kV voltage at room temperature. For the WAXD analysis, small sections of the injection molded samples were scanned using a rate of 0.017° s^{-1} and step time of 10 s in the 2θ angle range of 5-60°.

6.3 Results and Discussion

The water absorption curves (variation of percent water uptake with square root of ageing immersion time) of hygrothermally aged neat PP and its blends with EVA-MA, EBA and EAO in DW and SW are presented in Figure 6.1. The absorption curves of neat PP

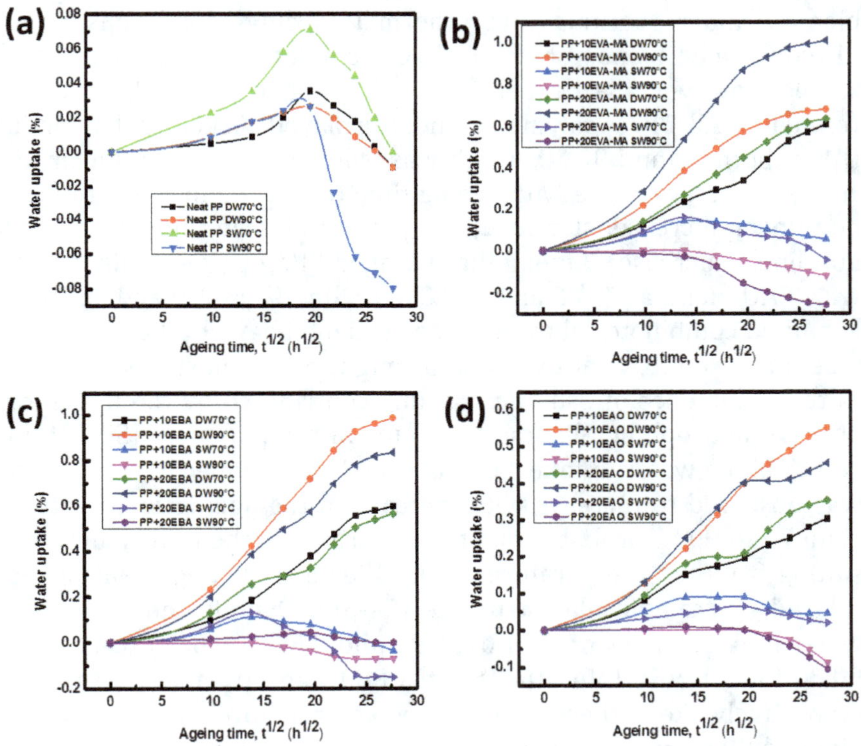

Figure 6.1 Water uptake curves of (a) neat PP, (b) EVA-MA blends, (c) EBA blends and (d) EAO blends.

(Figure 6.1a) indicated two sections, the first section exhibiting the Fickian behavior up to √h=19.6 observed by a non-linear water uptake increase with ageing time, whereas the section after √h=19.6 indicating the non-Fickian behavior observed by a decreasing water uptake tendency with ageing time. The observed Fickian behavior was related to the solvent diffusion in neat PP, and the optimum % values of water uptake were ca. 0.036 and 0.027 for DW aged samples at 70 °C and 90 °C and ca. 0.071 and 0.027 for respective SW aged samples at 70 °C and 90 °C. Thus, a higher water uptake value for SW aged neat PP at 70 °C was observed among hydrothermally aged PP samples under various conditions. The decreasing water uptake tendency in the second section of the sorption curves of the aged neat PP samples is probably associated with the ageing induced macromolecular chain reorganization, surface crosslinking, washing effect, etc. [38,50,51]. Figures 6.1b, 6.1c and 6.1d illustrate the water uptake behaviors of the respective EVA-MA, EBA and EAO blends under various hygrothermal ageing conditions. The PP/copolymer blends showed almost similar water uptake upon ageing, however, the behavior was different in DW and SW environments. All PP/copolymer blends displayed Fickian behavior in DW at both 70 and 90 °C, which was evidenced by a non-linear increase of water uptake with ageing time (or a combination of initial non-linear increase, succeeded by pseudo-plateau and afterwards non-linear increase). Among the DW aged PP/copolymer blends, 20 wt% EAO blend at 70 °C and 90 °C and 20 wt% EBA blend at 70 °C exhibited combinatorial three stage Fickian behavior. The water uptake increase in DW at extended ageing times is most possibly due to the occurrence of osmotic cracking contributed by the influence of more hydrophilic oligomeric groups in the polymers [52,53]. In addition, the water uptake trend of the PP/copolymer blends in DW was observed to increase with increase in the ageing temperature from 70 to 90 °C probably due to an increase in the free volume resulting from the temperature induced faster rearrangement of the relaxed macromolecular structure [54-56]. Overall, the maximum water uptake values of ca. 1.01, 0.99 and 0.55% were discerned at 90 °C for 20 wt% EVA-MA, 10 wt% EBA and 10 wt% EAO blends respectively. Regardless of the copolymer nature, the copolymer blended PP samples displayed almost similar water sorption behavior upon hygrothermal ageing in SW. At 70 °C, an initial increase in the water uptake followed by a decrease in the weight were noticed. However, at 90 °C, the PP/copolymer blend samples displayed pseudo-plateau behavior nearly close to zero initially, followed by weight decreasing tendency to negative values. In comparison to DW, the observed water diffusion rate was very low in SW due to the influence of salt and the resultant reduction of the chemical potential [51]. Here, the discerned maximum water sorption value of only ca. 0.07% for neat PP supports its persuasive barrier performance which is necessary for protective applications. Compared to PP blends with EVA-MA and EBA, EAO blends displayed higher water uptake resistance, and a maximum value of only 0.55% was noted, whereas the maximum water uptake values of EVA-MA and EBA blends were in the range of 1%. This underlines the superior barrier performance of the EAO toughened PP blends.

To study the structural changes owing to the hygrothermal ageing, FTIR analysis of the unaged and aged samples was carried out in transmission mode, as shown in Figures 6.2 and 6.3. As depicted in

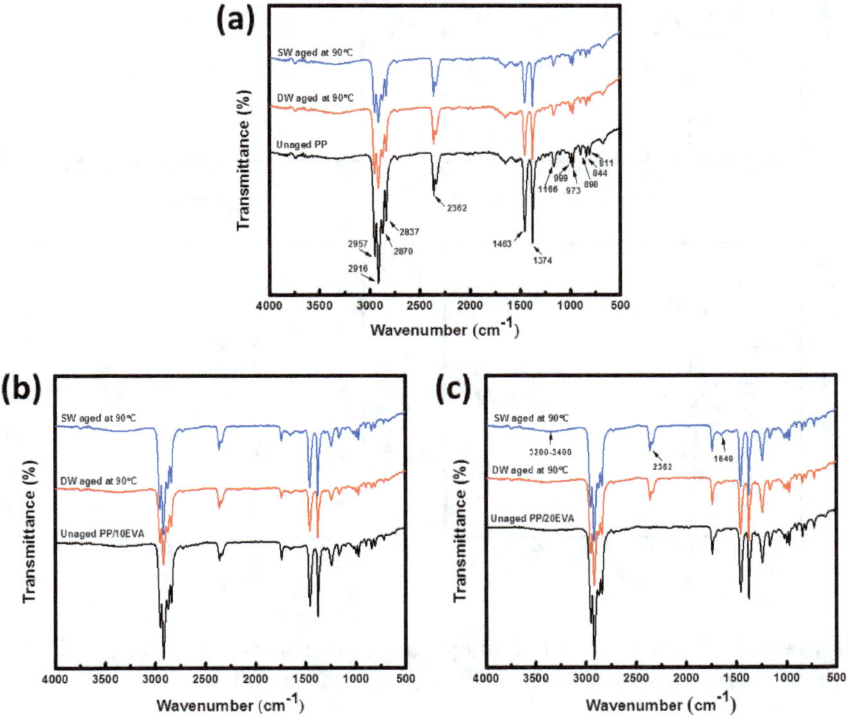

Figure 6.2 FTIR spectra of the unaged and aged samples: (a) neat PP, (b) 10EVA-MA blend and (c) 20EVA-MA blend.

Figure 6.2a, various absorption bands corresponding to CH_3 associated asymmetric and symmetric stretching vibrations at 2957 and 2870 cm^{-1} respectively, CH_2 associated symmetric stretching vibrations at 2916 and 2837 cm^{-1}, C-H associated bending vibration at 2362 cm^{-1}, CH_3 associated asymmetric bending vibration around 1463 cm^{-1}, CH_3 associated symmetric bending vibration around 1374 cm^{-1}, tertiary methyl skeleton associated deformation at 999, 973, 844 and 811 cm^{-1}, stretch direction induced perpendicular absorption nearby 898 cm^{-1}, etc., were observed for neat PP [57]. The discerned FTIR characteristic bands of EVA-MA, EBA and EAO blends also presented good resemblance with neat PP more likely

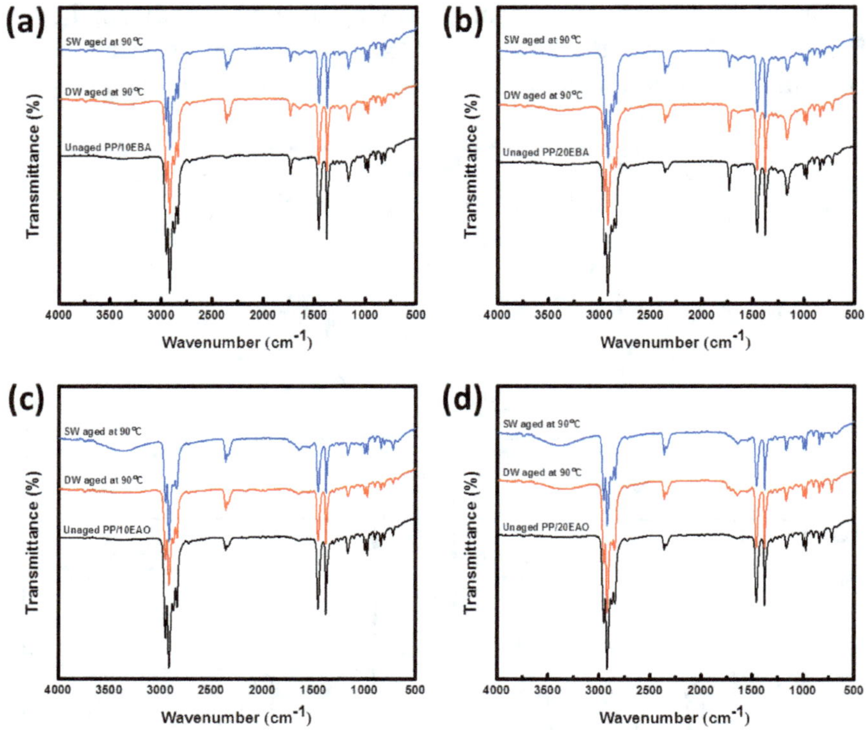

Figure 6.3 FTIR spectra of the unaged and aged samples: (a) 10EBA blend, (b) 20EBA blend (c) 10EAO blend and (d) 20EAO blend.

due to the compatible blend structures. Following hygrothermal ageing in both DW and SW, the samples exhibited characteristic IR bands observed for the unaged specimens, however, the structural changes could be assessed with the help of intensity variations of these bands [58,59]. For instance, the characteristic FTIR bands of the hygrothermally aged neat PP samples exhibited reduced intensity (Figure 6.2a), which is considered to be an indication for the initiation of the ageing induced macromolecular chain scission of neat PP [55]. For PP blends with EVA-MA, EBA and EAO copolymers, the discerned intensity reduction was higher for the blends with copolymer content of 10 wt% in comparison with the blends with 20 wt% copolymer. In addition, the FTIR spectra of the aged PP/copolymer blend samples displayed enhanced intensity for the absorption bands corresponding to absorbed water associated asymmetric bending in the broad range of 3200-3400 cm^{-1}, water induced C-H bending vibration nearby 2362 cm^{-1} and water induced OH stretching nearby 1640 cm^{-1} [47,58].

The WAXD plots of the samples, shown in Figures 6.4 and 6.5, provide insights into the hydrothermal ageing induced crystallinity variations. As depicted in Figure 6.4a, the unaged neat PP displayed

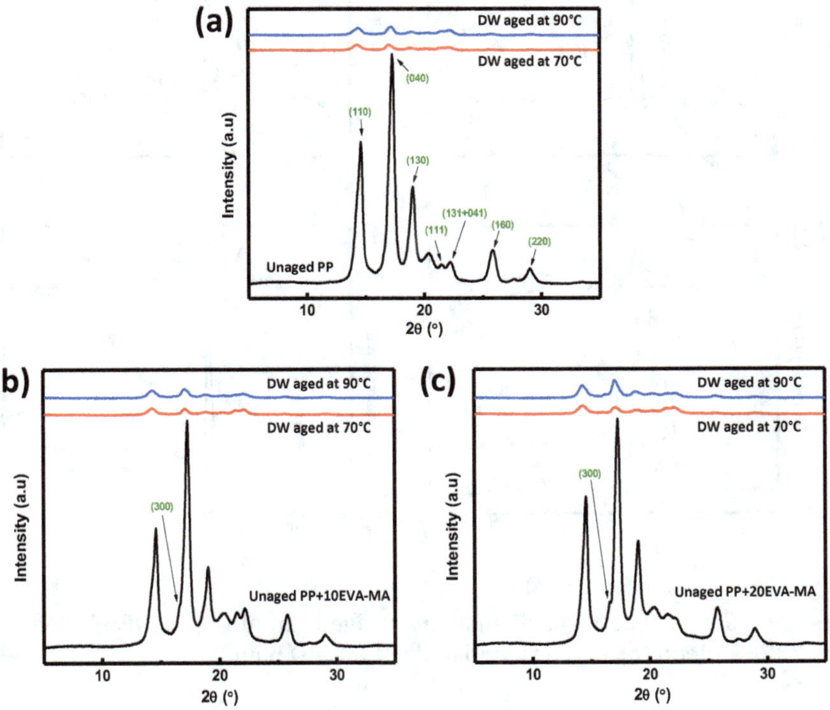

Figure 6.4 WAXD patterns of unaged and aged samples: (a) neat PP, (b) 10EVA-MA blend and (c) 20EVA-MA blend.

seven characteristic diffraction peaks associated with α crystalline planes (110) at 14.55°, (040) at 17.26°, (130) at 19.03°, (111) at 21.47°, (131+041) at 22.18°, (160) at 25.76° and (220) at 29.01° [60-62]. Also, the inter-planar distance (d) corresponding to these peaks was estimated using the Bragg's relation, $d = \lambda/(2\sin\theta_{max})$, where λ implies the X-ray wavelength (1.5406 Å) applied for the analysis. Hence, the estimated inter-planar distance values were 6.08 Å for (110), 5.13 Å for (040), 4.66 Å for (130), 4.14 Å for (111), 4.00Å for (131+041), 3.46Å for (160) and 3.08 Å for (220). After hygrothermal ageing in DW at 70 and 90 °C, the appearance of same characteristic crystalline peaks was observed, differing only in intensity. In actual, the intensity of all crystalline peaks was observed

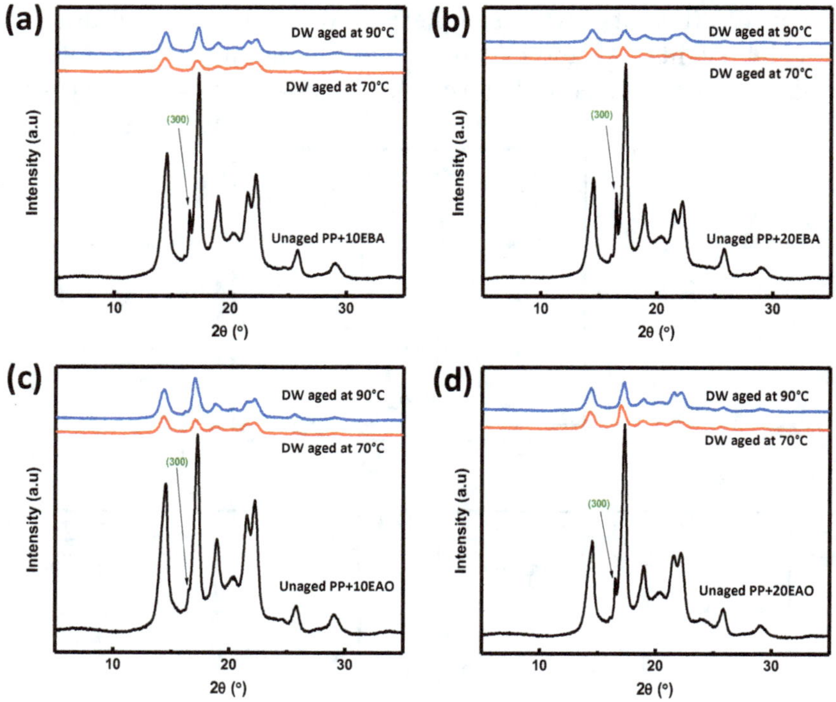

Figure 6.5 WAXD patterns of unaged and aged samples: (a) 10EBA blend, (b) 20EBA blend (c) 10EAO blend and (d) 20EAO blend.

to considerably decrease for aged neat PP, owing to the ageing induced macromolecular chain scission and corresponding crystallinity reduction [44]. The observed intensity reduction was higher for the neat PP aged in DW at 70 °C in comparison to 90 °C, suggesting the ageing induced macromolecular reorganization at 90° C. The WAXD plots of EVA-MA, EBA and EAO blends revealed the appearance of same α crystalline peaks as neat PP at the same diffraction angles, together with an additional β crystalline peak with a plane of (300) at 16.53° and the corresponding inter-planar distance of 5.36 Å. For PP/copolymer blends, similar crystalline peak intensity reduction was observed after hygrothermal ageing in DW at 70 and 90 °C. Among the blends, the crystalline peak intensity reduction was observed to be lower for EAO blends.

For a material, impact strength indicates its ability to absorb sudden force, thus, resulting in effective plastic deformation without any fracture development. Table 6.2 compares the impact strength

Table 6.2 Mechanical properties of unaged and aged samples

Sample	Impact strength (KJ/m^2)	Tensile strength (MPa)	Tensile modulus (MPa)	Breaking extension (%)
Unaged PP	7	39	1011	24
DW aged PP at 70 °C	51	36	815	23
DW aged PP at 90 °C	46	35	808	21
SW aged PP at 70 °C	45	36	809	21
SW aged PP at 90 °C	46	35	801	21
10EVA-MA	26	33	1095	38
DW aged 10EVA-MA at 70 °C	32	31	837	30
DW aged 10EVA-MA at 90 °C	33	30	764	29
SW aged 10EVA-MA at 70 °C	29	29	720	26
SW aged 10EVA-MA at 90 °C	39	28	705	26
20EVA-MA	42	30	1143	43
DW aged 20EVA-MA at 70 °C	35	27	718	28
DW aged 20EVA-MA at 90 °C	38	26	659	26
SW aged 20EVA-MA at 70 °C	31	27	710	26
SW aged 20EVA-MA at 90 °C	41	26	651	25
10EBA	54	32	1115	44
DW aged 10EBA at 70 °C	60	31	767	34
DW aged 10EBA at 90 °C	61	31	684	32
SW aged 10EBA at 70 °C	58	31	756	31
SW aged 10EBA at 90 °C	60	31	681	30
20EBA	74	26	1036	299
DW aged 20EBA at 70 °C	79	26	567	81
DW aged 20EBA at 90 °C	78	25	495	80
SW aged 20EBA at 70 °C	78	26	566	77
SW aged 20EBA at 90 °C	77	25	490	61
10EAO	19	35	1244	28
DW aged 10EAO at 70 °C	74	34	837	27
DW aged 10EAO at 90 °C	70	32	767	26
SW aged 10EAO at 70 °C	52	32	710	26
SW aged 10EAO at 90 °C	100	29	628	25
20EAO	96	30	1084	263
DW aged 20EAO at 70 °C	101	28	673	47
DW aged 20EAO at 90 °C	88	28	627	56
SW aged 20EAO at 70 °C	98	28	664	43
SW aged 20EAO at 90 °C	59	28	620	32

of the samples before and after ageing in both DW and SW at 70 and 90 °C. It can be observed that the addition of 10 and 20 wt% of copolymers was useful in appreciably improving the impact strength of neat PP. In the PP/copolymer blends, the copolymeric dispersed phase is expected to help in generating micro-voids in the matrix by undergoing crazing as well as shear yielding through effective absorption of the majority of the impact energy by acting as stress accumulator. Thus, the resultant PP/copolymer blends show favorable dissipation of the impact energy by undergoing fracture-free plastic deformation [31,63,64]. The impact strength of neat PP was observed to increase significantly in all hygrothermal ageing conditions by virtue of the water induced plasticization effect [41,65]. Therefore, the appearance of a highly appreciable impact strength retention rate of more than 550% was noted for the aged neat PP samples (Figure 6.6). Hygrothermally aged PP/copolymer blends also exhibited an enhancement of the impact strength except PP

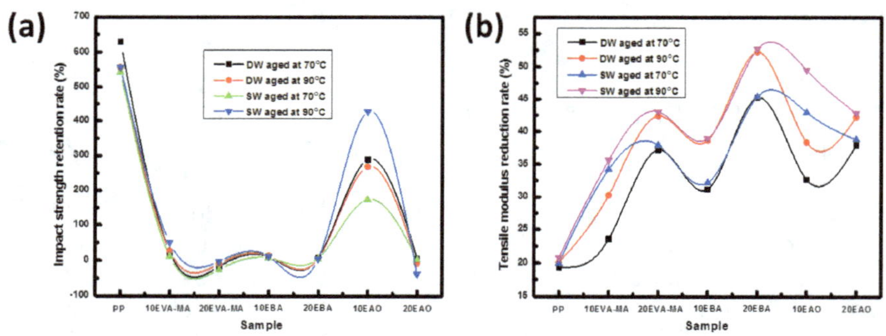

Figure 6.6 Variation of (a) impact strength retention rate and (b) tensile modulus reduction rate of neat PP, EVA-MA blends, EBA blends and EAO blends in various hygrothermal conditions.

blends with 20 wt% EVA-MA in all ageing conditions and 20 wt% EAO in both DW and SW at 90 °C (Figure 6.6). The drop in the impact strength of such PP/copolymer blends is most probably related to the structural integrity loss in hygrothermal conditions [44]. Among various PP/copolymer blend systems, the blend with 10 wt% EAO displayed higher rate of impact strength retention, observed in the range of 174-426% (Table 6.2). The observed impact strength values also revealed superior impact strength retention rate for 10 wt% copolymer blends than the blends with 20 wt% copolymer in hygrothermal conditions. In addition, the impact strength retention capability of a majority of the PP/copolymer blend samples was observed to increase with temperature in both DW and SW. Such increasing tendency of the impact strength retention rate at higher temperature relates to the increased degree of water induced plasticization effect at the blend interface and, thus, easier generation of micro-voids [41,44].

The mechanical properties such as strength, modulus and breaking extension of unaged and hygrothermally aged neat PP, EVA-MA blends, EBA blends and EAO blends are presented in Table 6.2. An improvement in the tensile modulus of PP was observed after blending with copolymers. The observed behavior indicates the ability of the blend constituents to proceed with stretching induced deformation simultaneously in accordance with the generation of defect-free interface owing to good phase compatibility [29]. On comparing the effect of copolymers, 20 wt% EVA-MA and 10 wt% EBA and EAO blended PP were observed to exhibit higher improvement in tensile modulus. The tensile modulus of the neat PP and PP/copolymer blends was observed to be negatively affected under hygrothermal ageing conditions. The modulus deteriorated proportionately with ageing temperature in both DW and SW. At the same time, the extent of tensile modulus deterioration was observed to be higher in SW as compared to DW. For instance, neat PP showed ca. 19-20% deterioration in DW and ca. 20-21% deterioration in SW (Figure 6.6). In the case of PP/copolymer blends, the tensile modulus deterioration rate was observed to increase with increase in the copolymer content. Overall, 10 wt% EVA-MA blend exhibited a lower rate of tensile modulus deterioration of ca. 24-30% and 34-36% in DW and SW respectively as compared to other blends. On the other hand, the observed tensile modulus deterioration in respective DW and SW were 37-42% and 38-43% for 20 wt% EVA-MA blend, 31-39% and 32-39% for 10 wt% EBA blend, 45-52% and 45-53% for 20 wt% EBA blend, 33-38% and 43-50% for 10 wt% EAO blend and 38-42% and 39-43% for 20 wt% EAO blend (Figure 6.6). Such deterioration in the tensile modulus might be due to the reduced material resistance resulting from the macromolecular breakdown induced by the absorbed water in hygrothermal conditions. Also, the extent of such structural changes was observed to increase with the ageing temperature as a result of increased water penetration tendency [55,66]. Thus, the immersion in SW caused higher tensile modulus deterioration as compared to DW due to the development of more micro-cracks in high pH environment [34,67].

The tensile strength of PP was observed to reduce after blending with copolymers and the extent of reduction was noticed to increase with the copolymer content. The observed reduction tendency after copolymer addition is possibly caused by the deterioration in load-carrying area. Similar to tensile modulus, hygrothermal ageing impacted the tensile strength negatively (Table 6.2). Overall, among the samples, the EBA blends exhibited superior resistance against the ageing induced tensile strength reduction under hygrothermal conditions. In addition, the elongation at break of neat PP was observed to enhance after copolymer addition. Upon tensile stretching, the dispersed copolymeric phase in the blend system can serve as effective stress absorber, thus, leading to successful macromolecular extension. The extension at break of the samples also deteriorated under hygrothermal conditions. Compared to neat PP, the observed breaking extension deterioration rate was higher for the PP/copolymer blends, which underlines the contribution of the amorphous portion for water penetration and macromolecular breakdown, as observed previously [58].

Table 6.3 presents the DSC data obtained from second heating and cooling scans shown in Figures 6.7-6.10 for unaged and hygrothermally aged samples. In order to estimate the X_c of the unaged and aged samples, the following equation was used [52,68]:

$$X_c = \frac{\Delta H}{\Delta H_0 w_{\mathrm{pp}}} \times 100$$

In this equation, the parameters ΔH_0, ΔH and w_{pp} represent theoretical melting enthalpy of fully crystalline PP (207.1 J/g) [68], DSC generated melting enthalpy of the studied samples and weight fraction of PP in the studied samples, respectively.

As observed from Table 6.3, the unaged neat PP exhibited the peak melting temperature of 168.5 °C. Also, the T_m values within a standard deviation of -1 °C were discerned after blending PP with 10 and 20 wt% of EVA-MA, EBA and EAO, which possibly indicates the blends with defect-free PP crystal structure [69]. After exposure to the accelerated hygrothermal ageing conditions, T_m of neat PP displayed slight reduction in DW, whereas it remained almost same in SW. A similar trend was observed for the PP blends with 10 wt% EVA-MA and EBA, though virtually similar T_m was observed for PP blends with 20 wt% EVA-MA and EBA. In the case of EAO blends, T_m remained same or slightly higher after hygrothermal exposure. In

Table 6.3 DSC data for unaged and aged samples (T_c: crystallization temperature; ΔH_c: crystallization enthalpy; Tm: melting temperature; ΔH_f: melting enthalpy; X_c: percentage crystallinity; ΔX_c: percentage crystallinity variation)

Sample	T_c (°C)	ΔH_c (J/g)	T_m (°C)	ΔH_f (J/g)	X_c (%)	ΔX_c (%)
Unaged PP	127.9	98.7	168.5	90.8	43.8	-
DW aged PP at 70 °C	126.3	106.2	166.9	112.2	54.2	10.4
DW aged PP at 90 °C	126.4	110.1	166.9	108.0	52.1	8.3
SW aged PP at 70 °C	126.2	111.2	168.2	106.5	51.4	7.6
SW aged PP at 90 °C	126.2	107.0	168.6	105.4	50.9	7.1
Unaged 10EVA-MA	127.9	90.5	168.3	87.3	46.8	-
DW aged 10EVA-MA (70 °C)	126.6	102.3	167.7	100.8	54.1	7.3
DW aged 10EVA-MA (90 °C)	126.6	98.3	167.7	96.8	51.9	5.1
SW aged 10EVA-MA (70 °C)	125.8	99.3	168.1	96.7	51.9	5.1
SW aged 10EVA-MA (90 °C)	125.8	95.7	168.5	93.7	50.3	3.5
Unaged 20EVA-MA	128.2	81.0	167.7	80.7	48.7	-
DW aged 20EVA-MA (70 °C)	125.0	86.9	167.7	88.5	53.5	4.8
DW aged 20EVA-MA (90 °C)	124.6	88.8	168.1	88.2	53.2	4.5
SW aged 20EVA-MA (70 °C)	125.0	84.0	169.2	81.8	49.4	0.7
SW aged 20EVA-MA (90 °C)	125.5	88.2	168.8	85.7	51.7	3.0
Unaged 10EBA	127.8	89.7	168.5	83.1	44.6	-
DW aged 10EBA (70 °C)	126.2	100.3	167.7	98.3	52.7	8.1
DW aged 10EBA (90 °C)	126.4	97.4	167.7	94.0	50.0	5.4
SW aged 10EBA (70 °C)	125.8	90.2	168.6	84.4	45.3	0.7
SW aged 10EBA (90 °C)	126.2	97.1	168.2	92.2	49.5	4.9
Unaged 20EBA	127.9	80.9	168.1	79.1	47.8	-
DW aged 20EBA (70 °C)	125.8	86.1	168.5	87.6	52.9	5.1
DW aged 20EBA (90 °C)	126.6	88.8	167.8	86.6	52.3	4.5
SW aged 20EBA (70 °C)	125.8	85.7	168.5	83.6	50.5	2.7
SW aged 20EBA (90 °C)	125.4	86.9	168.5	83.6	50.5	2.7
Unaged 10EAO	127.8	91.4	167.7	87.9	47.2	-
DW aged 10EAO (70 °C)	126.0	100.7	168.5	99.5	53.4	6.2
DW aged 10EAO (90 °C)	126.7	98.7	167.7	96.5	51.8	4.6
SW aged 10EAO (70 °C)	125.6	93.3	168.5	91.4	49.0	1.8
SW aged 10EAO (90 °C)	126.2	91.8	168.5	89.9	48.2	1.0
Unaged 20EAO	128.2	82.6	167.7	78.6	47.5	-
DW aged 20EAO (70 °C)	126.6	87.7	167.7	88.2	53.2	5.7
DW aged 20EAO (90 °C)	125.8	82.5	168.1	81.2	48.0	0.5
SW aged 20EAO (70 °C)	125.6	84.5	168.5	81.4	49.1	1.6
SW aged 20EAO (90 °C)	125.4	88.0	168.8	85.1	51.9	4.4

addition, the samples aged in SW displayed virtually same or slightly higher value of T_m in comparison to the DW aged samples exhibiting a slightly reduced T_m in majority of the cases (except EAO based blends). From these observations, the capability of the EAO based blend samples to withstand the hygrothermal ageing induced T_m

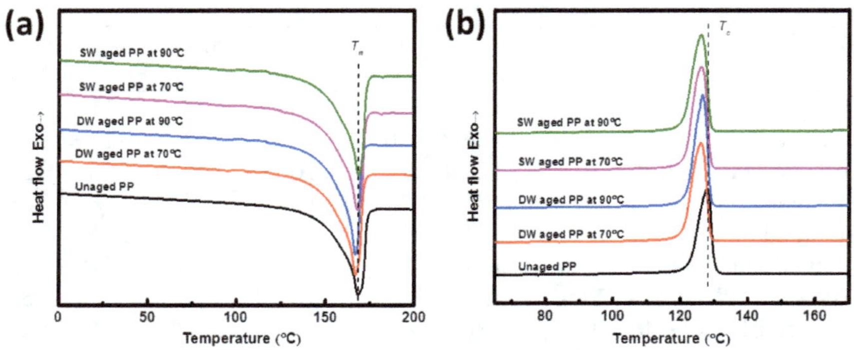

Figure 6.7 DSC (a) heating and (b) cooling curves of unaged and hygrothermally aged neat PP samples.

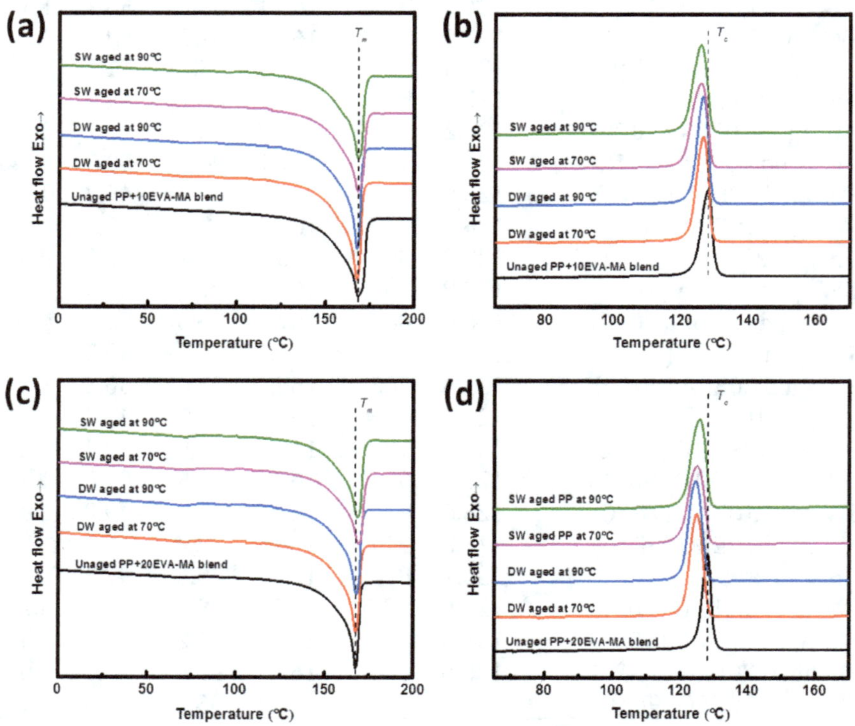

Figure 6.8 DSC heating and cooling curves of unaged and aged samples: (a,b) 10EVA-MA blend and (c,d) 20EVA-MA blend.

reduction in both DW and SW at 70 and 90 °C can be confirmed. This is in good agreement with the observations related to the water uptake discussed earlier.

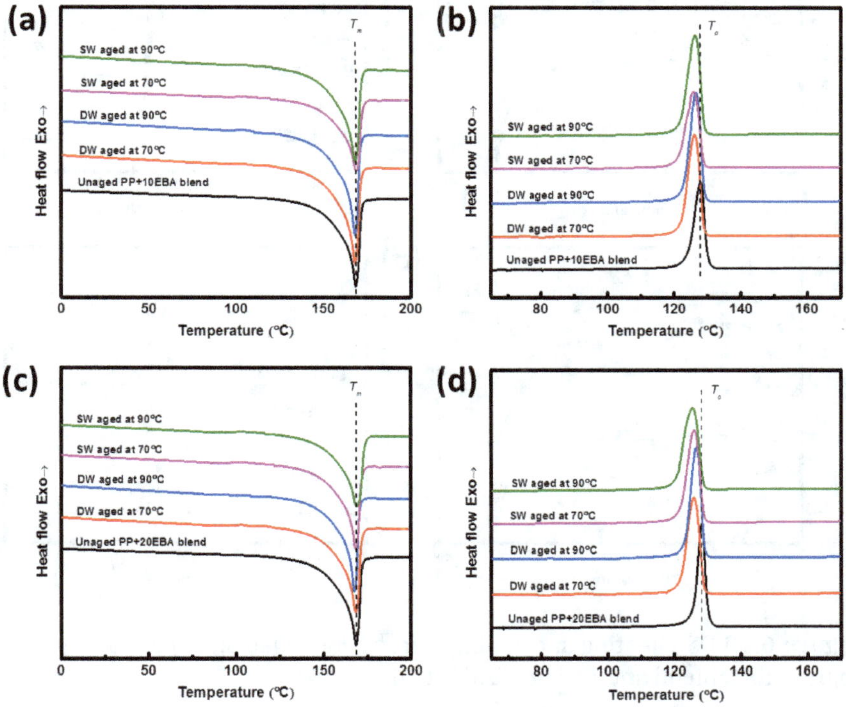

Figure 6.9 DSC heating and cooling curves of unaged and aged samples: (a,b) 10EBA blend and (c,d) 20EBA blend.

Concerning T_c, the unaged neat PP displayed a value of 127.9 °C and the blends with 10 and 20 wt% EVA-MA, EBA and EAO displayed virtually similar values. It indicated that the arrangement of PP crystallites remained undisturbed in the presence of copolymers [64]. The T_c of hygrothermally aged neat PP and PP/copolymer blends was observed to decrease, due to the influence of moisture, thus, altering the crystal structure perfection.

The observed X_c for the unaged neat PP was 43.8%, which was found to increase after blending with copolymers owing to a probable copolymer mediated ordering of PP crystallites [70]. X_c was also observed to increase for all samples after hygrothermal ageing (Table 6.3). For instance, the aged neat PP exhibited an increase of crystallinity in the range of 7.1-10.4% under various ageing conditions.

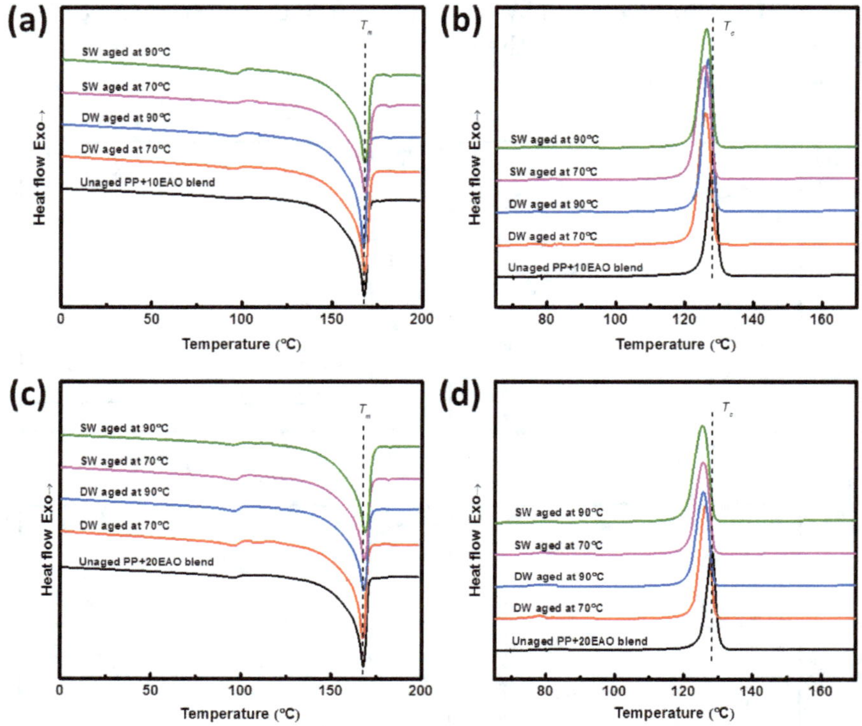

Figure 6.10 DSC heating and cooling curves of unaged and aged samples: (a,b) 10EAO blend and (c,d) 20EAO blend.

In addition, the observed ΔX_c of the blend samples was in the range of 3.5-7.3%, 0.7-4.8%, 0.7-8.1%, 2.7-5.1%, 1.0-6.2% and 0.5-5.7% for the blends with 10 wt% EVA-MA, 20 wt% EVA-MA, 10 wt% EBA, 20 wt% EBA, 10 wt% EAO and 20 wt% EAO respectively. Indeed, the noticed positive ΔX_c trend was lower for the PP blends with 20 wt% copolymers as compared to the PP blends with 10 wt% copolymer content. Also, the samples aged in DW exhibited higher enhancement in X_c as compared to SW, which also correlates with the observations pertaining to water uptake. Among various blends, the EAO blends demonstrated a lower degree of X_c increase, which indicated some degree of resistance against the ageing induced macromolecular chain scission under hygrothermal conditions.

The micro-morphology of the unaged and aged samples was studied using SEM of the gold sputter-coated impact fractured surfaces, as presented in Figures 6.11-6.13. Comparing the images of the unaged PP and PP/copolymer blend samples, a large number of

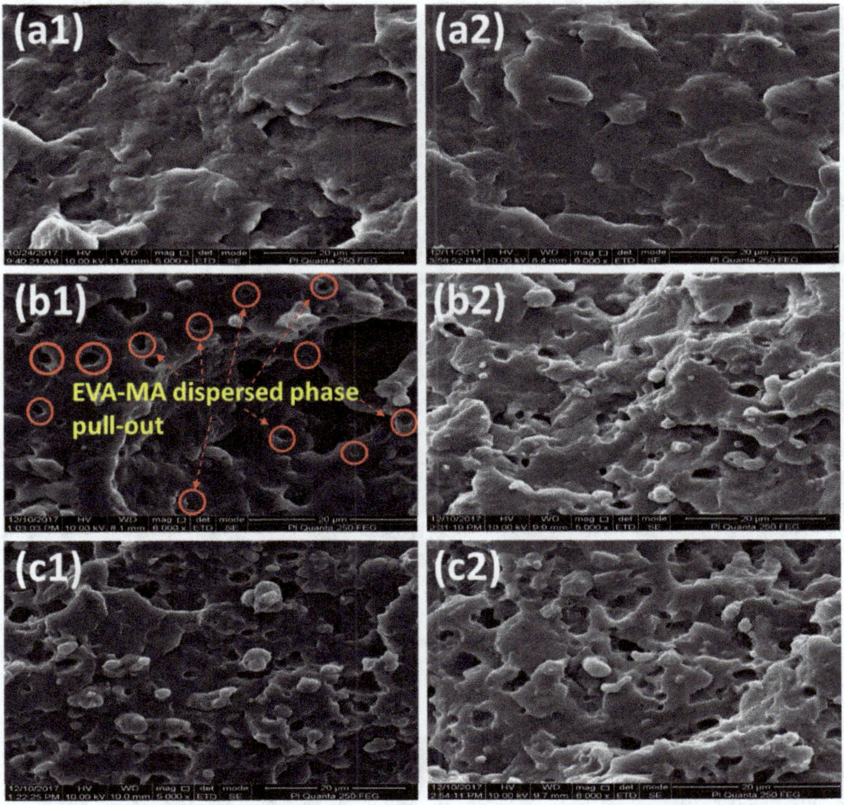

Figure 6.11 SEM images of the impact fractured surface for unaged and aged samples of ((a1) and (a2)) neat PP, ((b1) and (b2)) 10EVA-MA blend and ((c1) and (c2)) 20EVA-MA blend respectively.

spherical shaped dispersed copolymer phase pull-outs, also named as micro-voids, were observed in the PP matrix in the PP/copolymer blends, which correspondingly resulted in improved impact strength. This observation underlines the successful debonding or cavitation at the blend interface upon impact loading, by absorbing the majority of the force and consequent generation of crazing and shear yielding. From the SEM fractographs of the unaged PP/copolymer blends, uniform copolymeric phase distribution could also be observed in the continuous PP phase. Indeed, on increasing the copolymer content, the extent of shock absorbing sites was observed to increase, along with enhanced dispersed copolymeric phase size. It could be observed from Figures 6.11a1 and 6.11a2 that the fracture morphology changed from brittle to ductile

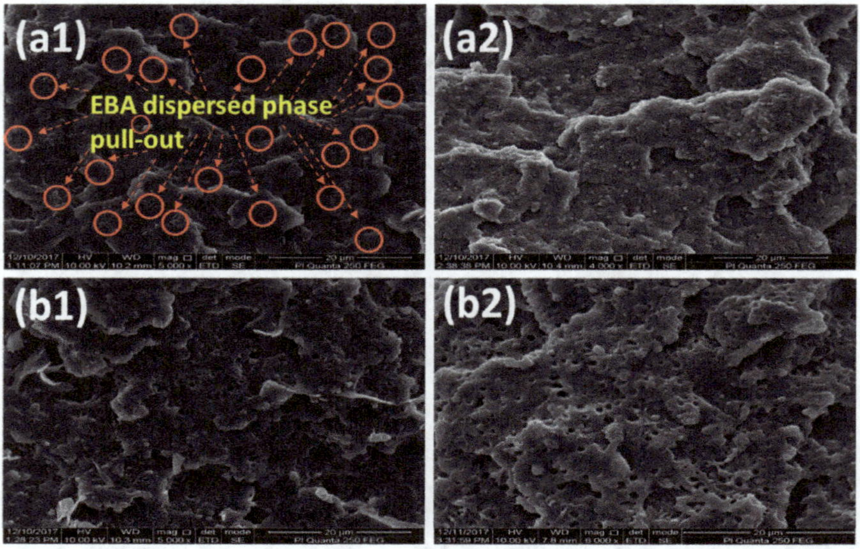

Figure 6.12 SEM images of the impact fractured surface for unaged and aged samples of ((a1) and (a2)) 10EBA blend and ((b1) and (b2)) 20EBA blend.

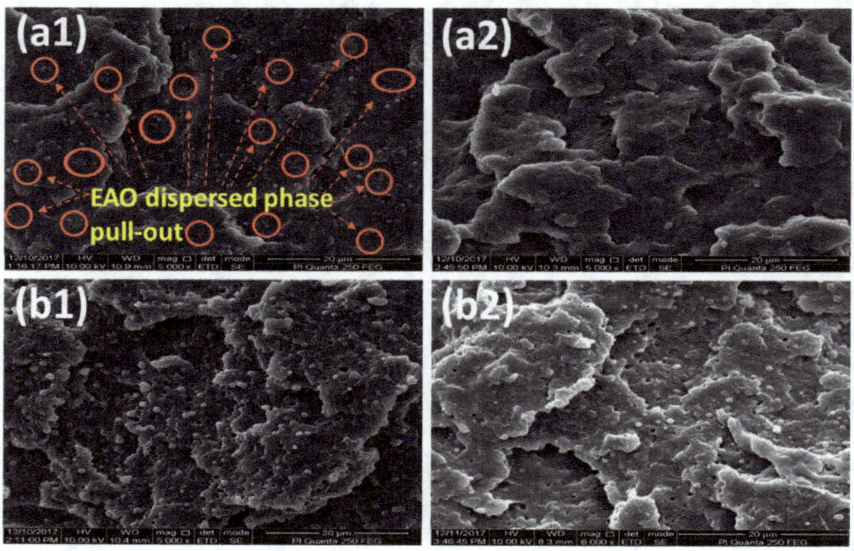

Figure 6.13 SEM images of the impact fractured surface for unaged and aged samples of ((a1) and (a2)) 10EAO blend and ((b1) and (b2)) 20EAO blend.

for neat PP after ageing (90 °C DW) due to the influence of the water induced plasticizing effect, which accordingly resulted in 557% enhancement in the impact strength. For the 90 °C DW aged 10 wt% copolymer blends (Figures 6.11b2, 6.12a2 and 6.13a2), the impact strength was also observed to exhibit an increasing trend due to the influence of the ageing induced water plasticizing effect, along with an improvement in the interfacial characteristics. Among the 10 wt% copolymer toughened PP samples, EAO blend displayed an optimal enhancement of ca. 268%. Considering 20 wt% copolymer toughened PP samples, ageing had no significant effect on the impact strength, as EVA-MA and EAO containing blends displayed a small decrease whereas EBA blend showed a small increase. This is due to the simultaneous presence of the ageing induced water plasticizing effect and decohesion of the copolymeric dispersed phase, as displayed in Figures 6.11c2, 6.12b2 and 6.13b2.

Figure 6.14 displays the optical micrographs of the 90 °C DW

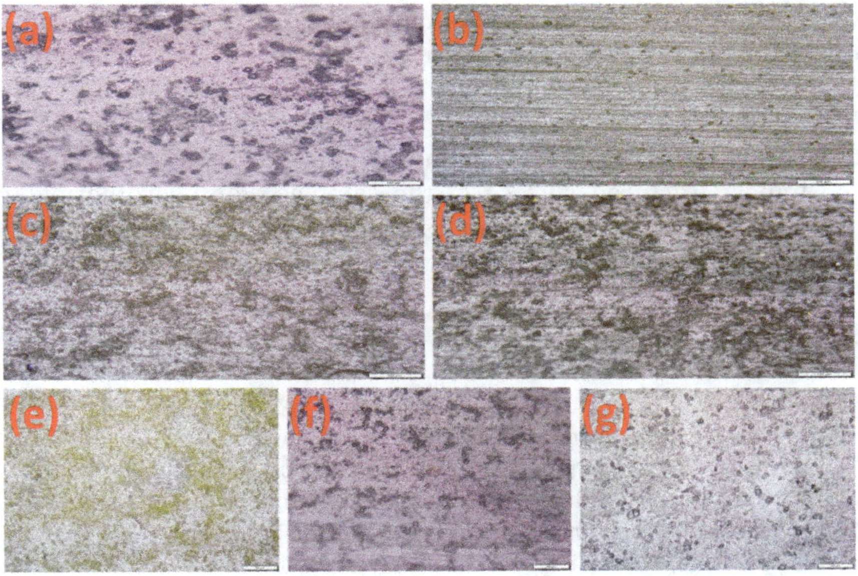

Figure 6.14 Surface micrographs of aged (a) neat PP, (b) 10EVA-MA blend, (c) 20EVA-MA blend, (d) 10EBA blend, (e) 20EBA blend, (f) 10EAO blend and (g) 10EAO blend.

aged samples. Surface deterioration trend through the loss of surface uniformity was observed in the samples. Also, the surface roughness of the aged samples was observed to increase, evidenced by the appearance of irregular pore like structures and small cracks on the surface. As highlighted in Figure 6.14a, neat PP experienced higher surface deterioration tendency compared to the PP/copolymer blend samples in hygrothermal ageing conditions. Overall, the PP blends with lower content (10 wt%) of EVA-MA and higher content (20 wt%) of EBA and EAO displayed superior resistance against surface deterioration.

6.4 Conclusions

In this study, PP/copolymer blends using EVA-MA, EBA and EAO copolymers in 10 and 20 wt% compositions were generated by direct melt blending in a twin screw extruder. Water uptake of the samples as a function of the hygrothermal ageing time exhibited a higher water uptake trend in DW as compared to SW. The impact strength was observed to increase due to the water plasticizing effect and improved interfacial properties in the blend systems, whereas the tensile strength, modulus and extension at break exhibited a decreasing trend due to the reduced material resistance resulting from the macromolecular breakdown caused by the combined effect of water and thermal ageing. The observations related to the impact strength were further supported by SEM analysis of the impact fractured surface of the samples. For hygrothermally aged samples, the DSC generated normalized X_c was also observed to increase by virtue of chemi-crystallization phenomenon.

References

1. Taktak, R., Guermazi, N., Derbeli, J., and Haddar, N. (2015) Effect of hygrothermal aging on the mechanical properties and ductile fracture of polyamide 6: Experimental and numerical approaches. *Engineering Fracture Mechanics*, **148**, 122-133.
2. Deroiné, M., Le Duigou, A., Corre, Y.-M., Le Gac, P.-Y., Davies, P., César, G., and Bruzaud, S. (2014). Accelerated ageing of polylactide in aqueous environments: comparative study between distilled water and seawater. *Polymer Degradation and Stability*, **108**, 319-329.
3. Eftekhari, M., and Fatemi, A. (2016) Tensile behavior of thermoplastic composites including temperature, moisture, and hygrothermal effects. *Polymer Testing*, **51**, 151-164.
4. Deroiné, M., Le Duigou, A., Corre, Y.-M., Le Gac, P.-Y., Davies, P.,

César, G., and Bruzaud, S. (2014) Accelerated ageing and lifetime prediction of poly (3-hydroxybutyrate-co-3-hydroxyvalerate) in distilled water. *Polymer Testing*, **39**, 70-78.

5. *Thermally STable and Flame Retardant Polymer Nanocomposites*, Mittal, V. (ed.), Cambridge University Press, UK (2011).
6. Wang, M., Xu, X., Ji, J., Yang, Y., Shen, J., and Ye, M. (2016) The hygrothermal aging process and mechanism of the novolac epoxy resin. *Composites, Part B: Engineering*, **107**, 1-8.
7. Bikiaris, D., Vassiliou, A., Chrissafis, K., Paraskevopoulos, K., Jannakoudakis, A., and Docoslis, A. (2008) Effect of acid treated multi-walled carbon nanotubes on the mechanical, permeability, thermal properties and thermo-oxidative stability of isotactic polypropylene. *Polymer Degradation and Stability*, **93**, 952-967.
8. Wang, Y., and Tsai, H. B. (2012) Thermal, dynamic-mechanical, and dielectric properties of surfactant intercalated graphite oxide filled maleated polypropylene nanocomposites. *Journal of Applied Polymer Science*, **123**, 3154-3163.
9. Xu, J., Mittal, V., and Bates, F. S. (2016) Toughened isotactic polypropylene: phase behavior and mechanical properties of blends with strategically designed random copolymer modifiers. *Macromolecules*, **49**, 6497-6506.
10. Ma, L.-F., Bao, R.-Y., Dou, R., Zheng, S.-D., Liu, Z.-Y., Zhang, R.-Y., Yang, M.-B., and Yang, W. (2016) Conductive thermoplastic vulcanizates (TPVs) based on polypropylene (PP)/ethylene-propylene-diene rubber (EPDM) blend: From strain sensor to highly stretchable conductor. *Composites Science and Technology*, **128**, 176-184.
11. Gao, Y., Li, J., Li, Y., Yuan, Y., Huang, S., and Du, B. (2017) Effect of Elastomer Type on Electrical and Mechanical Properties of Polypropylene/Elastomer Blends. *Electrical Insulating Materials (ISEIM) Symposium*, pp. 574-577.
12. Zhou, Y., He, J., Hu, J., Huang, X., and Jiang, P. (2015) Evaluation of polypropylene/polyolefin elastomer blends for potential recyclable HVDC cable insulation applications. *IEEE Transactions on Dielectrics and Electrical Insulation*, **22**, 673-681.
13. Yang, C.-j., Huang, T., Yang, J.-h., Zhang, N., Wang, Y., and Zhou, Z.-w. (2017) Carbon nanotubes induced brittle-ductile transition behavior of the polypropylene/ethylene-propylene-diene terpolymer blends. *Composites Science and Technology*, **139**, 109-116.
14. Hong, C. H., Lee, Y. B., Bae, J. W., Jho, J. Y., Nam, B. U., and Hwang, T. W. (2005) Preparation and mechanical properties of polypropylene/clay nanocomposites for automotive parts application. *Journal of Applied Polymer Science*, **98**, 427-433.
15. Zhang, X., Yan, X., He, Q., Wei, H., Long, J., Guo, J., Gu, H., Yu, J., Liu, J., and Ding, D. (2015) Electrically conductive polypropylene nanocomposites with negative permittivity at low carbon nanotube loa-

ding levels. *ACS Applied Materials and Interfaces*, **7**, 6125-6138.

16. Qiu, F., Hao, Y., Li, X., Wang, B., and Wang, M. (2015) Functionalized graphene sheets filled isotactic polypropylene nanocomposites. *Composites, Part B: Engineering*, **71**, 175-183.
17. *Functional Polymer Blends: Synthesis, Properties and Performance*, Mittal, V. (ed.), CRC Press, USA (2012).
18. *Characterization of Polymer Blends*, Thomas, S., Grohens, Y., and Jyotishkumar, P. (eds.), WIly VCH, Germany (2014).
19. Ibrahim, B. A., and Kadum, K. M. (2010) Influence of polymer blending on mechanical and thermal properties. *Modern Applied Science*, **4**, 157.
20. Cook, R. F., Koester, K. J., Macosko, C. W., and Ajbani, M. (2005) Rheological and mechanical behavior of blends of styrene-butadiene rubber with polypropylene. *Polymer Engineering and Science*, **45**, 1487-1497.
21. Abreu, F., Forte, M., and Liberman, S. (2005) SBS and SEBS block copolymers as impact modifiers for polypropylene compounds. *Journal of Applied Polymer Science*, **95**, 254-263.
22. Das, V., Gautam, S. S., and Pandey, A. K. (2011) Effect of SBS content on low temperature impact strength, morphology and rheology of PP-cp/SBS blends. *Polymer-Plastics Technology and Engineering*, **50**, 825-832.
23. Balkan, O., Demirer, H., and Kayali, E. S. (2011) Effects of deformation rates on mechanical properties of PP/SEBS blends. *Journal of Achievements in Materials and Manufacturing Engineering*, **47**, 26-33.
24. Mae, H., Omiya, M., and Kishimoto, K. (2008) Material ductility and toughening mechanism of polypropylene blended with bimodal distributed particle size of styrene–ethylene–butadiene–styrene triblock copolymer at high strain rate. *Journal of Applied Polymer Science*, **110**, 3941-3953.
25. Zebarjad, S., Bagheri, R., Reihani, S., and Lazzeri, A. (2003) Deformation, yield and fracture of elastomer-modified polypropylene. *Journal of Applied Polymer Science*, **90**, 3767-3779.
26. Kim, M. H., Alamo, R. G., and Lin, J. (1999) The cocrystallization behavior of binary blends of isotactic polypropylene and propylene-ethylene random copolymers. *Polymer Engineering and Science*, **39**, 2117-2131.
27. Balaji, A. B., Ratnam, C. T., Khalid, M., and Walvekar, R. (2017) Effect of electron beam irradiation on thermal and crystallization behavior of PP/EPDM blend. *Radiation Physics and Chemistry*, **141**, 179-189.
28. Bouchart, V., Bhatnagar, N., Brieu, M., Ghosh, A., and Kondo, D. (2008) Study of EPDM/PP polymeric blends: mechanical behavior and effects of compatibilization. *Comptes Rendus Mecanique*, **336**,

714-721.

29. Zhu, L., Fan, H.-N., Yang, Z.-Q., and Xu, X.-H. (2010) Evaluation of phase morphology, rheological, and mechanical properties based on polypropylene toughened with poly (ethylene-co-octene). *Polymer-Plastics Technology and Engineering*, **49**, 208-217.
30. Wang, J., Guo, J., Li, C., Yang, S., Wu, H., and Guo, S. (2014) Crystallization kinetics behavior, molecular interaction, and impact-induced morphological evolution of polypropylene/poly (ethylene-co-octene) blends: insight into toughening mechanism. *Journal of Polymer Research*, **21**, 618.
31. Liu, G., and Qiu, G. (2013) Study on the mechanical and morphological properties of toughened polypropylene blends for automobile bumpers. *Polymer Bulletin*, **70**, 849-857.
32. Yamaguchi, M., and Nitta, K. H. (1999) Optical and acoustical investigation for plastic deformation of isotactic polypropylene/ethylene-1-hexene copolymer blends. *Polymer Engineering and Science*, **39**, 833-840.
33. Ramírez-Vargas, E., Navarro-Rodríguez, D., Huerta-Martínez, B., Medellín-rodríguez, F., and Lin, J. (2000) Morphological and mechanical properties of polypropylene [PP]/poly (ethylene vinyl acetate)[EVA] blends. I. Homopolymer PP/EVA systems. *Polymer Engineering and Science*, **40**, 2241-2250.
34. Larbi, S., Bensaada, R., Bilek, A., and Djebali, S. (2015) Hygrothermal ageing effect on mechanical properties of FRP laminates. *AIP Conference Proceedings*, **1653**, 020066.
35. Han, M.-H., and Nairn, J. A. (2003) Hygrothermal aging of polyimide matrix composite laminates. *Composites, Part A: Applied Science and Manufacturing*, **34**, 979-986.
36. Hu, Y., Li, X., Lang, A. W., Zhang, Y., and Nutt, S. R. (2016) Water immersion aging of polydicyclopentadiene resin and glass fiber composites. *Polymer Degradation and Stability*, **124**, 35-42.
37. Retegi, A., Arbelaiz, A., Alvarez, P., Llano-Ponte, R., Labidi, J., and Mondragon, I. (2006) Effects of hygrothermal ageing on mechanical properties of flax pulps and their polypropylene matrix composites. *Journal of Applied Polymer Science*, **102**, 3438-3445.
38. Guermazi, N., Tarjem, A. B., Ksouri, I., and Ayedi, H. F. (2016) On the durability of FRP composites for aircraft structures in hygrothermal conditioning. *Composites, Part B: Engineering*, **85**, 294-304.
39. Le Gac, P.-Y., Arhant, M., Le Gall, M., and Davies, P. (2017) Yield stress changes induced by water in polyamide 6: characterization and modeling. *Polymer Degradation and Stability*, **137**, 272-280.
40. Mourad, A.-H., Fouad, H., and Elleithy, R. (2009) Impact of some environmental conditions on the tensile, creep-recovery, relaxation, melting and crystallinity behaviour of UHMWPE-GUR 410-

medical grade. *Materials and Design*, **30**, 4112-4119.

41. He, W., Liu, N., Chen, X., Guo, J., and Wei, T. (2016) The influence of hygrothermal ageing on the mechanical properties and thermal degradation kinetics of long glass fibre reinforced polyamide 6 composites filled with sepiolite. *RSC Advances*, **6**, 36689-36697.
42. Chen, Y., Davalos, J. F., Ray, I., and Kim, H.-Y. (2007) Accelerated aging tests for evaluations of durability performance of FRP reinforcing bars for concrete structures. *Composite Structures*, **78**, 101-111.
43. Ishak, Z. M., Ishiaku, U., and Karger-Kocsis, J. (2000) Hygrothermal aging and fracture behavior of short-glass-fiber-reinforced rubber-toughened poly (butylene terephthalate) composites. *Composites Science and Technology*, **60**, 803-815.
44. Islam, M. S., Pickering, K. L., and Foreman, N. J. (2010) Influence of hygrothermal ageing on the physico-mechanical properties of alkali treated industrial hemp fibre reinforced polylactic acid composites. *Journal of Polymers and the Environment*, **18**, 696-704.
45. Arhant, M., Le Gac, P.-Y., Le Gall, M., Burtin, C., Briançon, C., and Davies, P. (2016) Effect of sea water and humidity on the tensile and compressive properties of carbon-polyamide 6 laminates. *Composites, Part A: Applied Science and Manufacturing*, **91**, 250-261.
46. Shi, S., Chen, G., Wang, Z., and Chen, X. (2013) Mechanical properties of Nafion 212 proton exchange membrane subjected to hygrothermal aging. *Journal of Power Sources*, **238**, 318-323.
47. Le Gac, P.-Y., Choqueuse, D., Paris, M., Recher, G., Zimmer, C., and Melot, D. (2013) Durability of polydicyclopentadiene under high temperature, high pressure and seawater (offshore oil production conditions). *Polymer Degradation and Stability*, **98**, 809-817.
48. Wan, Y., Wang, Y., Huang, Y., He, B., and Han, K. (2005) Hygrothermal aging behaviour of VARTMed three-dimensional braided carbon-epoxy composites under external stresses. *Composites, Part A: Applied Science and Manufacturing*, **36**, 1102-1109.
49. Silva, L., Tognana, S., and Salgueiro, W. (2013) Study of the water absorption and its influence on the Young's modulus in a commercial polyamide. *Polymer Testing*, **32**, 158-164.
50. Guadagno, L., Fontanella, C., Vittoria, V., and Longo, P. (1999) Physical aging of syndiotactic polypropylene. *Journal of Polymer Science, Part B: Polymer Physics*, **37**, 173-180.
51. Guermazi, N., Elleuch, K., Ayedi, H., and Kapsa, P. (2008) Aging effect on thermal, mechanical and tribological behaviour of polymeric coatings used for pipeline application. *Journal of Materials Processing Technology*, **203**, 404-410.
52. Berthé, V., Ferry, L., Bénézet, J., and Bergeret, A. (2010) Ageing of different biodegradable polyesters blends mechanical and hygrothermal behavior. *Polymer Degradation and Stability*, **95**, 262-269.

53. Gautier, L., Mortaigne, B., Bellenger, V., and Verdu, J. (2000) Osmotic cracking nucleation in hydrothermal-aged polyester matrix. *Polymer*, **41**, 2481-2490.
54. Boubakri, A., Haddar, N., Elleuch, K., and Bienvenu, Y. (2010) Impact of aging conditions on mechanical properties of thermoplastic polyurethane. *Materials and Design*, **31**, 4194-4201.
55. Boubakri, A., Elleuch, K., Guermazi, N., and Ayedi, H. (2009) Investigations on hygrothermal aging of thermoplastic polyurethane material. *Materials and Design*, **30**, 3958-3965.
56. Mondal, S., Hu, J., and Yong, Z. (2006) Free volume and water vapor permeability of dense segmented polyurethane membrane. *Journal of Membrane Science*, **280**, 427-432.
57. Deshmane, C., Yuan, Q., Perkins, R., and Misra, R. (2007) On striking variation in impact toughness of polyethylene–clay and polypropylene–clay nanocomposite systems: the effect of clay–polymer interaction. *Materials Science and Engineering: A*, **458**, 150-157.
58. Guermazi, N., Elleuch, K., and Ayedi, H. (2009) The effect of time and aging temperature on structural and mechanical properties of pipeline coating. *Materials and Design*, **30**, 2006-2010.
59. *Handbook of Plastics Analysis*, Lobo, H., and Bonilla, J. V. (eds.), Taylor & Francis, USA (2003).
60. Zhao, S., Chen, F., Zhao, C., Huang, Y., Dong, J.-Y., and Han, C. C. (2013) Interpenetrating network formation in isotactic polypropylene/graphene composites. *Polymer*, **54**, 3680-3690.
61. Li, C.-Q., Zha, J.-W., Long, H.-Q., Wang, S.-J., Zhang, D.-L., and Dang, Z.-M. (2017) Mechanical and dielectric properties of graphene incorporated polypropylene nanocomposites using polypropylene-graft-maleic anhydride as a compatibilizer. *Composites Science and Technology*, **153**, 111-118.
62. Gopakumar, T., and Pagé, D. (2004) Polypropylene/graphite nanocomposites by thermo-kinetic mixing. *Polymer Engineering and Science*, **44**, 1162-1169.
63. Panda, B. P., Mohanty, S., and Nayak, S. K. (2015) Mechanism of toughening in rubber toughened polyolefin - A review. *Polymer-Plastics Technology and Engineering*, **54**, 462-473.
64. Bu, H., Qiu, W., Tan, Z., Li, Q., and Rong, J. (2015) Study on toughening of poly (4-methyl-1-pentene) with various thermoplastic elastomers. *Journal of Thermoplastic Composite Materials*, **28**, 1334-1342.
65. Chow, C., Xing, X., and Li, R. (2007) Moisture absorption studies of sisal fibre reinforced polypropylene composites. *Composites Science and Technology*, **67**, 306-313.
66. Zanni-Deffarges, M., and Shanahan, M. (1995) Diffusion of water into an epoxy adhesive: comparison between bulk behaviour and adhesive joints. International *Journal of Adhesion and Adhesives*,

15, 137-142.

67. Chu, W., Wu, L., and Karbhari, V. M. (2004) Durability evaluation of moderate temperature cured E-glass/vinyl ester systems. *Composite Structures*, **66**, 367-376.
68. Parameswaranpillai, J., Joseph, G., Shinu, K., Sreejesh, P., Jose, S., Salim, N. V., and Hameed, N. (2015) The role of SEBS in tailoring the interface between the polymer matrix and exfoliated graphene nanoplatelets in hybrid composites. *Materials Chemistry and Physics*, **163**, 182-189.
69. Ferrer, G. G., Sánchez, M. S., Sánchez, E. V., Colomer, F. R., and Ribelles, J. L. G. (2000) Blends of styrene–butadiene–styrene triblock copolymer and isotactic polypropylene: morphology and thermomechanical properties. *Polymer International*, **49**, 853-859.
70. Liao, C. Z., and Tjong, S. C. (2011) Effects of carbon nanofibers on the fracture, mechanical, and thermal properties of PP/SEBS-g-MA blends. *Polymer Engineering and Science*, **51**, 948-958.

7

Hygrothermal Ageing of Isotactic Polypropylene and Impact Polypropylene Copolymers: A Comparison of Mechanical, Thermal, Structural and Morphological Properties

7.1 Introduction

Thermoplastics are a promising class of polymers which can be reprocessed and reused by thermal means for a myriad of applications spanning over both engineering and non-engineering sectors owing to their tailorable properties, ease of processing, light weight, low cost, corrosion resistance, etc. [1-4]. Specifically, polypropylene (PP) is a versatile polyolefin thermoplastic material available in various forms like homopolymer, copolymer, reactor-blends, etc. As compared to other thermoplastics, the properties of PP can be relatively easily tuned through simple modifications in its structural arrangement [5-8].

Isotactic polypropylene (iPP), a homopolymer PP with same side methyl group alignment, is the most common PP grade accounting for more than 20 vol% of total PP production. In addition to usual benefits of PP such as low price and density as well as high resistance against corrosion and thermal degradation, iPP exhibits good balance of mechanical and physical characteristics along with superior recyclability due to the combinatorial effect of its isotactic nature and narrow distribution of molecular weight [9,10]. However, iPP displays poor impact performance on subjecting to sudden mechanical force due to its low crack growth resistance, especially at sub-ambient temperatures [11,12]. The introduction of various macromolecular materials such as elastomers, thermoplastic elastomers, copolymers, etc., to iPP has been reported to result in enhanced impact performance, thereby, paving the way for use in engineering applications [13-25]. Recently, we have also reported simultaneously improved impact strength and tensile modulus for iPP through the use of a combined system of maleic anhydride-*graft*-poly(ethylene-vinyl acetate) and graphene [23].

Anish Varghese and Vikas Mittal
Khalifa University of Science and Technology, Abu Dhabi, UAE

Impact polypropylene copolymers (IPC), also named as heterophasic copolymers, have been developed to overcome the poor impact performance associated with the iPP matrix. For this purpose, the reactor induced copolymerization of iPP is carried out with ethylene monomer *in-situ.* Borstar and Spheripol are the two widely accepted processes for this purpose, which comprise iPP bulk polymerization and subsequent propylene-ethylene copolymerization in the gas phase [26-30]. IPC comprise of a multiphase system with iPP crystalline homopolymer, ethylene-propylene semi-crystalline block copolymer (EPBC) and ethylene-propylene amorphous random copolymer (EPRC) [31-33]. Among these, EPRC can act as effective stress concentrator on the iPP surface, whereas EPBC is responsible for strengthening the interfacial performance between the other two by serving as a robust compatibilizer. On exposing IPC to the impact force, the stress concentrator phase, EPRC, is observed to consume a majority of the force, leading to the formation of micro-voids on the iPP surface by means of crazing and shear yielding, thus, resulting in plastic deformation and resistance to crack propagation. The ultimate impact performance of IPC greatly depends on the size, shape and distribution of EPRC, along with its miscibility with iPP phase under the influence of EPBC [32,34-36].

During their service life, the materials are occasionally subjected to harsh environmental conditions such as temperature and pressure variations, humidity, sunlight, marine environment, chemicals and pollutants, mechanical impact, etc. [37-39]. In order to estimate the endurance of the materials under harsh environmental conditions, laboratory accelerated ageing environments imitating the real-life harsh conditions can be employed, thus, avoiding long-term testing [40,41]. Among different types of ageing, hygrothermal ageing leads to structural changes in the polymer structure, which negatively affect the service life owing to the reduction in the mechanical and thermal characteristics [42-44]. The extent and severity of hygrothermal ageing induced changes in the polymers is closely related to the ageing conditions such as medium, temperature and time [45-51].

In this study, the behavior of three PP grades was studied under the influence of hygrothermal environment. The test specimens were prepared by injection molding and were subsequently immersed in both distilled water (DW) and saline water (SW) for 32 days at 70 and 90 °C. In order to assess the water uptake with ageing time, the sample weight was measured periodically after 4 days. Overall, hygrothermal ageing induced changes in the mechanical, thermal, structural and morphological properties of the polymer grades were analyzed.

7.2 Experimental

7.2.1 Materials

Isotactic polypropylene grade and two impact polypropylene copolymer grades (named as IPC1 and IPC2) were provided by Abu Dhabi Polymers Company Limited, UAE (in pellet form) and were used as received without any subsequent treatment. Table 7.1 presents further details of these materials.

Table 7.1 Material data for iPP, IPC1 and IPC2

Material	Nature/ Constituents of the Polymers	Appearance	Details
Isotactic polypropylene (iPP)	Homopolymer/iPP		**Density:** 900-910 $kg.m^{-3}$ **Melt flow rate:** 8 g/10 min (230 °C, 2.16 kg) **Melting point:** 168.5 °C
Impact polypropylene copolymer 1 (IPC1)	Heterophasic copolymer/iPP, EPRC and EPBC		**Density:** 900-910 $kg.m^{-3}$ **Melt flow rate:** 38 g/10 min (230 °C, 2.16 kg) **Melting point:** 166.5 °C
Impact polypropylene copolymer 2 (IPC2)	Same as for IPC1		**Density:** 900-910 $kg.m^{-3}$ **Melt flow rate:** 7 g/10 min (190 °C, 2.16 kg) **Melting point:** 167.7 °C

7.2.2 Preparation of Samples

In order to proceed with hygrothermal ageing, the PP samples were processed into dumbbell and rectangular bar shaped tensile and impact test specimens. For this purpose, a lab-scale plunger type injection molding machine HAAKE MiniJet PRO (Thermo Scientific, Germany) was employed. The test specimens were prepared by using the conditions: 180 °C cylinder temperature with a holding time of 5 min to obtain uniform PP melt, 430 bar injection pressure for 10 s to uniformly fill the mold volume and 500 bar holding pressure for 6 s to prevent backflow. The mold was maintained at a temperature of 125 °C.

7.2.3 Hygrothermal Ageing of Samples

To simulate the harsh environmental conditions, accelerated hygrothermal ageing in distilled water as well as synthetic saline water was carried out, in accordance with ASTM D570-90. Specifically, the accelerated ageing in DW and SW simulated humid and saline environments respectively. For this purpose, the samples were immersed in the media for 32 days at 70 and 90 °C. The concentration of NaCl in SW was fixed at 3%.

7.2.4 Characterization of Unaged and Aged Samples

Water Uptake

Water uptake, also termed as weight gain ratio (W_g), of the hygrothermally immersed PP samples was estimated using the unaged sample weight (W_0) and the weight of the aged sample (W_t) measured with a gap of 4 days, as per the following equation:

$$W_g = \frac{W_t - W_0}{W_0} \times 100$$

For measuring W_t, the samples were taken out from the immersed medium and weighed after wiping away the surface moisture using a neat cotton towel. The reported W_0 and W_t values constitute the average of at least five samples with less than 10% standard deviation.

Fourier-transform Infrared (FTIR) Spectroscopy

Ageing induced structural changes in the PP samples were analyzed using Fourier-transform infrared spectra, generated in transmission mode by employing BRUKER TENSOR II series spectrometer. With the aid of a diamond attenuated total reflectance (ATR) crystal, thin surface portion of the injection molded PP samples were examined by collecting 32 scans. The wavenumber range of 4000-400 cm^{-1} and a resolution of 4 cm^{-1} were used.

Wide-angle X-ray Diffraction (WAXD)

X'Pert PRO Panalytical powder diffractometer was used to study the diffraction patterns of the injection molded PP samples (at 40 mA and 45 kV). 2-theta angle range of 5-60° was scanned with a speed of 0.017° s^{-1} and a step time of 10 s. The analysis was carried out under ambient conditions using Cu-Kα irradiation (λ = 1.5406 Å).

Tensile Properties

Instron 3345 universal testing machine (UTM), attached with a load cell of 50 kN, was employed to study the impact of hygrothermal ageing on the tensile performance of the PP samples under ambient conditions (crosshead speed of 10 $mm.min^{-1}$). The dumbbell-shaped test specimens with dimensions 75 mm x 5 mm x 2 mm and a span length of 35 mm were used according to ISO 527 standard, and the reported data constitutes the mean of values from five specimens.

Izod Impact Strength

Resil impactor (Ceast) was employed to study the Izod impact strength of the unaged and aged samples at room temperature. The bar shaped test specimens with dimensions 80 mm x 10 mm x 4 mm were used according to ISO 180 standard. The hammer speed used for the analysis was 3.64 $m.s^{-1}$, and the reported data represents the average of five specimens.

Differential Scanning Calorimetry

Differential scanning calorimeter from TA Instruments was used to study the peak melting temperature (T_m), peak crystallization temperature and (T_c), melting enthalpy (ΔH_f), crystallization enthalpy (ΔH_c) and percentage crystallinity (X_c). 3-8 mg sample weight was examined in the temperature range of (-50, +200 °C) by applying two heating and cooling cycles using a dry nitrogen flow of 50 mL.min^{-1}. Second heating-cooling cycles were used to record the calorimetric behavior of the samples.

Scanning Electron and Optical Microscopy

Scanning electron microscope (SEM) FEI Quanta (FEG250, USA) was used to analyze the micro-morphology of the impact fractured surface of the PP samples. The images were taken at 10 kV after sputtering a thin layer of gold on the fracture surfaces. The samples were affixed on the aluminum stubs for analysis using conductive carbon tape. Optical microscope OLYMPUS BX51M (Japan) was also used to examine the surface changes induced by hygrothermal conditions (eyepiece magnification of 10x).

7.3 Results and Discussion

Figure 7.1 depicts the change in the extent of water uptake as a function of hygrothermal ageing time for iPP, IPC1 and IPC2, at 70 and

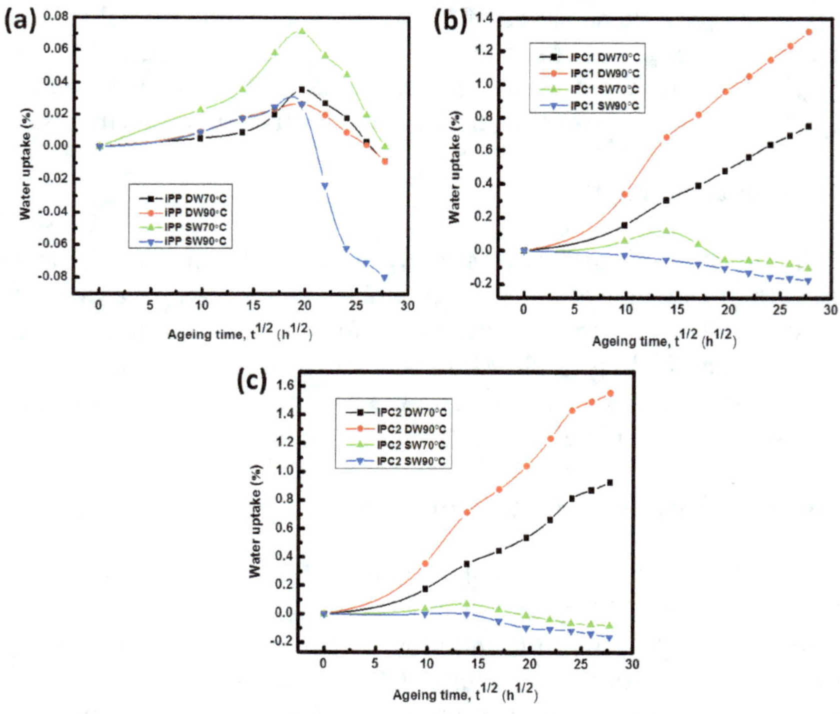

Figure 7.1 Water uptake curves of (a) iPP, (b) IPC1 and (c) IPC2.

90 °C in DW and SW. The water uptake behavior was observed to be different for iPP and IPC. Water uptake curves of iPP (Figure 7.1a) displayed two distinct parts, which supported the presence of combinatorial Fickian and non-Fickian behaviors regardless of the ageing conditions. The Fickian behavior, represented by an increasing trend in water uptake with time in a non-linear manner upon hygrothermal ageing, was visible until ageing time ($\sqrt{h}$) = 19.6. Following this, a decreasing trend of water uptake with time represented the non-Fickian behavior. Comparing different hygrothermal conditions, iPP aged at 70 °C in SW displayed the highest water uptake, i.e., ca. 0.071%. The Fickian behavior is associated with the moisture diffusion in the samples under hygrothermal conditions, whereas the non-Fickian behavior might result from surface crosslinking, PP chain rearrangement and washing effect [52,53]. Both IPC samples (Figures 7.1b and 7.1c) demonstrated similar behavior upon hygrothermal immersion, though the observed behavior differed in DW and SW. In DW, the water uptake increased with ageing time nonlinearly throughout the studied ageing time range regardless of the IPC type owing to the osmotic cracking effect induced by the hydrophilic oligomer moieties [54,55]. Also, such effect was observed to increase on enhancing the temperature from 70 to 90 °C due to the rapid reorganization of the loose macromolecules and, thus, the appearance of increased free volume [56,57]. A higher water uptake value of 1.55% was observed for IPC2 in DW at 90 °C as compared to 1.32% for IPC1. In SW, the observed water uptake values at 70 °C were extremely low for both IPC1 and IPC2 samples in comparison with the samples aged in DW. On the other hand, at 90 °C, the observed water uptake values were negative. The appearance of such behavior in SW is related to the reduction in the chemical potential contributed by salt [53]. In SW at 70 °C, the water uptake values increased only until $\sqrt{h}$ = 13.85 for both IPC1 and IPC2, followed by a decreasing trend which even reached negative values after $\sqrt{h}$ = 19.59. Comparing the two IPC samples, IPC2 demonstrated lower water uptake in SW. From these observations, it can be seen that iPP offered superior barrier performance in DW, whereas IPC1 and IPC2 were better in SW.

FTIR spectra were collected for samples before and after simulated hygrothermal ageing in DW and SW to assess the structural changes, as shown in Figure 7.2. For unaged iPP, characteristic peaks at 2957, 2916, 2870, 2837, 2362, 1463, 1374, 999-973-844-811 and 898 cm^{-1} could be attributed to CH_3 asymmetric stretching

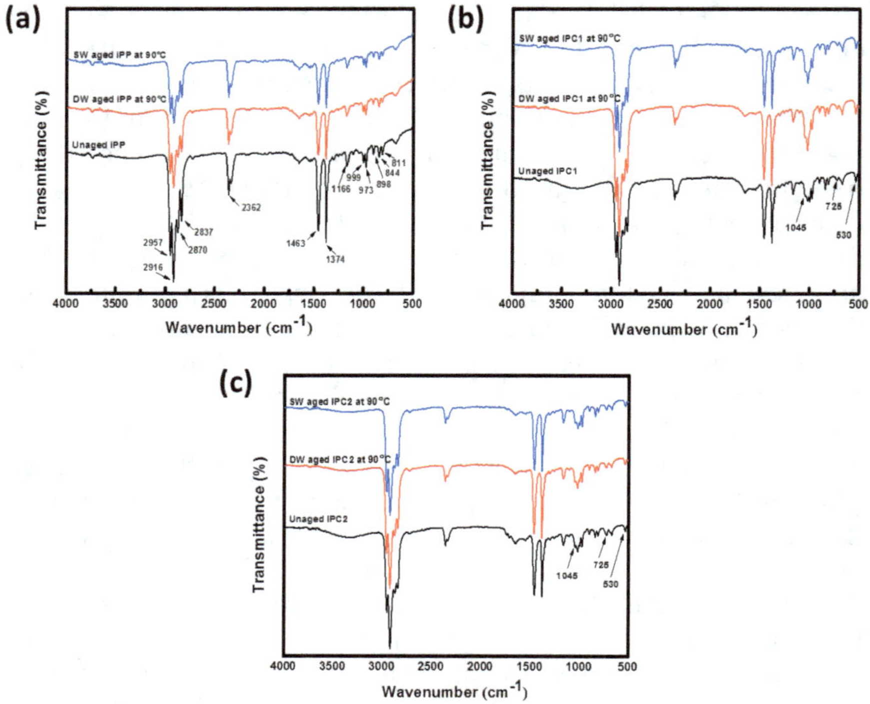

Figure 7.2 FTIR spectra for unaged and aged samples: (a) iPP, (b) IPC1 and (c) IPC2.

vibration, CH_2 symmetric stretching vibration, CH_3 symmetric stretching vibration, CH_2 symmetric stretching vibration, C-H bending vibration, CH_3 asymmetric bending vibration, CH_3 symmetric bending vibration, tertiary methyl skeleton deformation and stretching caused perpendicular absorption respectively [58]. FTIR spectra of 90 °C DW and SW aged iPP samples also displayed same characteristic peaks as unaged iPP, though the intensity of each peak was observed to reduce. Such intensity reduction indicates the initiation of chain scission under hygrothermal conditions. In the FTIR spectra of IPC1 (Figure 7.2b) and IPC2 (Figure 7.2c), three additional peaks at 1045, 725 and 530 cm^{-1} were observed along with the aforesaid characteristic iPP peaks. The additional peaks were contributed by the amorphous polyethylene composition in the IPC1 and IPC2 samples, which were attributed to CH_2 gauche configuration, CH_2 rocking vibration and CH_2 stretching vibration respectively [28,59-61]. After hygrothermal ageing at 90 °C, the characteristic peaks of IPC1 and IPC2 were observed to appear with increased intensity in both DW and SW. Specifically, DW aged IPC samples exhibited higher increase in the peak intensities as compared to SW. According to previous studies, thermal oxidation is the cause of such increase in the peak intensities after hygrothermal exposure [62,63]. Correspondingly, in the current study, a higher thermal oxidation tendency was confirmed in DW than SW for both IPC1 and IPC2 samples. Also, the observed higher extent of peak intensity increase in IPC1 underlines its greater tendency for thermal oxidation than IPC2.

Wide angle X-ray diffractograms of unaged and aged samples are depicted in Figure 7.3. For unaged iPP sample, the diffraction peaks

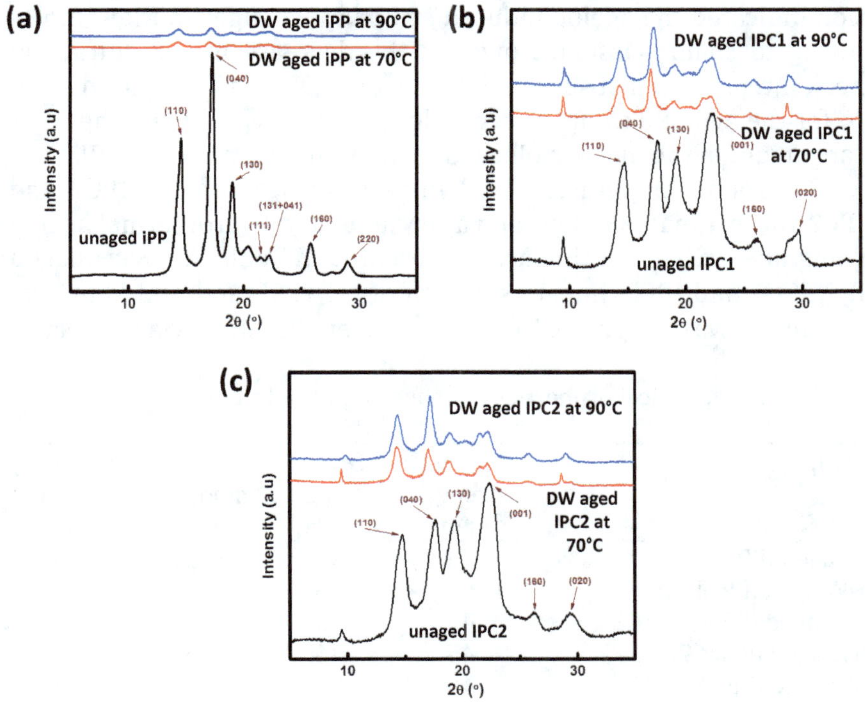

Figure 7.3 WAXD patterns for unaged and aged samples: (a) iPP, (b) IPC1 and (c) IPC2.

at 14.55°, 17.26°, 19.03°, 21.47°, 22.18°, 25.76° and 29.01° could be attributed to respective (110), (040), (130), (111), (131+041), (160) and (220) α-crystalline planes [23,64,65]. To estimate the interplanar distance (d) of each diffraction peak, Bragg's equation ($d = \lambda/(2\sin\theta_{max})$ where λ is the *X*-ray wavelength (1.5406 Å), was used.

Correspondingly, *d* values of 6.08, 5.13, 4.66, 4.14, 4.00, 3.46 and 3.08 Å respectively were obtained for the aforesaid α-crystalline planes. The unaged IPC1 and IPC2 exhibited α-crystalline peaks of iPP and PE attributed to respective (110), (040), (130), (160), (020) and (001) planes. These peaks were observed at 14.67°, 17.60°, 19.19°, 26.11°, 29.69° and 22.33° respectively for IPC1 and 14.77°, 17.72°, 19.29°, 26.22°, 29.36° and 22.33° respectively for IPC2. Accordingly, the obtained *d* values of these peaks were 6.03, 5.04, 4.62, 3.41, 3.01 and 3.98 Å respectively for IPC1 and 5.99, 5.00, 4.60, 3.39, 3.04 and 3.98 Å respectively for IPC2. Similar characteristic diffraction peaks were observed in the hygrothermally aged samples, however, with reduced intensity. The observed crystallinity reduction indicated the ageing induced structural changes in the samples owing to chain scission. However, the decrease in the diffraction peak intensity was observed to be lower at 90 °C as compared at 70 °C for the aged PP samples, regardless of their type, due to the rearrangement of the macromolecular chain structure at 90 °C [48].

The room temperature Izod impact strength of iPP, IPC1 and IPC2 was estimated before and after simulated hygrothermal ageing using un-notched specimens, as presented in Table 7.2. For unaged iPP, IPC1 and IPC2, impact strength of ca. 7, 53 and 103 kJ.m^{-2} respectively was observed. The impact strength of iPP was observed

Table 7.2 Mechanical properties of unaged and aged samples

Sample	Impact strength (kJ.m^2)	Tensile strength (MPa)	Tensile modulus (MPa)	Breaking extension (%)
Unaged iPP	7	39	1011	24
DW aged iPP at 70 °C	51	36	815	23
DW aged iPP at 90 °C	45	35	808	21
SW aged iPP at 70 °C	45	36	809	21
SW aged iPP at 90 °C	46	35	801	21
Unaged IPC1	53	22	1018	35
DW aged IPC1 at 70 °C	68	21	551	14
DW aged IPC1 at 90 °C	92	20	471	15
SW aged IPC1 at 70 °C	75	21	486	18
SW aged IPC1 at 90 °C	89	21	481	21
Unaged IPC2	103	21	864	299
DW aged IPC2 at 70 °C	101	19	270	120
DW aged IPC2 at 90 °C	101	19	205	163
SW aged IPC2 at 70 °C	102	18	259	113
SW aged IPC2 at 90 °C	102	18	220	154

to markedly enhance after ageing under hygrothermal conditions owing to the plasticization effect [43,66]. Accordingly, the observed impact strength retention rate of iPP stayed above 550% (Figure 7.4a) regardless of the ageing conditions. The impact strength of the

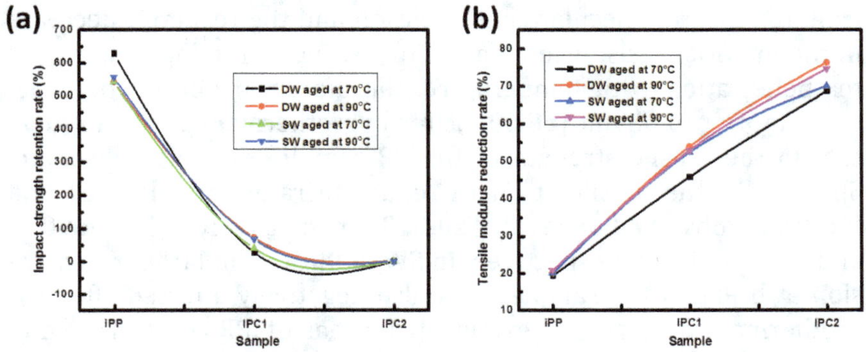

Figure 7.4 Effect of hygrothermal conditions on (a) impact strength retention rate and (b) tensile modulus reduction rate.

aged IPC1 also exhibited an increasing trend, which was also observed to increase with increase in temperature from 70 to 90°C in both DW and SW. For instance, the impact strength retention rates of ca. 74 and 68% were discerned for IPC1 after hygrothermal ageing at 90 °C in DW and SW respectively. The observed increase in the impact strength retention at higher temperature is related to the swift micro-voids formation at the iPP-EPRC interface in the IPC1 sample owing to the increasing degree of water plasticization [43,48]. The impact strength of IPC2 was observed to be retained after hygrothermal ageing. Thus, the impact strength values of 101 and 102 kJ.m^{-2} were discerned for IPC2 after hygrothermal ageing at 90 °C in DW and SW respectively, as compared to 103 kJ.m^{-2} for the unaged IPC2. Slight reduction in the impact strength indicates the beginning of the structural integrity loss under the influence harsh hygrothermal conditions [48].

The tensile modulus values of 1011, 1018, and 864 MPa were measured for unaged iPP, IPC1 and IPC2 respectively. After hygrothermal ageing, the tensile modulus of the PP samples was observed to decrease. In addition, the tensile modulus reduction rate increased with temperature in both DW and SW. For iPP, the observed tensile modulus reduction rate was higher in SW at both 70 and 90 °C as compared to DW due to the influence of high pH water medium and increased crack formation tendency [38,67]. An optimum reduction rate of ca. 21% (Figure 7.4b) was recorded for 90 °C SW aged iPP. For aged IPC1 and IPC2, higher reduction rates were observed at 70 °C in SW and at 90 °C in DW. Under hygrothermal ageing conditions, the reduction in the tensile modulus probably results from the macromolecular chain scission and the resultant decrease in the material resistance. In addition, owing to the increase in water penetration with temperature, the aforesaid effect would become more prominent [68,69]. After hygrothermal ageing, a reduction in the tensile strength of the PP samples was also observed. Specifically, the tensile strength reduction rates for iPP, IPC1 and IPC2 were observed to be 10, 9 and 10% respectively in DW at 90 °C and 10, 5 and 14% respectively in SW at 90 °C. Similarly, the extension at break of the samples was also negatively affected after hygrothermal exposure. The extension at break of iPP was observed to reduce with temperature in DW, though the reduction rate remained unchanged in SW at both 70 and 90 °C (Table 7.2). For IPC1 and IPC2, hygrothermal ageing had more negative effect on the extension at 70 °C in both DW and SW than 90 °C. Overall, the maximal extension reduction rates of 13, 60 and 62% were noticed for iPP, IPC1 and IPC2 respectively. From these observations, lower resistance of IPC samples against extension reduction in hygrothermal conditions as compared to iPP was confirmed due to the influence of amorphous content and, thus, higher water penetration tendency and molecular scission [70].

The calorimetric properties of the samples are depicted in Table 7.3. The second heating and cooling curves of the samples are also demonstrated in Figures 7.5 and 7.6. The peak melting temperature of iPP, IPC1 and IPC2 was observed to be 168.5, 166.5 and 167.7 °C respectively. For aged iPP, the observed T_m was slightly lower in DW and almost same in SW, as compared to the unaged polymer. In the case of aged IPC1, T_m was slightly higher in all hygrothermal conditions as compared to the unaged sample. For IPC2, a slight decrease in T_m was observed with temperature in both DW and SW. These observations indicated that the PP samples resisted severe changes in T_m upon hygrothermal exposure. The peak crystallization temperatures of 127.9, 130.4 and 128.9 °C were measured for unaged iPP, IPC1 and IPC2. For aged IPC1, T_c was slightly increased in DW with temperature, whereas it remained unchanged in SW, though still higher than unaged IPC1. For aged IPC2, T_c displayed a slight reduction of ≤0.5 °C in both DW and SW.

Table 7.3 Calorimetric properties of unaged and aged iPP, IPC1 and IPC2

Sample	T_c (°C)	ΔH_c (J/g)	T_m (°C)	ΔH_f (J/g)	X_c (%)	ΔX_c (%)
Unaged iPP	127.9	98.7	168.5	90.8	43.8	-
DW aged iPP at 70 °C	126.3	106.2	166.9	112.2	54.2	10.4
DW aged iPP at 90 °C	126.4	110.1	166.9	108.0	52.1	8.3
SW aged iPP at 70 °C	126.2	111.2	168.2	106.5	51.4	7.6
SW aged iPP at 90 °C	126.2	107.0	168.6	105.4	50.9	7.1
Unaged IPC1	130.4	84.3	166.5	82.8	40.0	-
DW aged IPC1 at 70 °C	130.9	88.9	166.9	93.8	45.3	6.3
DW aged IPC1 at 90 °C	131.2	92.0	166.6	91.6	44.2	5.2
SW aged IPC1 at 70 °C	130.9	91.7	166.7	91.1	44.0	4.0
SW aged IPC1 at 90 °C	130.9	90.3	166.9	90.8	43.8	3.8
Unaged IPC2	128.9	74.5	167.7	72.9	35.2	-
DW aged IPC2 at 70 °C	128.5	78.2	167.3	79.0	38.2	3.0
DW aged IPC2 at 90 °C	128.6	79.8	166.7	78.5	37.9	2.7
SW aged IPC2 at 70 °C	128.4	79.5	167.5	77.8	37.5	2.3
SW aged IPC2 at 90 °C	128.4	79.5	167.1	76.6	37.0	1.8

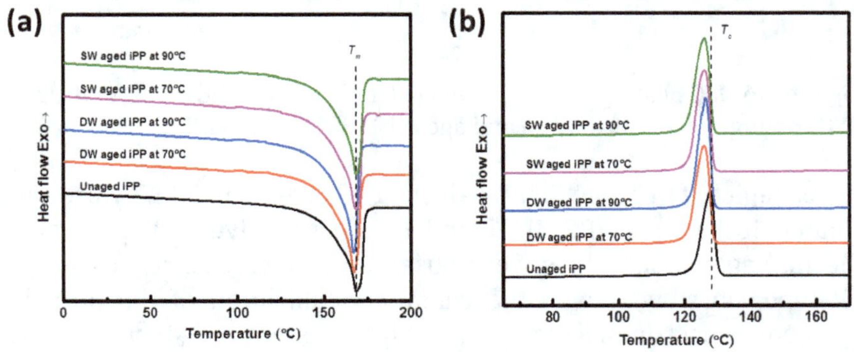

Figure 7.5 DSC (a) heating and (b) cooling thermograms of unaged and hygrothermally aged iPP.

For estimating X_c, the following equation was used [24,55]:

$$X_c = \frac{\Delta H}{\Delta H_0} \times 100$$

In this equation, ΔH_0, and ΔH correspond to theoretically calculated ΔH_f (207.1 J/g) for totally crystalline PP [24] and ΔH_f for the analyzed samples from DSC.

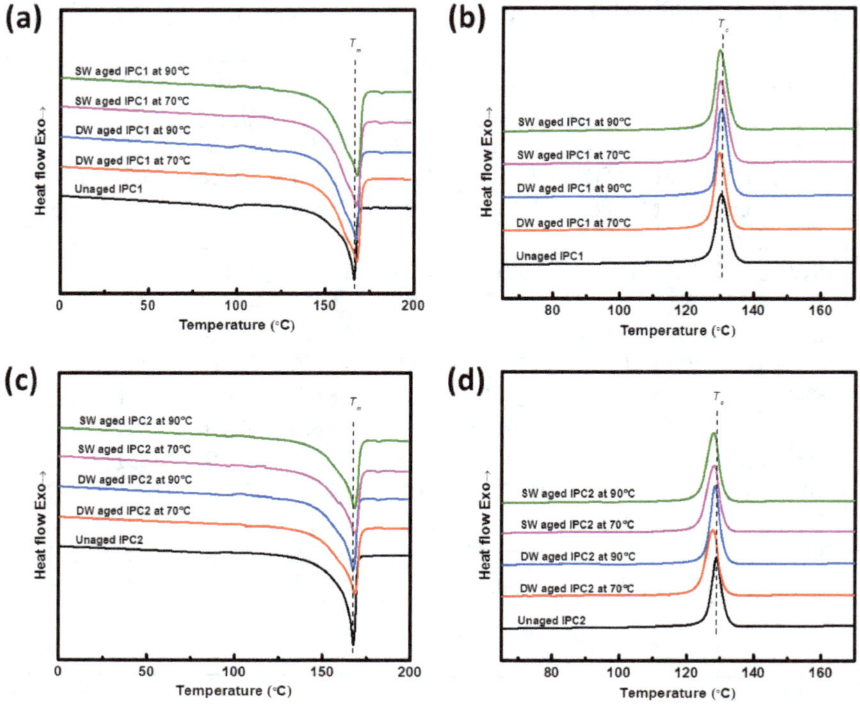

Figure 7.6 DSC heating curves for unaged and aged (a) IPC1 and (c) IPC2; DSC cooling curves for unaged and aged (b) IPC1 and (d) IPC2.

As depicted in Table 7.3, X_c values of 43.8, 40.0 and 35.2 were calculated for unaged iPP, IPC1 and IPC2 respectively. After hygrothermal ageing, X_c values were observed to increase, regardless of PP type, possibly due to the chemi-crystallization of the initially scissioned macromolecules [55]. A higher extent of X_c increase was observed at 70 °C than 90 °C in both DW and SW. In addition, the samples exhibited higher rate of increase in DW as compared to SW. Specifically, the extent of X_c increase was aligned in the range of 7.1-10.4%, 2.8-6.3% and 1.8-3.0% for iPP, IPC1 and IPC2 respectively.

Figure 7.7 presents the SEM images for unaged and aged iPP, IPC1 and IPC2 samples. The SEM images of the unaged iPP, 70 °C DW aged iPP and 90 °C DW aged iPP (Figure 7.7a-c) supported the transition of the iPP fracture morphology from brittle to ductile upon hygrothermal exposure due to the effect of the water plasticizing effect. These observations also supported the earlier observed impact properties of the samples. The SEM images of the unaged IPC1 and IPC2 (Figures 7.7d and 7.7g) exhibited several round-shaped micro-voids in the continuous phase. The observed micro-voids confirmed the effective debonding or cavitation at the EPRC-iPP

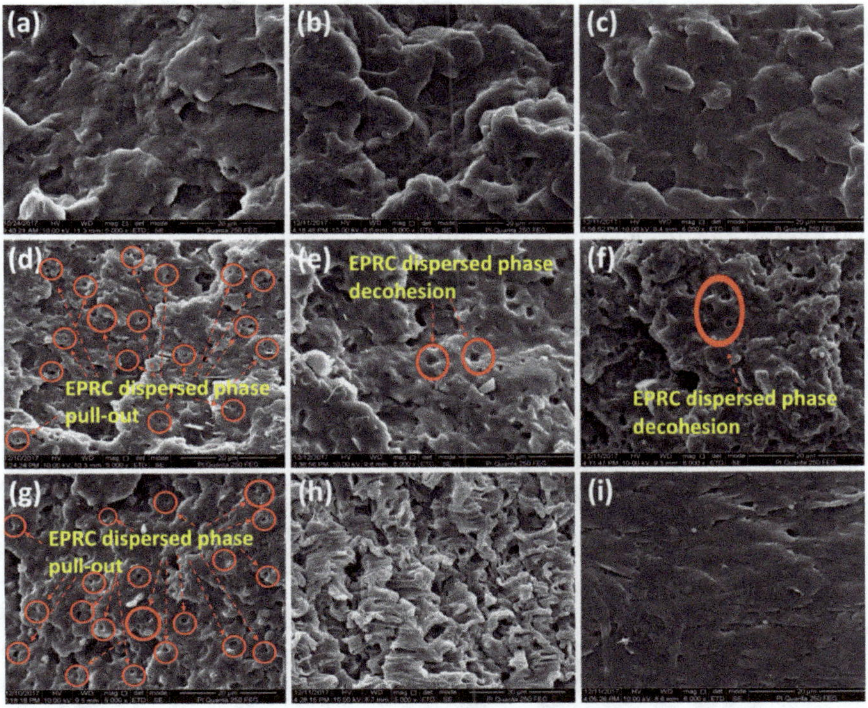

Figure 7.7 SEM images of the impact fractured surface for unaged (a) iPP, (d) IPC1 and (g) IPC2; 70 °C DW aged (b) iPP (e) IPC1 and (h) IPC2; and 90 °C DW aged (c) iPP, (f) IPC1 and (i) IPC2.

interface owing to shear yielding and crazing upon impact. Accordingly, the increased impact strength values of 53 and 103 $kJ.m^{-2}$ were observed for the corresponding IPC1 and IPC2 samples. By virtue of better EPRC-iPP interfacial performance under the influence of EPBC, IPC2 exhibited higher impact strength than IPC1. Figure 7.7e-f demonstrate the impact fracture morphology of DW aged IPC1 at 70 and 90 °C respectively. Decohesion of the EPRC phase in the continuous matrix was evident after hygrothermal exposure at both 70 and 90 °C. As a result, slightly bigger round-shaped EPRC phase pull-outs were observed to appear. Figure 7.7h-i show the impact fracture morphology of the hygrothermally aged IPC2 at 70 and 90 °C respectively. The initiation of the phase damage of the dispersed EPRC phase was observed after hygrothermal exposure.

Accordingly, slightly reduced impact strength of 101 kJ.m^{-2} was observed at both 70 and 90 °C for the aged IPC2 sample.

Optical micrographs of the hygrothermally aged samples at 70 and 90 °C in DW are depicted in Figure 7.8. Surface uniformity loss and resultant deterioration were observed for all samples. Accordingly, the occurrence of pore analogous structure with irregular shapes and minor surface cracks was observed, which underlined the increasing trend of surface roughness with hygrothermal ageing. Among the studied samples, IPC1 suffered more surface damage upon ageing as compared to IPC2 and iPP.

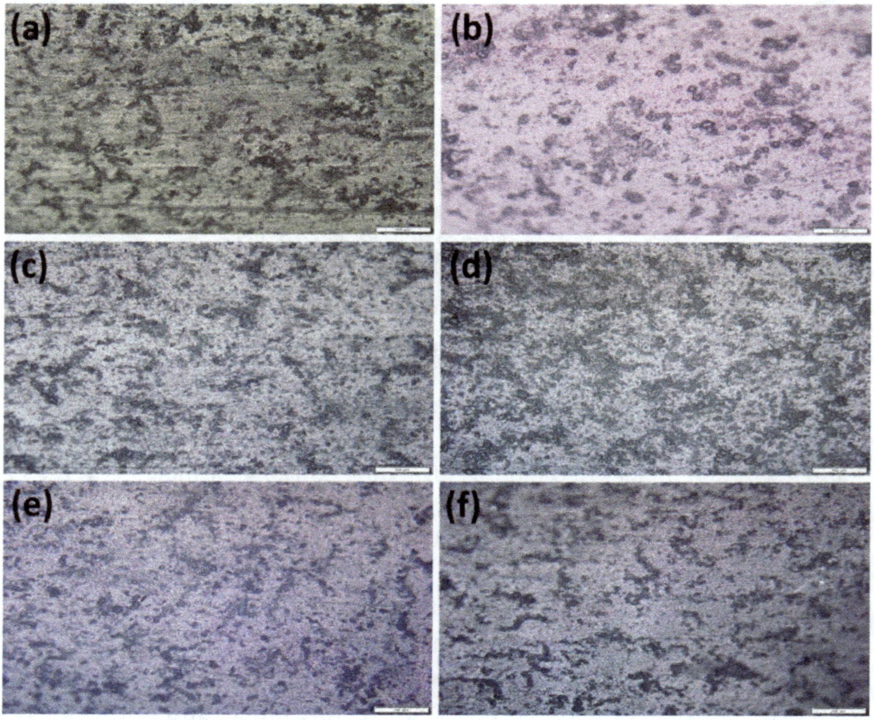

Figure 7.8 Optical micrographs of 70 °C DW aged (a) iPP, (c) IPC1 and (e) IPC2; and 90 °C DW aged (b) iPP, (d) IPC1 and (f) IPC2.

7.4 Conclusions

In this study, hygrothermal ageing analysis of isotactic polypropylene (iPP) and two different impact polypropylene copolymers (IPC1 and IPC2) was carried out using different ageing conditions. The iPP polymer displayed superior barrier performance in DW, though IPC samples were better in SW. Significantly enhanced values of impact strength were recorded after hygrothermal exposure for iPP and IPC1 due to the water plasticizing effect, whereas a slight reduction was noted for IPC2. The impact fracture morphology of the samples also supported the observations from the impact analysis. Owing to the combinatorial thermal and water ageing effect and the resultant molecular chain scission, the tensile modulus, strength and elongation at break of the aged samples were observed to decrease. As a result of the chemi-crystallization phenomenon, increased percentage crystallinity values were discerned for the PP samples.

References

1. Mortazavian, S., Fatemi, A., Mellott, S. R., and Khosrovaneh, A. (2015) Effect of cycling frequency and self-heating on fatigue behavior of reinforced and unreinforced thermoplastic polymers. *Polymer Engineering and Science*, **55**, 2355-2367.
2. Teoh, S., Tang, Z., and Hastings, G. W. (2016) Thermoplastic polymers in biomedical applications: Structures, properties and processing. In: *Handbook of Biomaterial Properties*, Black, J., and Hastings, G. (eds.), Springer, USA, pp. 261-290.
3. Paoletti, A., Lambiase, F., and Di Ilio, A. (2016) Analysis of forces and temperatures in friction spot stir welding of thermoplastic polymers. *The International Journal of Advanced Manufacturing Technology*, **83**, 1395-1407.
4. Eftekhari, M., and Fatemi, A. (2016) Tensile behavior of thermoplastic composites including temperature, moisture, and hygrothermal effects. *Polymer Testing*, **51**, 151-164.
5. Maddah, H. A. (2016) Polypropylene as a promising plastic: A review. *American Journal of Polymer Science*, **6**, 1-11.
6. Gahleitner, M., and Paulik, C. (2017) Polypropylene and other polyolefins. In: *Brydson's Plastics Materials*, 8th edition, Elsevier, USA, pp. 279-309.
7. Tripathi, D. (2002) *Practical Guide to Polypropylene*, iSmithers Rapra Publishing, USA.
8. *Handbook of Polypropylene and Polypropylene Composites*, Karian, H. G. (ed.), CRC Press, USA (2003).
9. Bikiaris, D., Vassiliou, A., Chrissafis, K., Paraskevopoulos, K., Jannakoudakis, A., and Docoslis, A. (2008) Effect of acid treated multiwalled carbon nanotubes on the mechanical, permeability, thermal properties and thermo-oxidative stability of isotactic polypropylene. *Polymer Degradation and Stability*, **93**, 952-967.

10. Wang, Y., and Tsai, H. B. (2012) Thermal, dynamic-mechanical, and dielectric properties of surfactant intercalated graphite oxide filled maleated polypropylene nanocomposites. *Journal of Applied Polymer Science*, **123**, 3154-3163.
11. Chen, F., Qiu, B., Wang, B., Shangguan, Y., and Zheng, Q. (2015) Balanced toughening and strengthening of ethylene–propylene rubber toughened isotactic polypropylene using a poly (styrene-b-ethylene–propylene) diblock copolymer. *RSC Advances*, **5**, 20831-20837.
12. Zebarjad, S., Bagheri, R., Reihani, S., and Lazzeri, A. (2003) Deformation, yield and fracture of elastomer-modified polypropylene. *Journal of Applied Polymer Science*, **90**, 3767-3779.
13. Wu, C. L., Zhang, M. Q., Rong, M. Z., and Friedrich, K. (2005) Silica nanoparticles filled polypropylene: effects of particle surface treatment, matrix ductility and particle species on mechanical performance of the composites. *Composites Science and Technology*, **65**, 635-645.
14. Kukaleva, N., Jollands, M., Cser, F., and Kosior, E. (2000) Influence of phase structure on impact toughening of isotactic polypropylene by metallocene-catalyzed linear low-density polyethylene. *Journal of Applied Polymer Science*, **76**, 1011-1018.
15. Xu, J., Mittal, V., and Bates, F. S. (2016) Toughened isotactic polypropylene: phase behavior and mechanical properties of blends with strategically designed random copolymer modifiers. *Macromolecules*, **49**, 6497-6506.
16. Keskin, R., and Adanur, S. (2011) Improving toughness of polypropylene with thermoplastic elastomers in injection molding. *Polymer-Plastics Technology and Engineering*, **50**, 20-28.
17. Kukackova, O., Dung, N. V., Abbrent, S., Urbanova, M., Kotek, J., and Brus, J. (2017) A novel insight into the origin of toughness in polypropylene-calcium carbonate microcomposites: Multivariate analysis of ss-NMR spectra. *Polymer*, **132**, 106-113.
18. Yang, C.-j., Huang, T., Yang, J.-h., Zhang, N., Wang, Y., and Zhou, Z.-w. (2017) Carbon nanotubes induced brittle-ductile transition behavior of the polypropylene/ethylene-propylene-diene terpolymer blends. *Composites Science and Technology*, **139**, 109-116.
19. Zhu, L., Fan, H.-N., Yang, Z.-Q., and Xu, X.-H. (2010) Evaluation of phase morphology, rheological, and mechanical properties based on polypropylene toughened with poly (ethylene-co-octene). *Polymer-Plastics Technology and Engineering*, **49**, 208-217.
20. Liang, J., and Li, R. (2000) Rubber toughening in polypropylene: A review. *Journal of Applied Polymer Science*, **77**, 409-417.
21. Zuiderduin, W., Westzaan, C., Huetink, J., and Gaymans, R. (2003) Toughening of polypropylene with calcium carbonate particles. *Polymer*, **44**, 261-275.

22. Tjong, S., Bao, S., and Liang, G. (2005) Polypropylene/montmorillonite nanocomposites toughened with SEBS-g-MA: Structure - property relationship. *Journal of Polymer Science, Part B: Polymer Physics*, **43**, 3112-3126.
23. Varghese, A. M., Vengatesan, M. R., Mun, S. C., Macosko, C. W., and Mittal, V. (2018) Effect of graphene on polypropylene/maleic anhydride grafted ethylene vinyl acetate (PP/EVA-g-MA) blend: Mechanical, thermal, morphological and rheological properties. *Industrial and Engineering Chemistry Research*, **57**(23), 7834-7845.
24. Parameswaranpillai, J., Joseph, G., Shinu, K., Sreejesh, P., Jose, S., Salim, N. V., and Hameed, N. (2015) The role of SEBS in tailoring the interface between the polymer matrix and exfoliated graphene nanoplatelets in hybrid composites. *Materials Chemistry and Physics*, **163**, 182-189.
25. Haghnegahdar, M., Naderi, G., and Ghoreishy, M. (2017) Fracture toughness and deformation mechanism of un-vulcanized and dynamically vulcanized polypropylene/ethylene propylene diene monomer/graphene nanocomposites. *Composites Science and Technology*, **141**, 83-98.
26. Li, R., Zhang, X., Zhao, Y., Hu, X., Zhao, X., and Wang, D. (2009) New polypropylene blends toughened by polypropylene/poly (ethylene-co-propylene) in-reactor alloy: Compositional and morphological influence on mechanical properties. *Polymer*, **50**, 5124-5133.
27. Gahleitner, M., Tranninger, C., and Doshev, P. (2013) Heterophasic copolymers of polypropylene: Development, design principles, and future challenges. *Journal of Applied Polymer Science*, **130**, 3028-3037.
28. Hongjun, C., Xiaolie, L., Dezhu, M., Jianmin, W., and Hongsheng, T. (1999) Structure and properties of impact copolymer polypropylene. I. Chain structure. *Journal of Applied Polymer Science*, **71**, 93-101.
29. Mirabella, Jr., F. M. (1993) Impact polypropylene copolymers: Fractionation and structural characterization. *Polymer*, **34**, 1729-1735.
30. Li, Y., Xu, J.-T., Dong, Q., Fu, Z.-S., and Fan, Z.-Q. (2009) Morphology of polypropylene/poly (ethylene-co-propylene) in-reactor alloys prepared by multi-stage sequential polymerization and two-stage polymerization. *Polymer*, **50**, 5134-5141.
31. Tan, H., Li, L., Chen, Z., Song, Y., and Zheng, Q. (2005) Phase morphology and impact toughness of impact polypropylene copolymer. *Polymer*, **46**, 3522-3527.
32. Zhang, C., Shangguan, Y., Chen, R., Wu, Y., Chen, F., Zheng, Q., and Hu, G. (2010) Morphology, microstructure and compatibility of impact polypropylene copolymer. *Polymer*, **51**, 4969-4977.

33. Zhu, H., Monrabal, B., Han, C. C., and Wang, D. (2008) Phase structure and crystallization behavior of polypropylene in-reactor alloys: insights from both inter-and intramolecular compositional heterogeneity. *Macromolecules*, **41**, 826-833.
34. Doshev, P., Lach, R., Lohse, G., Heuvelsland, A., Grellmann, W., and Radusch, H.-J. (2005) Fracture characteristics and deformation behavior of heterophasic ethylene–propylene copolymers as a function of the dispersed phase composition. *Polymer*, **46**, 9411-9422.
35. Starke, J., Michler, G., Grellmann, W., Seidler, S., Gahleitner, M., Fiebig, J., and Nezbedova, E. (1998) Fracture toughness of polypropylene copolymers: Influence of interparticle distance and temperature. *Polymer*, **39**, 75-82.
36. Grein, C., Bernreitner, K., Hauer, A., Gahleitner, M., and Neißl, W. (2003) Impact modified isotatic polypropylene with controlled rubber intrinsic viscosities: some new aspects about morphology and fracture. *Journal of Applied Polymer Science*, **87**, 1702-1712.
37. Han, M.-H., and Nairn, J. A. (2003) Hygrothermal aging of polyimide matrix composite laminates. *Composites, Part A: Applied Science and Manufacturing*, **34**, 979-986.
38. Larbi, S., Bensaada, R., Bilek, A., and Djebali, S. (2015) Hygrothermal ageing effect on mechanical properties of FRP laminates. *AIP Conference Proceedings*, **1653**, 020066.
39. Hu, Y., Li, X., Lang, A. W., Zhang, Y., and Nutt, S. R. (2016) Water immersion aging of polydicyclopentadiene resin and glass fiber composites. *Polymer Degradation and Stability*, **124**, 35-42.
40. Guermazi, N., Tarjem, A. B., Ksouri, I., and Ayedi, H. F. (2016) On the durability of FRP composites for aircraft structures in hygrothermal conditioning. *Composites, Part B: Engineering*, **85**, 294-304.
41. Retegi, A., Arbelaiz, A., Alvarez, P., Llano-Ponte, R., Labidi, J., and Mondragon, I. (2006) Effects of hygrothermal ageing on mechanical properties of flax pulps and their polypropylene matrix composites. *Journal of Applied Polymer Science*, **102**, 3438-3445.
42. Mourad, A.-H., Fouad, H., and Elleithy, R. (2009) Impact of some environmental conditions on the tensile, creep-recovery, relaxation, melting and crystallinity behaviour of UHMWPE-GUR 410-medical grade. *Materials and Design*, **30**, 4112-4119.
43. He, W., Liu, N., Chen, X., Guo, J., and Wei, T. (2016) The influence of hygrothermal ageing on the mechanical properties and thermal degradation kinetics of long glass fibre reinforced polyamide 6 composites filled with sepiolite. *RSC Advances*, **6**, 36689-36697.
44. Le Gac, P.-Y., Arhant, M., Le Gall, M., and Davies, P. (2017) Yield stress changes induced by water in polyamide 6: characterization and modeling. *Polymer Degradation and Stability*, **137**, 272-280.
45. Ishak, Z. M., Ishiaku, U., and Karger-Kocsis, J. (2000) Hygrothermal

aging and fracture behavior of short-glass-fiber-reinforced rubber-toughened poly (butylene terephthalate) composites. *Composites Science and Technology*, **60**, 803-815.

46. Chen, Y., Davalos, J. F., Ray, I., and Kim, H.-Y. (2007) Accelerated aging tests for evaluations of durability performance of FRP reinforcing bars for concrete structures. *Composite Structures*, **78**, 101-111.
47. Wang, M., Xu, X., Ji, J., Yang, Y., Shen, J., and Ye, M. (2016) The hygrothermal aging process and mechanism of the novolac epoxy resin. *Composites, Part B: Engineering*, **107**, 1-8.
48. Islam, M. S., Pickering, K. L., and Foreman, N. J. (2010) Influence of hygrothermal ageing on the physico-mechanical properties of alkali treated industrial hemp fibre reinforced polylactic acid composites. *Journal of Polymers and the Environment*, **18**, 696-704.
49. Arhant, M., Le Gac, P.-Y., Le Gall, M., Burtin, C., Briançon, C., and Davies, P. (2016) Effect of sea water and humidity on the tensile and compressive properties of carbon-polyamide 6 laminates. *Composites, Part A: Applied Science and Manufacturing*, **91**, 250-261.
50. Silva, L., Tognana, S., and Salgueiro, W. (2013) Study of the water absorption and its influence on the Young's modulus in a commercial polyamide. *Polymer Testing*, **32**, 158-164.
51. Le Gac, P.-Y., Choqueuse, D., Paris, M., Recher, G., Zimmer, C., and Melot, D. (2013) Durability of polydicyclopentadiene under high temperature, high pressure and seawater (offshore oil production conditions). *Polymer Degradation and Stability*, **98**, 809-817.
52. Guadagno, L., Fontanella, C., Vittoria, V., and Longo, P. (1999) Physical aging of syndiotactic polypropylene. *Journal of Polymer Science, Part B: Polymer Physics*, **37**, 173-180.
53. Guermazi, N., Elleuch, K., Ayedi, H., and Kapsa, P. (2008) Aging effect on thermal, mechanical and tribological behaviour of polymeric coatings used for pipeline application. *Journal of Materials Processing Technology*, **203**, 404-410.
54. Gautier, L., Mortaigne, B., Bellenger, V., and Verdu, J. (2000) Osmotic cracking nucleation in hydrothermal-aged polyester matrix. *Polymer*, **41**, 2481-2490.
55. Berthé, V., Ferry, L., Bénézet, J., and Bergeret, A. (2010) Ageing of different biodegradable polyesters blends mechanical and hygrothermal behavior. *Polymer Degradation and Stability*, **95**, 262-269.
56. Boubakri, A., Haddar, N., Elleuch, K., and Bienvenu, Y. (2010) Impact of aging conditions on mechanical properties of thermoplastic polyurethane. *Materials and Design*, **31**, 4194-4201.
57. Mondal, S., Hu, J., and Yong, Z. (2006) Free volume and water vapor permeability of dense segmented polyurethane membrane. *Journal of Membrane Science*, **280**, 427-432.
58. Deshmane, C., Yuan, Q., Perkins, R., and Misra, R. (2007) On striking

variation in impact toughness of polyethylene–clay and polypropylene–clay nanocomposite systems: the effect of clay–polymer interaction. *Materials Science and Engineering A*, **458**, 150-157.
59. Hagemann, H., Snyder, R., Peacock, A., and Mandelkern, L. (1989) Quantitative infrared methods for the measurement of crystallinity and its temperature dependence: polyethylene. *Macromolecules*, **22**, 3600-3606.
60. Liang, C., and Krimm, S. (1959) Infrared spectra of high polymers: Part IX. Polyethylene terephthalate. *Journal of Molecular Spectroscopy*, **3**, 554-574.
61. Krimm, S., Liang, C., and Sutherland, G. (1956) Infrared spectra of high polymers. II. Polyethylene. *The Journal of Chemical Physics*, **25**, 549-562.
62. Ivanova, K. I., Pethrick, R. A., and Affrossman, S. (2001) Hygrothermal aging of rubber modified and mineral filled dicyandiamide cured digylcidyl ether of bisphenol A epoxy resin. I. Diffusion behavior. *Journal of Applied Polymer Science*, **82**, 3468-3476.
63. DeNeve, B., and Shanahan, M. E. R. (1993) Water absorption by an epoxy resin and its effect on the mechanical properties and infrared spectra. *Polymer*, **34**(24), 5099-5105.
64. Zhao, S., Chen, F., Zhao, C., Huang, Y., Dong, J.-Y., and Han, C. C. (2013) Interpenetrating network formation in isotactic polypropylene/graphene composites. *Polymer*, **54**, 3680-3690.
65. Li, C.-Q., Zha, J.-W., Long, H.-Q., Wang, S.-J., Zhang, D.-L., and Dang, Z.-M. (2017) Mechanical and dielectric properties of graphene incorporated polypropylene nanocomposites using polypropylene-graft-maleic anhydride as a compatibilizer. *Composites Science and Technology*, **153**, 111-118.
66. Chow, C., Xing, X., and Li, R. (2007) Moisture absorption studies of sisal fibre reinforced polypropylene composites. *Composites Science and Technology*, **67**, 306-313.
67. Chu, W., Wu, L., and Karbhari, V. M. (2004) Durability evaluation of moderate temperature cured E-glass/vinylester systems. *Composite Structures*, **66**, 367-376.
68. Boubakri, A., Elleuch, K., Guermazi, N., and Ayedi, H. (2009) Investigations on hygrothermal aging of thermoplastic polyurethane material. *Materials and Design*, **30**, 3958-3965.
69. Zanni-Deffarges, M., and Shanahan, M. (1995) Diffusion of water into an epoxy adhesive: comparison between bulk behaviour and adhesive joints. *International Journal of Adhesion and Adhesives*, **15**, 137-142.
70. Guermazi, N., Elleuch, K., and Ayedi, H. (2009) The effect of time and aging temperature on structural and mechanical properties of pipeline coating. *Materials and Design*, **30**, 2006-2010.

8

Effect of Accelerated UV Weathering on the Properties of Isotactic Polypropylene and Impact Polypropylene Copolymers

8.1 Introduction

Owing to tunable mechanical and structural properties, light weight structures, ease of processing and low cost, polymer based materials have found widespread application in engineering, non-engineering and specialty fields [1-4]. In their lifespan, the polymeric materials have to sustain highly active environmental conditions, which can have detrimental effects on their performance. In brief, continuously exposing the polymeric materials to severe conditions like sunlight, UV radiation, temperature, pressure, humidity, oxygen, pollutants, chemicals, mechanical force, biological media, marine conditions, etc., results in the molecular degradation and irreversible negative effects on the physical and chemical properties of the polymer matrices [5-9]. Thus, an in-depth evaluation of the material performance in such harsh environmental conditions is very useful to estimate the life time as well as replacement or repair needs [10,11], along with an efficient design of materials. In this respect, accelerated ageing studies, which suitably mimic the environmental conditions, are preferred over natural ageing because of the much longer time span needed to obtain sufficient insights about the materials in the case of natural ageing [12,13].

In laboratory, the UV radiation induced accelerated ageing of the polymeric materials can be conducted with the help of fluorescent tube, xenon long arc, mercury arc, metal halide or carbon arc lamp as effective light source [10,14], with xenon long arc representing the most preferred source. When polymer based materials are in contact with a UV source, the process of photo-initiation takes place through chromophore excitation, followed by photo-degradation by undergoing either chain scission or crosslinking [14-17]. Accordingly, color fading is the first indication of ageing, with subsequent loss in physico-chemical and mechanical properties [14,18,19]. Indeed,

Anish Varghese and Vikas Mittal
Khalifa University of Science and Technology, Abu Dhabi, UAE

the influence of temperature, humidity, pollutants, etc., has an accelerating effect on the degree of UV radiation induced photodegradation [4]. One of the crucial factors deciding the degree of degradation is the nature of interaction of the UV radiations with the polymer surface, which, in turn, is related to the polymer nature [4,9]. The process of chain scission is related to the amorphous part of the polymer, which results in reduced molecular weight and increased crystallinity, whereas crosslinking relates to the imperfect crystalline parts, which results in increased molecular weight with unaltered crystallinity [15,20]. The basic reactions during photodegradation via chain scission are classified as Norrish Type I or Norrish Type II reactions [21]. Norrish Type I reaction is the photochemical breakdown of aldehydes or ketones and subsequent generation of active free radical sites, while Norrish Type II reaction is the excitation of the polymer-oxygen complex through the generation of carbonyl or vinyl groups [15,21,22]. Both reaction types result in the breakdown of highly entangled macromolecular structure into smaller crystallites, which correspondingly leads to reduced molecular weight, decreased mechanical properties and increased crystallinity [15,20,21,23].

Polypropylene (PP) is a versatile polyolefin thermoplastic material available in various forms like homopolymer, copolymer, reactor-blends, etc. [24-27]. PP homopolymer, isotactic polypropylene (iPP), exhibits narrow molecular weight distribution and exceptional clarity, along with balanced physical and mechanical properties, recyclability, low production cost, corrosion resistance and thermal stability [28-30]. However, iPP suffers from poor crack propagation resistance [31,32]. The addition of elastomers, thermoplastics, thermoplastic elastomers, copolymers, etc., has been reported to be beneficial for strengthening the impact properties of iPP [30,33-44]. As an alternative, impact polypropylene copolymers (IPC), also known as heterophasic copolymers, have been developed by *in-situ* copolymerization of iPP and ethylene in the polymerization reactor using Borstar or Spheripol processes [45-48]. Generally, these processes include the bulk polymerization of iPP, followed by gas phase propylene-ethylene copolymerization [49]. The structural analysis of IPC reveals the presence of three phases such as semi-crystalline iPP phase, semi-crystalline phase of ethylene-propylene block copolymer (EPBC) and amorphous phase of ethylene-propylene random copolymer (EPRC). In other words, IPC structure can be visualized as the stress concentrator EPRC phase distributed in the iPP continuous phase under the influence of compatibilizer EPBC phase [49-51]. On impact loading, IPC leads to effective micro-voids generation via crazing or shear yielding through the absorption of impact energy by EPRC, thereby, resulting in fracture-free plastic deformation [50,52]. The factors like EPRC phase size, shape and distribution in iPP phase as well as EPRC-iPP interaction have strong influence on the overall performance [53-55].

In the current study, iPP and two IPC grades were analyzed for the effect of accelerated UV weathering conditions. For this, injection molded specimens were exposed to a fluorescent light source for two different time periods, i.e., 400 h and 700 h, together with other accelerating factors.

8.2 Experimental

8.2.1 Materials

Commercial iPP grade HD915CF and two impact polypropylene copolymer grades (termed as IPC1 and IPC2) were procured from Abu Dhabi Polymers Company Limited, UAE. The reported melt flow index (MFI) values at 230 °C and 2.16 kg were 8 and 38 g/10 min for iPP and IPC1 respectively, whereas the MFI of IPC2 was 7 g/10 min at 190 °C and 2.16 kg. The density of the PP grades was measured to be in the range 900-910 kg.m^{-3}. The polymers were used for analysis as received.

8.2.2 Sample Preparation

Dumbbell and rectangular bar shaped specimens were prepared using a plunger type lab-scale injection molding machine HAAKE MiniJet PRO (Thermo Scientific, Germany). For this purpose, PP pellets were melted by applying a temperature of 180 °C for 5 min, followed by injection into a mold at a temperature of 125 °C by applying a pressure of 430 bar for 10 s. In order to prevent the backflow of the injected sample, a holding pressure of 500 bar was applied for 6 s.

8.2.3 Photo-oxidation

Artificial photo-oxidation studies were carried out in a BGD 856 UV light accelerated weathering cabinet from Buiged Laboratory Instruments (Guangzhou), equipped with 4 fluorescent UV lamps (UVA-340) oriented horizontally on both sides in accordance with ASTM D 4329. The samples were exposed to accelerated UV conditions for 400 h and 700 h. Specifically, the accelerated weathering cycle was as follows: UV exposure with an irradiance of 0.76 W.m^{-2} at 50 °C for 8 h, combined water spray and UV exposure for 0.25 h, followed by combined condensation at 50 °C and UV exposure for 1.75 h. In this cycle, UV exposure imitates the role of sunlight, condensation imitates rain and water spray imitates dew.

8.2.4 Characterization

Fourier-transform infrared (FTIR) spectrometer (BRUKER TENSOR II Series) was used for the structural analysis of unweathered and UV weathered PP samples. FTIR spectra were generated by collecting 32 scans from the thin sections of the injection molded samples in transmission mode. A diamond attenuated total reflectance (ATR) crystal was used to obtain the spectra in the 4000-400 cm^{-1} wavenumber region with a resolution of 4 cm^{-1}.

X'Pert PRO Panalytical powder diffractometer was used to study the wide angle X-ray diffraction (WAXD) of the unweathered and UV weathered PP samples. Thin sections from the injection molded samples were scanned in the 5-60° 2-theta region using Cu-Kα irradiation with a wavelength (λ) of 1.5406 Å at 45 kV, 40 mA and room temperature. A step size of 0.017° s^{-1} and a step time of 10 s were employed.

Tensile properties of the PP samples were measured at room temperature with the aid of a 50 kN load cell universal testing machine (Instron 3345, USA). The cross-head speed used for the analysis was 10 mm.min^{-1}. The reported tensile data represents the mean of values from five dumbbell shape specimens with size of 75 mm x 5 mm x 2 mm and span length of 35 mm, in accordance with ISO 527. The effect of UV weathering on the impact strength of iPP, IPC1 and IPC2 was estimated at room temperature using Resil impactor from Ceast, USA (4 J hammer energy). For the analysis, the hammer speed of 3.64 m.s^{-1} was used. The reported impact strength represents the mean of values from five un-notched rectangular bar shaped specimens with size of 80 mm x 10 mm x 4 mm, in agreement with ISO 180.

UV weathering induced changes in the thermal properties (melting temperature and enthalpy (T_m and ΔH_f), crystallization temperature and enthalpy (T_c and ΔH_c)), percentage crystallinity (X_c), etc.) were assessed by using differential scanning calorimeter (DSC) from TA instruments, USA. 3-8 mg sample weight was examined in the temperature range of (-50, +200 °C) by applying two heating and cooling cycles using a dry nitrogen flow of 50 mL.min^{-1}. Second heating-cooling cycles were used to record the calorimetric behavior of the samples.

The morphology of the impact fractured surface of the PP samples was analyzed in a scanning electron microscope (SEM) (FEI Quanta, FEG250, USA). An accelerating voltage of 10 kV was used for the analysis. Prior to analysis, gold sputter coated samples were affixed on aluminum stubs with the help of carbon conductive adhesive tape. In addition, energy dispersive X-ray analysis (EDX) for the UV weathered PP samples was carried out by focusing on the area near the surface in order to confirm the presence of photo-oxidation products. For the examination of surface changes, micrographs were also obtained using an optical microscope Olympus BX51M, Japan, operated with an ocular magnification of 10x.

8.3 Results and Discussion

Figure 8.1 presents the FTIR spectra of unweathered and UV weathered iPP, IPC1 and IPC2 samples. For unweathered iPP, the presence of FTIR characteristic bands at 2957 cm^{-1}, 2916 cm^{-1}, 2870 cm^{-1}, 2837 cm^{-1}, 2362 cm^{-1}, 1463 cm^{-1}, 1374 cm^{-1}, 999-973-844-811 cm^{-1} and 898 cm^{-1} could be assigned to asymmetric CH_3 stretching, symmetric CH_2 stretching, symmetric CH_3 stretching, symmetric CH_2 stretching, C-H bending, asymmetric CH_3 bending, symmetric CH_3 bending, tertiary methyl skeleton deformation and stretching induced perpendicular absorption respectively [56-58]. Owing to photo-oxidation and thermal oxidation reactions on the iPP surface upon continuous UV weathering, the appearance of hydroxyl band at 3660-3200 cm^{-1}, alkyne band at 2150-2025 cm^{-1}, carbonyl band at 1850-1650 cm^{-1}, vinyl band at 1650-1560 cm^{-1} and C-O-C bonds at 1085-1055 cm^{-1} was observed for 400 h UV weathered iPP sample, as shown in Figure 8.1a [18,59-65]. Carbonyl band comprises γ-lactones, peresters, esters, conjugated ketones, aldehyde and carboxylic acids as oxidation products, whereas hydroxyl band comprises alcohols, hydroperoxides and some carboxylic acid groups [60,61]. Overall, the oxidation products evolved due to the reaction of the generated macromolecular radicals and oxygen in continuous

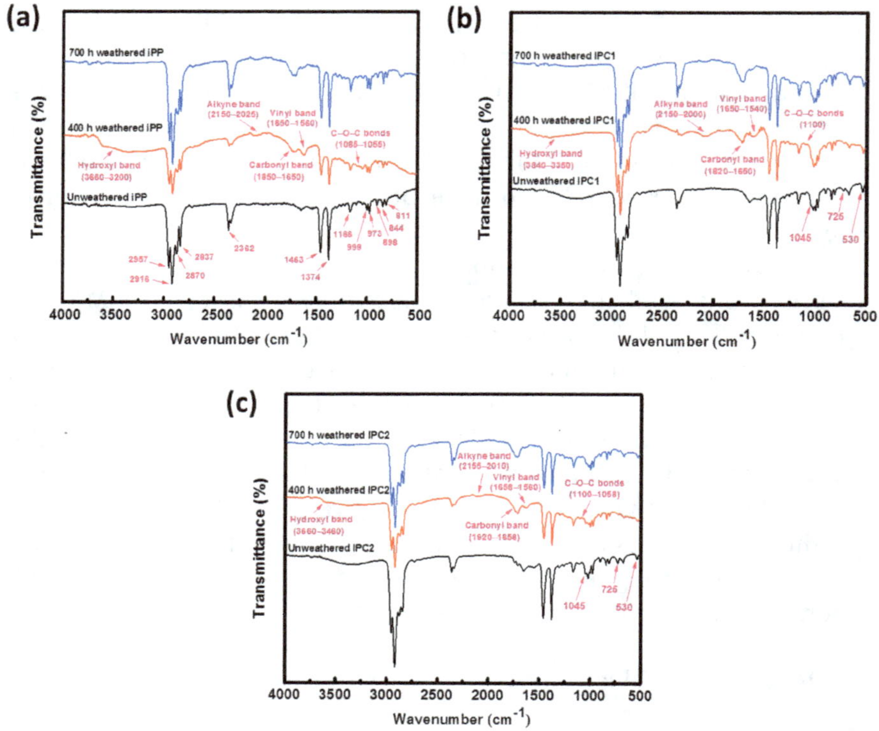

Figure 8.1 FTIR spectra of unweathered and UV weathered samples (400 h and 700 h exposure): (a) iPP, (b) IPC1 and (c) IPC2.

UV weathering conditions. For both unweathered IPC1 (Figure 8.1b) and IPC2 (Figure 8.1c), three amorphous polyethylene bands at 1045 cm^{-1}, 725 cm^{-1} and 530 cm^{-1}, assigned to CH_2 gauche configuration, CH_2 rocking and CH_2 stretching, were noticed in addition to the iPP characteristic bands [47,66-68]. Comparing the spectra of 400 h UV weathered IPC1 and IPC2 with iPP, appearance of the similar bands corresponding to the oxidation products was observed, however, the band positions differed to some extent. Accordingly, bands attributed to hydroxyl, alkyne, carbonyl and vinyl groups as well as C-O-C bonds were discerned at 3840-3350, 2150-2000, 1820-1650, 1650-1540 and 1100 cm^{-1} respectively for 400 h weathered IPC1 and 3660-3460, 2155-2010, 1920-1658, 1658-1560 and 1100-1058 cm^{-1} respectively for 400 h weathered IPC2. In addition, the intensity of the iPP, IPC1 and IPC2 bands was observed to decrease after 400 h UV weathering due to the effect of macromolecular degradation. Considering the FTIR bands of 700 h UV weathered samples, the intensity of the bands assigned to the oxidation products (except carbonyl groups) was observed to decrease probably due to the tendency to decompose [59] and/or leach under the influence of water [19]. In addition, the intensity of the iPP, IPC1 and IPC2 characteristic bands displayed an increasing trend on enhancing the exposure time from 400 to 700 h.

Figure 8.2 demonstrates the WAXD patterns of PP samples. iPP

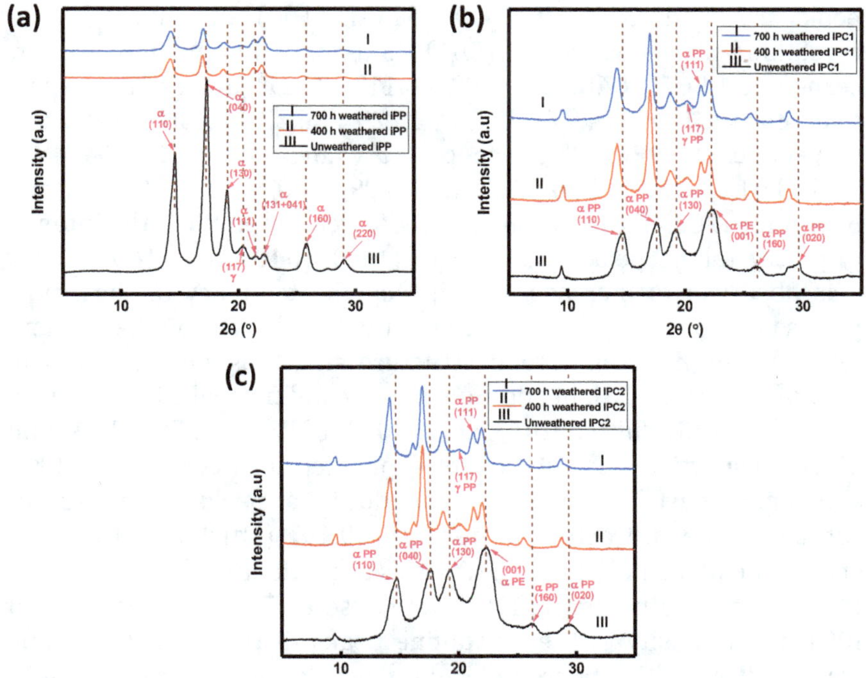

Figure 8.2 WAXD patterns of unweathered and UV weathered samples: (a) iPP, (b) IPC1 and (c) IPC2.

exhibited characteristic peaks corresponding to (110), (040), (130), (111), (131+041), (160) and (220) α-crystalline planes as well as (007) γ-crystalline plane at 14.55°, 17.26°, 19.03°, 21.47°, 22.18°, 25.76°, 29.01° and 20.41° respectively [30,69-71]. The related interplanar distance (d) was estimated by applying the Bragg's relation $d = \lambda/(2\sin\theta_{max})$, where λ (X-ray wavelength) = 1.5406 Å. As a result, the respective d values of 6.08, 5.13, 4.66, 4.14, 4.00, 3.46, 3.08 and 4.35 Å were obtained. For iPP after 400 h of UV weathering, the characteristic diffraction peaks attributed to (110), (040), (130),

(111), (131+041), (160) and (007) crystalline planes were observed to slightly shift to 14.29°, 16.97°, 18.75°, 21.39°, 22.06°, 25.62° and 20.18°, along with a reduction in the peak intensity due to significant macromolecular degradation. Accordingly, the increased *d* values of 6.19, 5.22, 4.73, 4.15, 4.03, 3.47 and 4.40 Å were obtained. For 700 h UV weathered iPP, the intensity of the characteristic peaks displayed an increasing trend, probably due to the chemi-crystallization effect. For unweathered IPC1 and IPC2, characteristic peaks assigned to (110), (040), (130), (160) and (020) iPP α-crystalline planes as well as (001) PE α-crystalline plane were discerned at 14.67°, 17.60°, 19.19°, 26.11°, 29.69° and 22.33° respectively for IPC1 and 14.77°, 17.72°, 19.29°, 26.22°, 29.36° and 22.33° respectively for IPC2. The respective *d* values were 6.03, 5.04, 4.62, 3.41, 3.01 and 3.98 Å for IPC1 and 5.99, 5.00, 4.60, 3.39, 3.04 and 3.98 Å for IPC2. In the case of 400 h UV weathered IPC1, the intensity of crystalline peaks assigned to (110), (040) and (160) planes was observed to increase. It indicated the tendency of the amorphous part to undergo degradation initially, followed by the reorientation of the macromolecular structure and, thus, enhancement in crystallinity [8,17,63]. In addition, the crystalline peaks for 400 h UV weathered IPC1 shifted to 14.22°, 17.04°, 18.75°, 25.57°, 28.84° and 22.13° respectively, which further supported the degradation of the amorphous part. Interestingly, two additional peaks were also observed in the diffraction pattern at 20.19° (attributed to PP γ-crystalline plane (117)) and 21.39° (as a shoulder peak attributed to PP α-crystalline plane (111)). In the case of 700 h UV weathered IPC1, the intensity of the crystalline peaks displayed a decreasing trend indicating the macromolecular degradation. In the case of 400 and 700 h UV weathered IPC2 samples, similar behavior as IPC1 was observed. For 400 h UV weathered IPC2, three additional peaks appeared at 16.23°, 20.13° and 21.33° corresponding to PP β-crystalline plane (300), PP γ-crystalline plane (117) and PP α-crystalline plane (111) respectively. For 700 h UV weathered IPC2, the intensity of the peaks decreased because of the effect of macromolecular degradation.

Table 8.1 depicts the impact strength values of the unweathered and accelerated UV weathered samples. The impact strength values of 7, 53, and 103 $kJ.m^{-2}$ were recorded for unweathered iPP, IPC1 and IPC2 respectively (Figure 8.3a). The significantly higher impact strength for IPC1 and IPC2 resulted due to the influence of multiphase structural composition, i.e., iPP, EPRC and EPBC [50,51]. Here,

Table 8.1 Mechanical properties of the PP samples

Sample	Impact strength (kJ.m^{-2})	Tensile strength (MPa)	Tensile modulus (MPa)	Extension at break (%)
Unweathered iPP	7	39	1011	24
400 h UV weathered iPP	3	15	675	8
700 h UV weathered iPP	2.5	8	315	7
Unweathered IPC1	53	22	1018	35
400 h UV weathered IPC1	3.7	10	569	7
700 h UV weathered IPC1	3	9	495	7
Unweathered IPC2	103	21	864	299
400 h UV weathered IPC2	6.1	9	345	8
700 h UV weathered IPC2	5.9	7	219	7

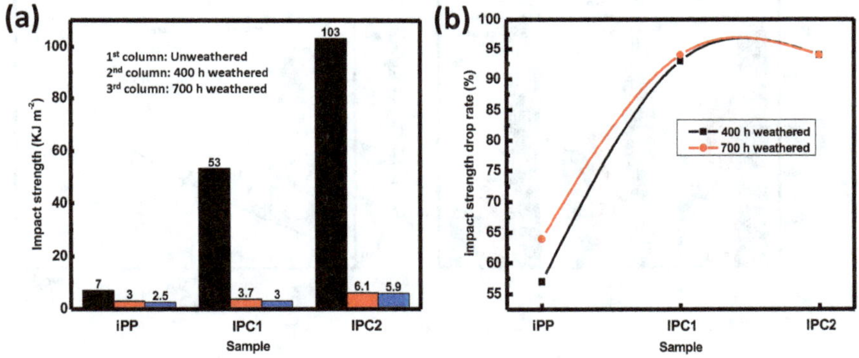

Figure 8.3 (a) Impact strength of unweathered and UV weathered samples; (b) impact strength drop rate of 400 and 700 h UV weathered samples.

EPRC can act as robust stress absorber, whereas EPBC can act as a compatibilizer to strengthen the interfacial performance between iPP and EPRC. Under impact load, EPRC can, thus, absorb a majority of the load and generate micro-voids through crazing and/or shear yielding, which provide improved stress propagation resistance [50,54]. In addition, the difference in the impact strength of IPC1 and IPC2 could have resulted due to their multiphase structural behavior, i.e., interaction between iPP and EPRC, along with distribution, size and shape of EPRC [52,53,72]. The impact strength of the samples was observed to decrease significantly after 400 h of UV weathering. For instance, ca. 57, 93 and 94% loss in the impact strength were recorded for iPP, IPC1 and IPC2 respectively (Figure 8.3b). The observed loss in the impact strength values for 700 h UV weathered samples was ca. 64, 94 and 94% respectively. Thus, a slowdown in the impact strength loss was observed on increasing the weathering time to 700 h from 400 h. The decrease in the impact strength can be attributed to the structural embrittlement resulting from the macromolecular degradation. Therefore, the weathered materials exhibited diminished capability to dissipate impact energy [17,73,74]. Also, the greater extent of impact strength loss for IPC1 and IPC2 as compared to iPP indicated the influence of amorphous part and its severe degradation in IPC1 and IPC2.

The tensile modulus of unweathered iPP, IPC1 and IPC2 was measured to be 1011, 1018 and 864 MPa respectively (Figure 8.4a).

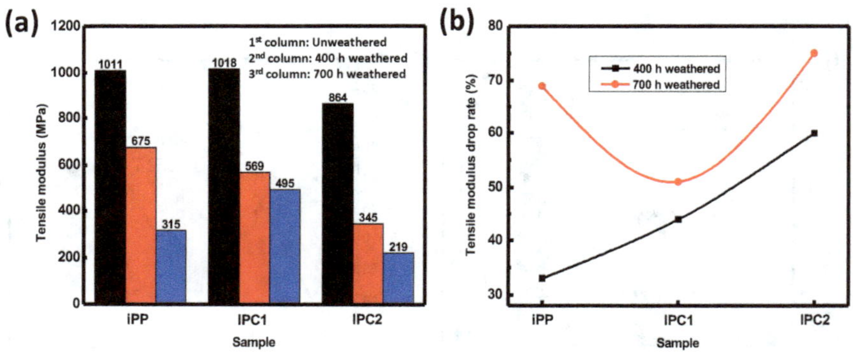

Figure 8.4 (a) Tensile modulus of unweathered and UV weathered samples; (b) tensile modulus drop rate of 400 and 700 h UV weathered samples.

The tensile modulus of iPP, IPC1 and IPC2 dropped by ca. 33, 44 and 60% respectively after 400 h UV weathering (Figure 8.4b). The observed trend of tensile modulus drop was also visible in the case of 700 h UV weathered samples, where the loss was ca. 69, 51 and 75% for iPP, IPC1 and IPC2 respectively. The significant drop in tensile modulus is probably caused by the combined effect of surface cracks and macromolecular degradation. Here, the macromolecular degradation is related to the molecular weight drop resulting due to the generation of oxidation products upon continuous UV weathering [8,75-55]. Furthermore, on enhancing the UV weathering time, the degree of surface crack formation would also increase [62,78]. Regardless of the polymer type, the PP grades also suffered a drop in the tensile strength after continuous UV weathering. For instance, ca. 62, 55 and 57% drop in the tensile strength were recorded for iPP, IPC1 and IPC2 respectively after 400 h UV weathering, whereas these values were ca. 79, 59 and 67% for 700 h UV weathered samples. Among the PP samples, iPP suffered a large tensile strength drop after 400 h and 700 h UV weathering, whereas IPC1 displayed relatively higher resistance against the loss of tensile strength. The observed significant drop in the tensile strength of the UV weathered PP samples can be correlated with the macromolecular degradation induced structural embrittlement upon continuous UV weathering [3,17,55]. Furthermore, the surface cracks are expected to have an additional effect on the tensile strength of the samples [16]. As observed for other tensile properties, the extension at break also suffered a reduction after UV weathering. For instance, after 400 h of UV weathering, the extension at break of iPP, IPC1 and IPC2 dropped by ca. 67, 80 and 97% respectively, whereas the drop was 71, 80 and 98% for 700 h UV weathered samples. Macromolecular degradation resulting from the significant drop in molecular weight can be attributed to the observed loss in elongation [78-80]. The appearance of nearly stabilized elongation in the 400-700 h UV weathering range may result due to the termination of the surface crack generation process once the photo-oxidation process has completed [8].

The calorimetric properties of the samples are presented in Table 8.2, and the second heating and cooling traces of the samples are plotted in Figures 8.5 and 8.6. The T_m values of 168.5, 166.5 and 167.7 °C were observed for unweathered iPP, IPC1 and IPC2. Due to the detrimental effect of UV weathering leading to macromolecular degradation and molecular weight reduction, the T_m values of the samples were reduced, regardless of their type [3,16,17,55,73,81]. Specifically, the T_m drop of 12.3, 6.4 and 12.6 °C was observed for iPP, IPC1 and IPC2 after 400 h UV weathering. The observed drop in T_m slowed down in the UV weathering period of 400-700 h. The T_c values of 127.9, 130.4 and 128.9 °C were observed for unweathered iPP, IPC1 and IPC2 respectively. T_c for iPP, IPC1 and IPC2 after 400 h of UV weathering was reduced by 4, 2.6 and 2.5 °C respectively. For 700 h of UV weathering, the recorded reduction in T_c was 4.5, 3 and 3.9 °C respectively. The observed drop in T_c of the PP grades indicated reduced molecular weight and an enhanced degree of chemical deformities [19,82]. Regardless of the type, ΔH_f and ΔH_c of the PP grades were observed to increase with UV weathering time, thus,

Table 8.2 Calorimetric properties of the samples

Sample	T_c (°C)	ΔT_c (°C)	ΔH_c (J/g)	T_m (°C)	ΔT_m (°C)	ΔH_f (J/g)	X_c (%)
Unweathered iPP	127.9	-	98.7	168.5	-	90.8	43.8
400 h UV weathered iPP	123.9	-4	102.1	156.2	-12.3	105.3	50.8
700 h UV weathered iPP	123.4	-4.5	107.8	154.7	-13.8	110.9	53.5
Unweathered IPC1	130.4	-	84.3	166.5	-	82.8	40.0
400 h UV weathered IPC1	127.8	-2.6	87.3	160.1	-6.4	89.9	43.3
700 h UV weathered IPC1	127.4	-3	88.2	157.8	-8.7	91.8	44.3
Unweathered IPC2	128.9	-	74.5	167.7	-	72.9	35.2
400 h UV weathered IPC2	126.4	-2.5	77.1	155.1	-12.6	79.4	38.3
700 h UV weathered IPC2	125.0	-3.9	77.9	153.2	-14.5	80.8	39.0

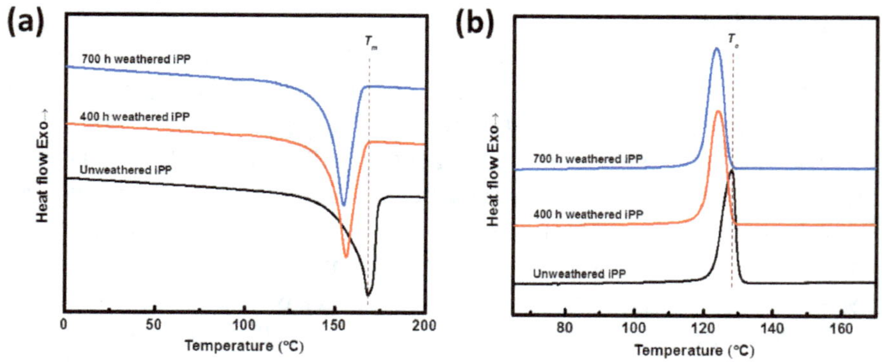

Figure 8.5 DSC (a) heating and (b) cooling thermograms of unweathered and UV weathered iPP.

indicating increased X_c. The following equation [30,43] was used for the estimation of X_c for the samples:

$$X_c = \frac{\Delta H}{\Delta H_0} \times 100$$

Here, ΔH_0 and ΔH correspond to theoretical ΔH_f of 100% crystalline PP (207.1 J/g) [30] and DSC measured ΔH_f of the samples. The

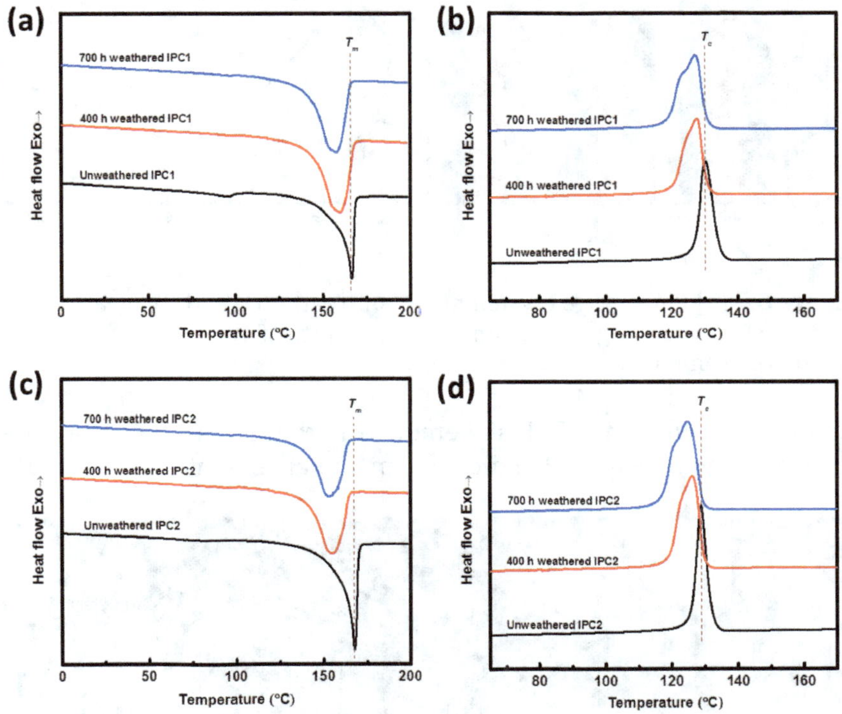

Figure 8.6 DSC heating thermograms of unweathered and UV weathered samples: (a) IPC1 and (c) IPC2; and cooling thermograms of unweathered and UV weathered samples: (b) IPC1 and (d) IPC2.

X_c values of 43.8, 40.0 and 35.2% were calculated for unweathered iPP, IPC1 and IPC2 respectively (Figure 8.7a). After continuous UV weathering, the X_c values of the PP grades were observed to increase, regardless of the type. Specifically, X_c increase of 9.7, 4.3 and 3.8% was recorded for 700 h UV weathered samples (Figure 8.7b). Such increase in X_c is opined to be the effect of chemi-crystallization which leads to the reorganization of the degraded non-crystalline sections into crystallites [8,63,73,83]. It is considered as an indirect effect of UV weathering. In accordance with such changes, the cracks would appear on the sample surface due to shrinkage and associated dimensional changes [17,84].

The analysis of the UV weathering induced changes in the micromorphology of the impact fractured surface of iPP, IPC1 and IPC2

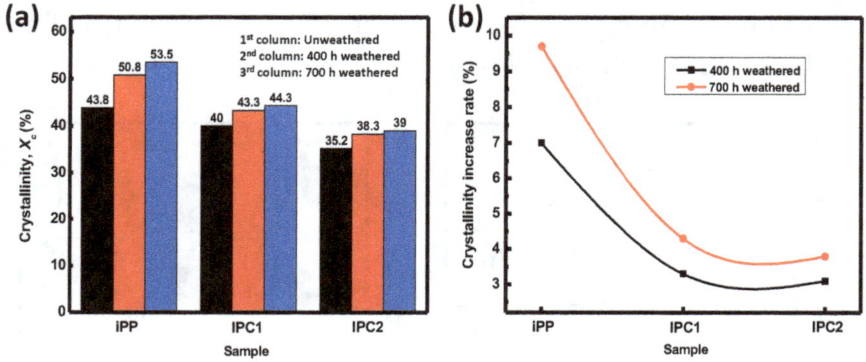

Figure 8.7 (a) Percentage crystallinity of unweathered and UV weathered samples; (b) percentage crystallinity increase rate of 400 and 700 h UV weathered samples.

was performed using SEM, as demonstrated in Figure 8.8. The fracture morphology of iPP indicated an enhancement in the brittle

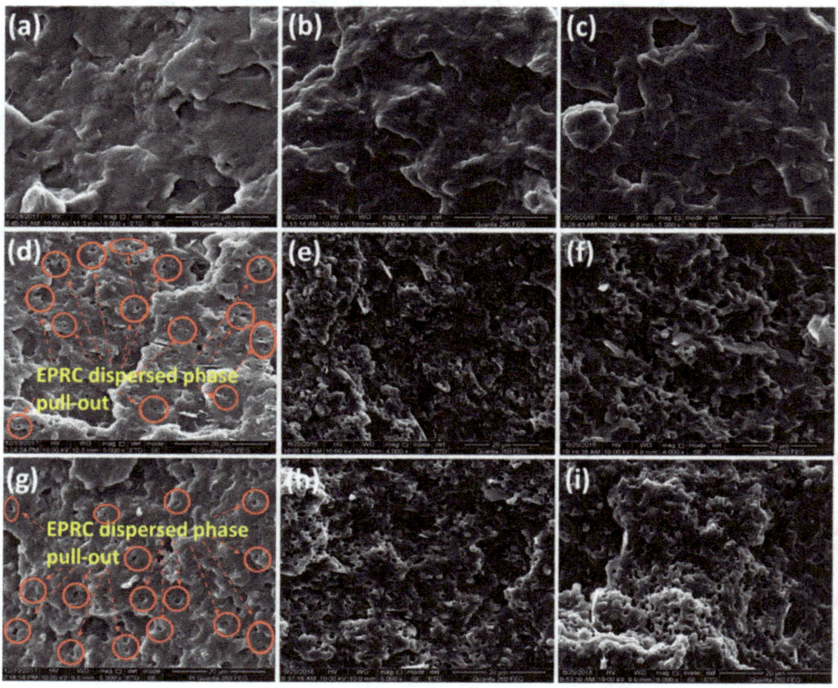

Figure 8.8 Micrographs of the impact fractured surface of unweathered (a) iPP, (d) IPC1 and (g) IPC2; 400 h UV weathered (b) iPP, (e) IPC1 and (h) IPC2; and 700 h UV weathered (c) iPP, (f) IPC1 and (i) IPC2.

nature with UV weathering time. The observed structural embrittlement can be expected to result from the photo-oxidation induced macromolecular degradation during UV weathering. These results coincided well with the impact properties of the unweathered and UV weathered iPP samples. Due to increased structural embrittlement and corresponding impact load dissipation inability, the impact strength values dropped for the UV weathered iPP samples. In the micrographs for IPC1 and IPC2, the observed micro-voids or dispersed EPRC phase pull-outs in the continuous phase evidenced the successful iPP-EPRC interface debonding or cavitation under impact loading, by means of either shear yielding or crazing. Also, the observed higher impact strength of IPC2 can be attributed to the influence of more desirable interfacial features between iPP and EPRC in the presence of EPBC. UV weathered IPC1 and IPC2 exhibited severe damage to the EPRC dispersed phase and iPP-EPRC interface. These observations underline the photo-degradation of the amorphous regions present in the IPC macromolecular structure and the corresponding increase in embrittlement when exposed to continuous UV weathering.

The EDX spectra in Figure 8.9 represent the elemental composition with respect to carbon (C) and oxygen (O) content. Table 8.3 also lists the recorded C and O content as well as the C/O ratio for 400 h and 700 h UV weathered iPP, IPC1 and IPC2. An increasing O content with weathering time was observed, which supported the evolution of photo-oxidation products. Specifically, enhanced O content of 18.86, 23.21 and 24.44 wt% was recorded for 700 h UV weathered iPP, IPC1 and IPC2 respectively. Correspondingly, the C/O ratio exhibited a decreasing trend with UV weathering time. After 700 h of UV weathering, the observed C/O ratio for iPP, IPC1 and IPC2 samples was 4.30, 3.31 and 3.09 respectively. Overall, the results indicated the presence of more oxidation products in the UV weathered IPC2 as compared to iPP.

The surface morphology of the UV weathered samples was also analyzed through optical analysis. The optical micrographs of 400 h and 700 h UV weathered iPP, IPC1 and IPC2 samples are depicted in Figure 8.10. UV weathering induced surface degradation was observed as surface cracks in the PP samples, regardless of their type. Specifically, the appearance of a transverse crack network was noticed on the UV weathered surface of the samples, which became more intense on increasing the weathering time. Careful analysis also revealed that the nature of the surface cracks was different for

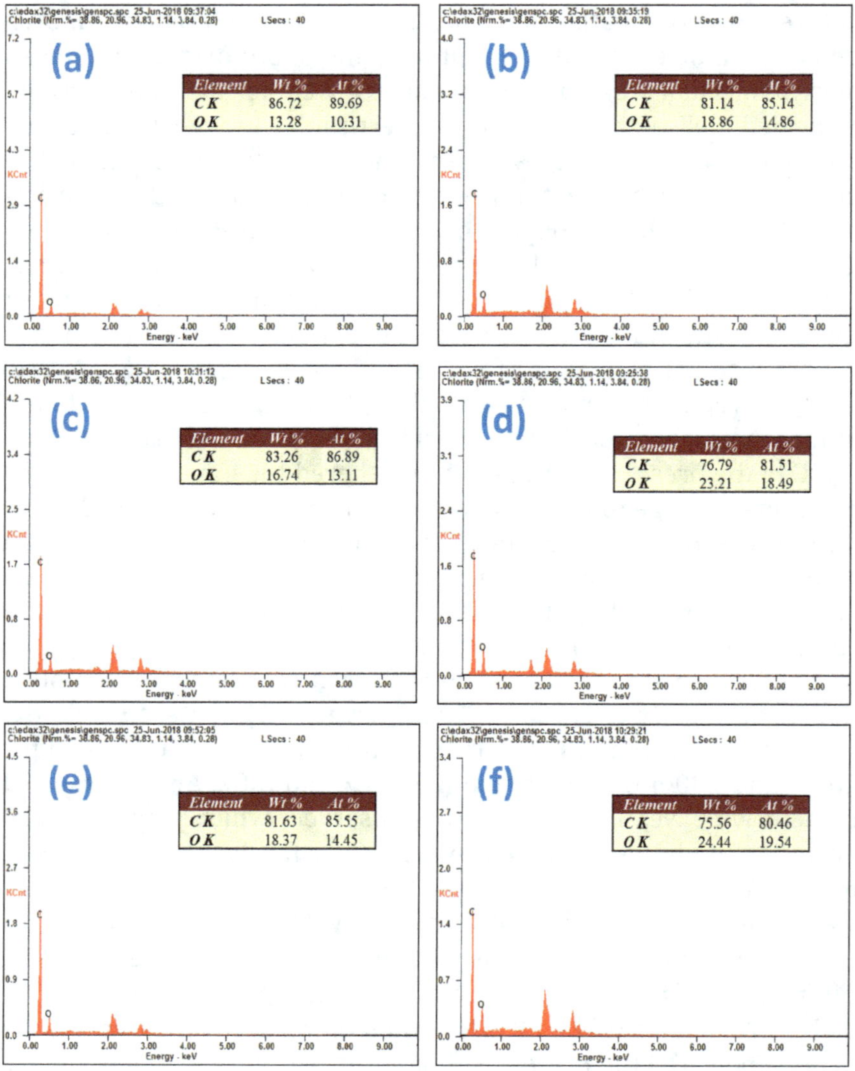

Figure 8.9 EDX spectra of 400 h UV weathered samples: (a) iPP, (c) IPC1 and (e) IPC2; and 700 h UV weathered samples: (b) iPP, (d) IPC1 and (f) IPC2.

each PP grade. For iPP, transverse cracks with smaller width were observed in large numbers after 700 h of UV weathering, whereas no surface crack was visible after 400 h. Large width transverse cracks in small numbers were observed for IPC1 after 400 h and 700 h of weathering in UV conditions. In the case of IPC2, medium width slightly tapered transverse cracks in large numbers were observed.

Table 8.3 EDX data for unweathered and UV weathered iPP, IPC1 and IPC2

Sample	Carbon (C)		Oxygen (O)		C/O ratio	
	Wt%	At%	Wt%	At%	Wt%	At%
400 h weathered iPP	86.72	89.69	13.28	10.31	6.53	8.70
700 h weathered iPP	81.14	85.14	18.86	14.86	4.30	5.73
400 h weathered IPC1	83.26	86.89	16.74	13.11	4.97	6.63
700 h weathered IPC1	76.79	81.51	23.21	18.49	3.31	4.41
400 h weathered IPC2	81.63	85.55	18.37	14.45	4.44	5.92
700 h weathered IPC2	75.56	80.46	24.44	19.54	3.09	4.12

The appearance of such surface cracks during continuous UV weathering is related to shrinkage and associated dimensional change resulting from chemi-crystallization of the photo-degraded non-crystalline sections [17,84].

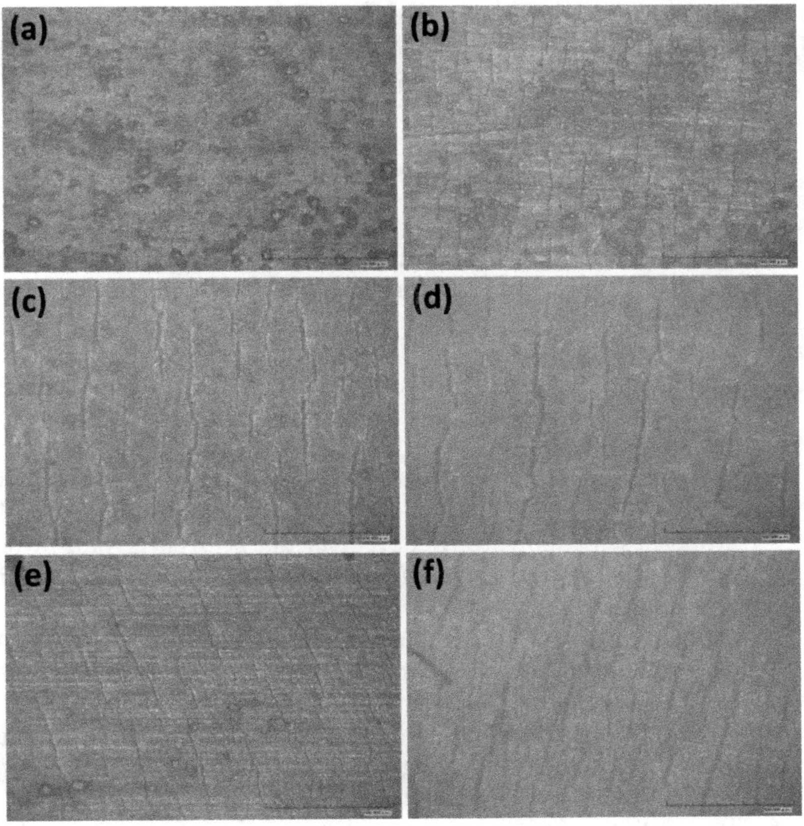

Figure 8.10 Surface micrographs of 400 h UV weathered (a) iPP, (c) IPC1 and (e) IPC2; and 700 h UV weathered (b) iPP, (d) IPC1 and (f) IPC2.

8.4 Conclusions

The objective of the study was to compare the photo-degradation of isotactic polypropylene (iPP) and impact polypropylene copolymers (IPC1 and IPC2) in continuous UV weathering conditions. The FTIR spectra of the unweathered and UV weathered samples indicated the formation of various oxidation products upon UV weathering. These observations were further supported by the EDX results associated with the elemental composition. Due to the combined effect of macromolecular degradation, structural embrittlement and surface cracks, the mechanical properties such as impact strength, tensile modulus, tensile strength and extension at break were significantly reduced. Similar to the mechanical properties, melting and crystallization temperatures, along with melting and crystallization enthalpy, were also decreased after UV weathering due to the aforementioned weathering induced changes. Regardless of the type, the crystallinity of the PP samples increased with UV weathering due to the chemi-crystallization phenomenon.

References

1. Brinson, H. F., and Brinson, L. C. (2015). Characteristics, applications and properties of polymers. In: *Polymer Engineering Science and Viscoelasticity*, Springer, USA, pp. 57-100.
2. Paoletti, A., Lambiase, F., and Di Ilio, A. (2016) Analysis of forces and temperatures in friction spot stir welding of thermoplastic polymers. *The International Journal of Advanced Manufacturing Technology*, **83**, 1395-1407.
3. Darie, R. N., Bodirlau, R., Teaca, C. A., Macyszyn, J., Kozlowski, M., and Spiridon, I. (2013) Influence of accelerated weathering on the properties of polypropylene/polylactic acid/eucalyptus wood composites. *International Journal of Polymer Analysis and Characterization*, **18**, 315-327.
4. Pandey, J. K., Reddy, K. R., Kumar, A. P., and Singh, R. (2005) An overview on the degradability of polymer nanocomposites. *Polymer Degradation and Stability*, **88**, 234-250.
5. Pickett, J., Gibson, D., and Gardner, M. (2008) Effects of irradiation conditions on the weathering of engineering thermoplastics. *Polymer Degradation and Stability*, **93**, 1597-1606.
6. Tan, K., White, C., Benatti, D., and Hunston, D. (2010) Effects of ultraviolet radiation, temperature and moisture on aging of coatings and sealants–A chemical and rheological study. *Polymer Degrada-*

tion and Stability, **95**, 1551-1556.

7. Tan, K., White, C., Benatti, D., and Hunston, D. (2008) Evaluating aging of coatings and sealants: Mechanisms. *Polymer Degradation and Stability*, **93**, 648-656.
8. Yakimets, I., Lai, D., and Guigon, M. (2004) Effect of photo-oxidation cracks on behaviour of thick polypropylene samples. *Polymer Degradation and Stability*, **86**, 59-67.
9. Kumar, A. P., Depan, D., Tomer, N. S., and Singh, R. P. (2009) Nanoscale particles for polymer degradation and stabilization - trends and future perspectives. *Progress in Polymer Science*, **34**, 479-515.
10. Heikkilä, A., Kärhä, P., Tanskanen, A., Kaunismaa, M., Koskela, T., Kaurola, J., Ture, T., and Syrjälä, S. (2009) Characterizing a UV chamber with mercury lamps for assessment of comparability to natural UV conditions. *Polymer Testing*, **28**, 57-65.
11. Larché, J.-F., Bussière, P.-O., Therias, S., and Gardette, J.-L. (2012) Photooxidation of polymers: Relating material properties to chemical changes. *Polymer Degradation and Stability*, **97**, 25-34.
12. Azwa, Z., Yousif, B., Manalo, A., and Karunasena, W. (2013) A review on the degradability of polymeric composites based on natural fibres. *Materials & Design*, **47**, 424-442.
13. Chen, Y., Davalos, J. F., Ray, I., and Kim, H.-Y. (2007) Accelerated aging tests for evaluations of durability performance of FRP reinforcing bars for concrete structures. *Composite Structures*, **78**, 101-111.
14. Tocháček, J., and Vrátníčková, Z. (2014) Polymer life-time prediction: The role of temperature in UV accelerated ageing of polypropylene and its copolymers. *Polymer Testing*, **36**, 82-87.
15. Stark, N. M., and Matuana, L. M. (2004) Surface chemistry changes of weathered HDPE/wood-flour composites studied by XPS and FTIR spectroscopy. *Polymer Degradation and Stability*, **86**, 1-9.
16. Rabello, M., and White, J. (1997) The role of physical structure and morphology in the photodegradation behaviour of polypropylene. *Polymer Degradation and Stability*, **56**, 55-73.
17. Beg, M. D. H., and Pickering, K. L. (2008) Accelerated weathering of unbleached and bleached Kraft wood fibre reinforced polypropylene composites. *Polymer Degradation and Stability*, ***93***, 1939-1946.
18. La Mantia, F., and Morreale, M. (2008) Accelerated weathering of polypropylene/wood flour composites. *Polymer Degradation and Stability*, **93**, 1252-1258.
19. Butylina, S., Hyvärinen, M., and Kärki, T. (2012) A study of surface changes of wood-polypropylene composites as the result of exterior weathering. *Polymer Degradation and Stability*, **97**, 337-345.
20. Jabarin, S. A., and Lofgren, E. A. (1994) Photooxidative effects on properties and structure of high-density polyethylene. *Journal of*

Applied Polymer Science, **53**, 411-423.

21. Fabiyi, J. S., McDonald, A. G., Wolcott, M. P., and Griffiths, P. R. (2008) Wood plastic composites weathering: Visual appearance and chemical changes. *Polymer Degradation and Stability*, **93**, 1405-1414.
22. *Handbook of Polymer Degradation*, Hamid, S. H. (ed.), CRC Press, USA (2000).
23. Philip, M., Attwood, J., Hulme, A., Williams, G., and Shipton, P. (2004) Evaluation of Weathering in Mixed Polyethylene and Polypropylene Products, The Waste and Resources Action Program, UK. Online: https://cms.esi.info/Media/documents/11653_13615 34798115.pdf [accessed 19th May 2019].
24. Tripathi, D. (2002) *Practical Guide to Polypropylene*, iSmithers Rapra Publishing, USA.
25. *Handbook of Polypropylene and Polypropylene Composites*, Karian, H. G. (ed.), CRC Press, USA (2003).
26. Maddah, H. A. (2016) Polypropylene as a promising plastic: A review. *American Journal of Polymer Science*, **6**, 1-11.
27. Gahleitner, M., and Paulik, C. (2017) Polypropylene and other polyolefins. In: *Brydson's Plastics Materials*, 8th edition, Elsevier, USA, pp. 279-309.
28. Bikiaris, D., Vassiliou, A., Chrissafis, K., Paraskevopoulos, K., Jannakoudakis, A., and Docoslis, A. (2008) Effect of acid treated multi-walled carbon nanotubes on the mechanical, permeability, thermal properties and thermo-oxidative stability of isotactic polypropylene. *Polymer Degradation and Stability*, **93**, 952-967.
29. Wang, Y., and Tsai, H.B. (2012) Thermal, dynamic-mechanical, and dielectric properties of surfactant intercalated graphite oxide filled maleated polypropylene nanocomposites. *Journal of Applied Polymer Science*, **123**, 3154-3163.
30. Varghese, A. M., Vengatesan, M. R., Mun, S. C., Macosko, C. W., and Mittal, V. (2018) Effect of graphene on polypropylene/maleic anhydride-graft-ethylene-vinyl acetate (PP/EVA-g-MA) blend: Mechanical, thermal, morphological, and rheological properties. *Industrial & Engineering Chemistry Research*, **57**, 7834-7845.
31. Chen, F., Qiu, B., Wang, B., Shangguan, Y., and Zheng, Q. (2015) Balanced toughening and strengthening of ethylene–propylene rubber toughened isotactic polypropylene using a poly (styrene-b-ethylene–propylene) diblock copolymer. *RSC Advances*, **5**, 20831-20837.
32. Zebarjad, S., Bagheri, R., Reihani, S.S., and Lazzeri, A. (2003) Deformation, yield and fracture of elastomer-modified polypropylene. *Journal of Applied Polymer Science*, **90**, 3767-3779.
33. Liang, J., and Li, R. (2000) Rubber toughening in polypropylene: A review. *Journal of Applied Polymer Science*, **77**, 409-417.

34. Kukaleva, N., Jollands, M., Cser, F., and Kosior, E. (2000) Influence of phase structure on impact toughening of isotactic polypropylene by metallocene-catalyzed linear low-density polyethylene. *Journal of Applied Polymer Science*, **76**, 1011-1018.
35. Keskin, R., and Adanur, S. (2011) Improving toughness of polypropylene with thermoplastic elastomers in injection molding. *Polymer-Plastics Technology and Engineering*, **50**, 20-28.
36. Zhu, L., Fan, H.-N., Yang, Z.-Q., and Xu, X.-H. (2010) Evaluation of phase morphology, rheological, and mechanical properties based on polypropylene toughened with poly (ethylene-co-octene). *Polymer-Plastics Technology and Engineering*, **49**, 208-217.
37. Xu, J., Mittal, V., and Bates, F. S. (2016) Toughened isotactic polypropylene: phase behavior and mechanical properties of blends with strategically designed random copolymer modifiers. *Macromolecules*, **49**, 6497-6506.
38. Wu, C. L., Zhang, M. Q., Rong, M. Z., and Friedrich, K. (2005) Silica nanoparticles filled polypropylene: effects of particle surface treatment, matrix ductility and particle species on mechanical performance of the composites. *Composites Science And Technology*, **65**, 635-645.
39. Zuiderduin, W., Westzaan, C., Huetink, J., and Gaymans, R. (2003) Toughening of polypropylene with calcium carbonate particles. *Polymer*, **44**, 261-275.
40. Kukackova, O., Dung, N. V., Abbrent, S., Urbanova, M., Kotek, J., and Brus, J. (2017) A novel insight into the origin of toughness in polypropylene–calcium carbonate microcomposites: Multivariate analysis of ss-NMR spectra. *Polymer*, **132**, 106-113.
41. Yang, C.-j., Huang, T., Yang, J.-h., Zhang, N., Wang, Y., and Zhou, Z.-w. (2017) Carbon nanotubes induced brittle-ductile transition behavior of the polypropylene/ethylene-propylene-diene terpolymer blends. *Composites Science and Technology*, **139**, 109-116.
42. Tjong, S., Bao, S., and Liang, G. (2005) Polypropylene/montmorillonite nanocomposites toughened with SEBS-g-MA: Structure–property relationship. *Journal of Polymer Science, Part B: Polymer Physics*, **43**, 3112-3126.
43. Parameswaranpillai, J., Joseph, G., Shinu, K., Sreejesh, P., Jose, S., Salim, N. V., and Hameed, N. (2015) The role of SEBS in tailoring the interface between the polymer matrix and exfoliated graphene nanoplatelets in hybrid composites. *Materials Chemistry and Physics*, **163**, 182-189.
44. Haghnegahdar, M., Naderi, G., and Ghoreishy, M. (2017) Fracture toughness and deformation mechanism of un-vulcanized and dynamically vulcanized polypropylene/ethylene propylene diene monomer/graphene nanocomposites. *Composites Science and Technology*, **141**, 83-98.

45. Li, R., Zhang, X., Zhao, Y., Hu, X., Zhao, X., and Wang, D. (2009) New polypropylene blends toughened by polypropylene/poly (ethylene-co-propylene) in-reactor alloy: Compositional and morphological influence on mechanical properties. *Polymer*, **50**, 5124-5133.
46. Gahleitner, M., Tranninger, C., and Doshev, P. (2013) Heterophasic copolymers of polypropylene: Development, design principles, and future challenges. *Journal of Applied Polymer Science*, **130**, 3028-3037.
47. Hongjun, C., Xiaolie, L., Dezhu, M., Jianmin, W., and Hongsheng, T. (1999) Structure and properties of impact copolymer polypropylene. I. Chain structure. *Journal of Applied Polymer Science*, **71**, 93-101.
48. Mirabella, Jr., F. M. (1993) Impact polypropylene copolymers: fractionation and structural characterization. *Polymer*, **34**, 1729-1735.
49. Li, Y., Xu, J.-T., Dong, Q., Fu, Z.-S., and Fan, Z.-Q. (2009) Morphology of polypropylene/poly (ethylene-co-propylene) in-reactor alloys prepared by multi-stage sequential polymerization and two-stage polymerization. *Polymer*, **50**, 5134-5141.
50. Zhang, C., Shangguan, Y., Chen, R., Wu, Y., Chen, F., Zheng, Q., and Hu, G. (2010) Morphology, microstructure and compatibility of impact polypropylene copolymer. *Polymer*, **51**, 4969-4977.
51. Tan, H., Li, L., Chen, Z., Song, Y., and Zheng, Q. (2005) Phase morphology and impact toughness of impact polypropylene copolymer. *Polymer*, **46**, 3522-3527.
52. Grein, C., Bernreitner, K., Hauer, A., Gahleitner, M., and Neißl, W. (2003) Impact modified isotatic polypropylene with controlled rubber intrinsic viscosities: Some new aspects about morphology and fracture. *Journal of Applied Polymer Science*, **87**, 1702-1712.
53. Doshev, P., Lach, R., Lohse, G., Heuvelsland, A., Grellmann, W., and Radusch, H.-J. (2005) Fracture characteristics and deformation behavior of heterophasic ethylene–propylene copolymers as a function of the dispersed phase composition. *Polymer*, **46**, 9411-9422.
54. Grellmann, W., Seidler, S., Gahleitner, M., and Nezbedova, E. (1998) Fracture toughness of polypropylene copolymers: influence of interparticle distance and temperature. *Polymer*, **39**, 75-82.
55. Joseph, P., Rabello, M. S., Mattoso, L., Joseph, K., and Thomas, S. (2002) Environmental effects on the degradation behaviour of sisal fibre reinforced polypropylene composites. *Composites Science and Technology*, **62**, 1357-1372.
56. Deshmane, C., Yuan, Q., Perkins, R., and Misra, R. (2007) On striking variation in impact toughness of polyethylene–clay and polypropylene–clay nanocomposite systems: the effect of clay–polymer interaction. *Materials Science and Engineering A*, **458**, 150-157.
57. Wu, Q., and Qu, B. (2001) Synergistic effects of silicotungistic acid

on intumescent flame-retardant polypropylene. *Polymer Degradation and Stability*, **74**, 255-261.

58. Butylina, S., Hyvärinen, M., and Kärki, T. (2012) Accelerated weathering of wood–polypropylene composites containing minerals. *Composites, Part A: Applied Science and Manufacturing*, **43**, 2087-2094.
59. Lv, Y., Huang, Y., Yang, J., Kong, M., Yang, H., Zhao, J., and Li, G. (2015) Outdoor and accelerated laboratory weathering of polypropylene: A comparison and correlation study. *Polymer Degradation and Stability*, **112**, 145-159.
60. Bocchini, S., Morlat-Therias, S., Gardette, J. L., and Camino, G. (2008) Influence of nanodispersed hydrotalcite on polypropylene photooxidation. *European Polymer Journal*, **44**, 3473-3481.
61. Gadioli, R., Morais, J. A., Waldman, W. R., and De Paoli, M. -A. (2014) The role of lignin in polypropylene composites with semi-bleached cellulose fibers: mechanical properties and its activity as antioxidant. *Polymer Degradation and Stability*, **108**, 23-34.
62. Goel, A., Chawla, K., Vaidya, U., Koopman, M., and Dean, D. (2008) Effect of UV exposure on the microstructure and mechanical properties of long fiber thermoplastic (LFT) composites. *Journal of Materials Science*, **43**, 4423-4432.
63. Grigoriadou, I., Paraskevopoulos, K., Karavasili, M., Karagiannis, G., Vasileiou, A., and Bikiaris, D. (2013) HDPE/Cu-nanofiber nanocomposites with enhanced mechanical and UV stability properties. *Composites, Part B: Engineering*, **55**, 407-420.
64. Guadagno, L., Naddeo, C., and Vittoria, V. (2004) Structural and morphological changes during UV irradiation of the crystalline helical form of syndiotactic polypropylene. *Macromolecules*, **37**, 9826-9834.
65. Vallejo-Montesinos, J., Martínez, J. L., Montejano-Carrizales, J., Perez, E., Pérez, J. B., Almendárez-Camarillo, A., and Gonzalez-Calderon, J. (2017) Passivation of titanium oxide in polyethylene matrices using polyelectrolytes as titanium dioxide surface coating. *Mechanics, Materials Science & Engineering Journal*, **8**, doi: 10.2412/mmse.96.48.950.
66. Krimm, S., Liang, C., and Sutherland, G. (1956) Infrared spectra of high polymers. II. Polyethylene. *The Journal of Chemical Physics*, **25**, 549-562.
67. Hagemann, H., Snyder, R., Peacock, A., and Mandelkern, L. (1989) Quantitative infrared methods for the measurement of crystallinity and its temperature dependence: polyethylene. *Macromolecules*, **22**, 3600-3606.
68. Liang, C., and Krimm, S. (1959) Infrared spectra of high polymers: Part IX. Polyethylene terephthalate. *Journal of Molecular Spectroscopy*, **3**, 554-574.

69. Li, C.-Q., Zha, J.-W., Long, H.-Q., Wang, S.-J., Zhang, D.-L., and Dang, Z.-M. (2017) Mechanical and dielectric properties of graphene incorporated polypropylene nanocomposites using polypropylene-graft-maleic anhydride as a compatibilizer. *Composites Science and Technology*, **153**, 111-118.
70. Zhao, S., Chen, F., Zhao, C., Huang, Y., Dong, J.-Y., and Han, C. C. (2013) Interpenetrating network formation in isotactic polypropylene/graphene composites. *Polymer*, **54**, 3680-3690.
71. Yang, C., Wang, M., Xing, Z., Zhao, Q., Wang, M., and Wu, G. (2018) A new promising nucleating agent for polymer foaming: effects of hollow molecular-sieve particles on polypropylene supercritical CO_2 microcellular foaming. *RSC Advances*, **8**, 20061-20067.
72. Zhu, H., Monrabal, B., Han, C. C., and Wang, D. (2008) Phase structure and crystallization behavior of polypropylene in-reactor alloys: insights from both inter-and intramolecular compositional heterogeneity. *Macromolecules*, **41**, 826-833.
73. Selden, R., Nyström, B., and Långström, R. (2004) UV aging of poly (propylene)/wood-fiber composites. *Polymer Composites*, **25**, 543-553.
74. *Developments in Polymer Stabilisation - 8*, Scott, G. (ed.), Elsevier, USA (1987).
75. Popa, M. I., Pernevan, S., Sirghie, C., Spiridon, I., Chambre, D., Copolovici, D. M., and Popa, N. (2013) Mechanical properties and weathering behavior of polypropylene-hemp shives composites. *Journal of Chemistry*, **2013**, Article ID 343068.
76. Onggo, H., and Pujiastuti, S. (2010) Effect of weathering on functional group and mechanical properties of polypropylene-kenaf composites. *Indonesian Journal of Materials Science*, **11**, 123-128.
77. Grigoriadou, I., Paraskevopoulos, K., Chrissafis, K., Pavlidou, E., Stamkopoulos, T.-G., and Bikiaris, D. (2011) Effect of different nanoparticles on HDPE UV stability. *Polymer Degradation and Stability*, **96**, 151-163.
78. Al-Shabanat, M. (2011) Study of the effect of weathering in natural environment on polypropylene and its composites: Morphological and mechanical properties. *International Journal of Chemistry*, **3**, 129.
79. Rajakumar, K., Sarasvathy, V., Chelvan, A. T., Chitra, R., and Vijayakumar, C. (2009) Natural weathering studies of polypropylene. *Journal of Polymers and the Environment*, **17**, 191.
80. Hoekstra, H., Spoormaker, J., and Breen, J. (1997) Mechanical and morphological properties of stabilized and non-stabilized HDPE films versus exposure time. *Macromolecular Materials and Engineering*, **247**, 91-110.
81. Rabello, M., and White, J. (1997) Crystallization and melting behaviour of photodegraded polypropylene-I. Chemi-crystallization. *Pol-*

ymer, **38**, 6379-6387.

82. Rabello, M., and White, J. (1997) Crystallization and melting behaviour of photodegraded polypropylene-II. Re-crystallization of degraded molecules. *Polymer*, **38**, 6389-6399.
83. Craig, I., White, J., and Kin, P.C. (2005) Crystallization and chemi-crystallization of recycled photo-degraded polypropylene. *Polymer*, **46**, 505-512.
84. Schoolenberg, G., and Meijer, H. (1991) Ultra-violet degradation of polypropylene: 2. Residual strength and failure mode in relation to the degraded surface layer. *Polymer*, **32**, 438-444.

9

Nanocomposites of Polypropylene, Polypropylene/PP-g-MA and Polypropylene/Polyamide Blend with Graphene and Boron Nitride: Thermal, Mechanical, Morphological and Electrical Studies

9.1 Introduction

In the last two decades, polymer nanocomposites have evolved into the materials of high commercial relevance due to the significant enhancement in the polymer properties by the incorporation of small fractions of functional nanofillers [1]. Due to the presence of filler particles at the nanoscale, a large number of polymer-filler interfacial contacts develop in the materials, thus, resulting in effective transfer of load as well as heat, leading to a synergistic improvement in the material performance.

Polypropylene (PP) is a stereo-regular polymer, with properties depending on the crystalline structure and various tacticities. Moreover, PP possesses various beneficial properties like low density, high thermal resistance, resistance to chemical attack (insoluble in almost all solvents at room temperature), ease of processing and recyclability. Interestingly, PP manifests mechanical profile closely to the engineering thermoplastics. Owing to this extraordinary set of properties, PP is a commonly used thermoplastic for structural applications [2]. A large variety of reinforcing fillers such as silica, clay, nanotubes and graphene have been explored as fillers in PP. Graphene, which consists of one atomic thick sheets of covalently sp^2-bonded carbon atoms in a hexagonal arrangement, has received immense attention of researchers for the generation of polymer nanocomposites [3,4]. Its choice as a filler is due to its excellent electrical and mechanical properties, which are significantly better than other inorganic filler materials. For instance, a single defect-free graphene layer has Young's modulus of ~1.0 TPa, intrinsic strength ~42 N/m, thermal conductivity ~4840-5300 W/(m.K), electron mobility exceeding 25,000 $cm^2/V.s$, excellent gas impermeability and specific surface area of ~2630 m^2/g [3]. In a recent study, Pawar *et al.* [5] reported functional

Mona Al Hosani, M. R. Vengatesan and Vikas Mittal
Khalifa University of Science and Technology, Abu Dhabi, UAE

PP/EVA-MA (ethylene vinyl acetate-*graft*-maleic anhydride) nanocomposites with amine functionalized graphite oxide, GrO-NH_2. As shown in Figure 9.1, the filler stacks were only a few layers thick and

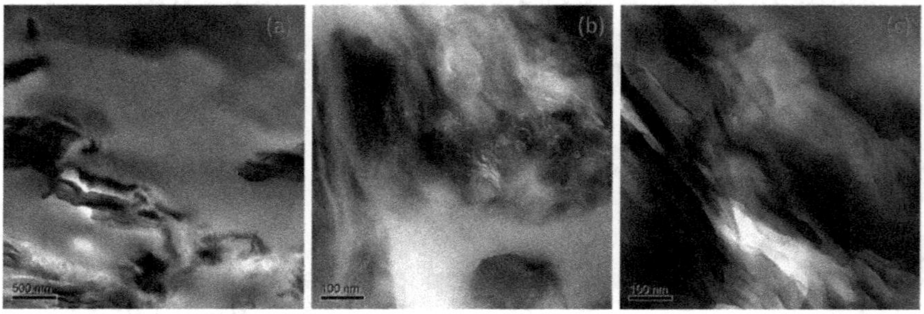

Figure 9.1 (a-c) Morphology of PP+EVA-MA+1%GrONH_2 composite. Reproduced from Reference 5.

were largely present in the EVA-MA phase. The storage modulus of the composites was also observed to be enhanced owing to the polymer-filler interactions and effective filler dispersion leading to higher stress transfer from the polymer chains to the filler platelets. In another study, Mittal *et al.* [6] reported blend nanocomposites of polypropylene and polystyrene with graphene and ethylene-co-acrylic acid (EAA) copolymer. The generated blend nanocomposites exhibited superior properties than the corresponding blends and are more useful alternatives than not only the parent materials but also other thermoplastics due to optimal properties. In addition, the morphology of the blend nanocomposites was also stable even after a few high temperature processing cycles, though the components were only kinetically mixed. Figure 9.2 demonstrates the effect of filler as well as compatibilizer on the tensile properties of the materials. Overall, even though the filler did not exfoliate extensively in the polymer matrix, the EAA incorporation still helped to achieve a superior composite performance by enhancing the overall extent of filler dispersion. In addition, the localization of filler in the PP phase also correspondingly affected the composite properties, as a function of blend composition.

Most of the polymers are insulators at macro-scale and usually have thermal conductivity below 0.5 W/(m.K). Generally, the polymers with higher degree of crystalline domains have higher thermal conductivity than the amorphous polymers. Similarly, the presence of side chains in the main chain also results in lower conductivity of the polymer. Specifically, in thermoplastic polymers, thermal conductivity is affected by the factors such as chain structure and orientation, crystallinity and interchain interactions. On the contrary, the factors which affect the thermal conductivity in thermoset polymers are mainly liquid crystal domain size and content, curing conditions

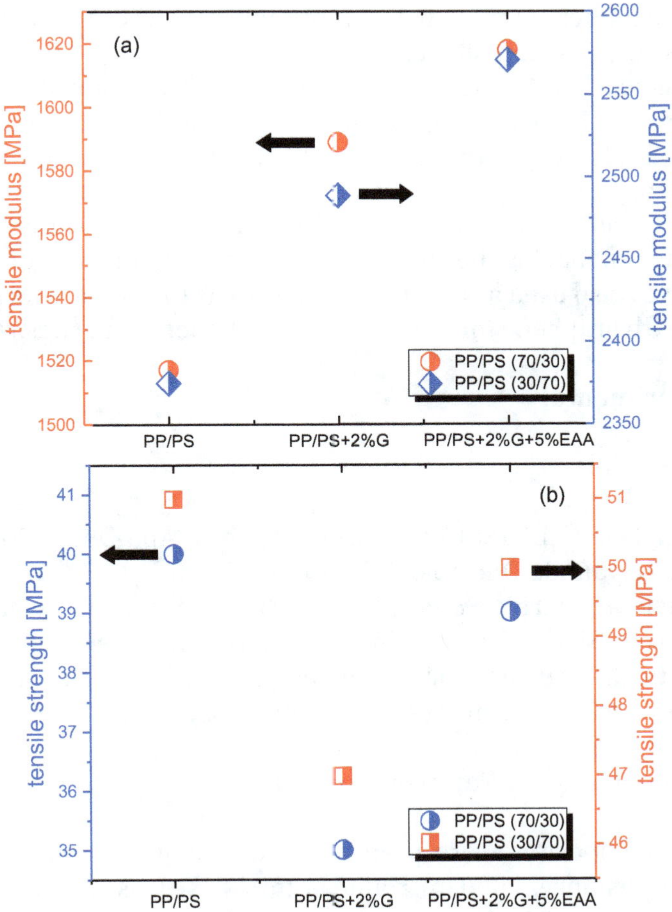

Figure 9.2 Tensile properties of (a) PP/PS (70/30) blend and its composites; (b) PP/PS (30/70) blend and its composites. Reproduced from Reference 6.

and orientation [7]. In addition to the mechanical and thermal properties, incorporation of fillers to the polymers is also beneficial to enhance the thermal and electrical conductivities. Usually, one dimensional fillers (1-D) develop well connected long conductive pathways in composite systems which result in significantly improved thermal conductivity. Examples of 1-D fillers are carbon nanotubes, carbon fibers, Si_3N_4 nanowires, boron nitride nanotubes, silver nanowires, copper nanowires, etc. [8-10]. Similarly, two dimensional (2-D) fillers, such as platelet fillers, are also considered as high aspect ratio fillers and can exhibit very high in-plane thermal conductivity, examples of such fillers with plate like morphology are boron nitride, graphene, Al_2O_3, TiB_2 and SiC [11].

In this work, PP nanocomposites have been developed with graphene as filler for achieving enhancement in mechanical and thermal properties, along with thermal and electrical conductivities. The use of polypropylene-*graft*-maleic anhydride (PP-g-MA) based compatibilizer as well as polyamide (PA) as blend component was employed for further enhancing the composite performance. In addition, boron nitride was also used as reinforcing filler so as to compare its performance with graphene for enhancing the polymer properties.

9.2 Experimental

9.2.1 Materials

Polypropylene (HD91SCF) was procured from Abu Dhabi Polymers Limited. Polyamide Durethan T40, PP-g-MA (M_w ~9,100, M_n ~3,900 and maleic anhydride % grafting 8-10%) and *o*-xylene were purchased from Sigma Aldrich. The materials were used as received, without further purification. Graphene (G) with a trade name N002-PDR was purchased from Angstron Materials, USA.

9.2.2 Preparation of Nanocomposites

The nanocomposites were prepared using solution blending technique. For instance, a 500 mL round bottom flask was added with 300 mL of xylene and 150 mg (3 wt%) of graphene. The suspension was kept for 1 h in ultra-sonication bath. 4.85 g of polypropylene was subsequently added to the flask. Afterwards, the flask was fitted with a reflux condenser in a silicon oil bath at 130 °C for 2 h. The solvent was subsequently evaporated at 100 °C in a vacuum atmosphere for around 12 h until the composite become fully dry. The prepared composite was then melted and mixed in the extruder for 15 min. The composites with PP-g-MA were generated by adding the PP-g-MA (fixed at 5 wt%) before the addition of PP. Similarly, the composites with 5 wt% graphene were generated. By following the same procedure, the composites with 5 and 10 wt% BN were also generated. For generating the blend composites with PA, a 50:50 PP/PA blend was used. For comparison, PP/PP-g-MA and PP/PA were also prepared using the same procedure. In addition, pure PP was also subjected to same processing protocols so as to obtain accurate property comparisons among the materials. The disc and dumbbell-shaped test samples were prepared by mini injection molding machine (HAAKE MiniJet, Germany) at a processing temperature of 190 °C. The injection pressure was 1000 bar for 6 s, whereas holding pressure was 400 bar for 3 s. The temperature of the mold was kept at 55 °C.

9.2.3 Characterization

Differential scanning calorimetric analysis of the materials was carried out using Discovery DSC from TA Instruments under nitrogen atmosphere. The scans were obtained from 30 to 200 °C (first heating), 200 to -50 °C (first cooling) and -50 to 200 °C (second hating) using heating and cooling rates of 10 °C/min and 5 °C/min, respectively. Second heating runs were used for the comparison purpose. Thermal degradation properties of the polymer, blends and nanocomposites were recorded using Discovery thermogravimetric analyzer (TGA) from TA Instruments. Nitrogen was used as a carrier gas and the scans were obtained from 50 to 700 °C at a heating rate of 10 °C/min.

Tensile analysis of the blends and composites was carried out using universal testing machine (Instron, USA) (ASTM D 638). The dumbbell-shaped samples were used for the analysis. A loading rate of 10 mm/min was used, and the tests were carried out at room temperature. The tensile strength and modulus of the composites were calculated using Win Test Analysis software. An average of five sample values was recorded. DHR-3 rheometer from TA Instruments was used to characterize the rheological properties such as storage modulus (G') and complex viscosity elasticity (η^*). Disc shaped samples of 25 mm diameter and 2 mm thickness were measured at 190 °C using a gap opening of 1.2 mm. Flow sweep experiment was carried out apriori and all experiments were performed in linear viscoelastic region (LVR) at 0.5% strain from $\omega = 0.1$ to 100 rad/s.

Wide angle X-ray diffraction (WAXRD) analysis of the materials was performed with analytical powder diffractometer (X'Pert PRO) using Cu Kα radiation ($\lambda = 1.5406$ Å) in reflection mode. A zero-background holder was used to minimize the noise. The samples were step-scanned from 2 θ = 5-60° at room temperature using a step size of 2θ = 0.02° and a step time of 10 s. The microstructure of the surface of the nanocomposites was analyzed using scanning electron microscope (SEM) (FEI Quanta, FEG250, USA) with elemental mapping at accelerating voltages of 10-20 kV. The sample surfaces were sputter-coated with 3 nm thick gold layer.

Laser flash method (using Netzsch LFA 447) was employed to measure the thermal conductivity of the samples at varying temperatures (25, 50, 75 and 100 °C). The thermal conductivity was measured according to ASTM E1461, using the measured heat capacity and thermal diffusivity, with separately entered density data. The disc-shaped samples of 12 mm diameter and 2 mm thickness were used for the analysis and vespel was used as the standard. For the electrical conductivity analysis, four-point probe method containing a four-point resistivity probing fixture was employed.

9.3 Results and Discussion

9.3.1 Boron Nitride Composites

Figure 9.3 demonstrates the DSC melting and crystallization thermograms of PP, PP/PP-g-MA, PP/PA and the composites with BN. Table 9.1 also details the calorimetric properties of these materials. Addition of BN to PP without PP-g-MA resulted in a small decrease in the peak melting point, whereas the peak crystallization temperature was observed to slightly enhance as a function of BN content. The melting (and crystallization) enthalpy (normalized to 100% polymer content) was observed to increase with BN indicating nucleation effect as well as increased crystallinity of the polymer. Addition of PP-g-MA to PP resulted in a small reduction in the melt enthalpy. The addition of BN to the PP/PP-g-MA exhibited similar trend as the PP+BN composites. Addition of amorphous PA to PP resulted in a significant reduction in the melt enthalpy of the PP/PA blend. On incorporation 5% and 10% BN, peak melting and crystallization temperatures as well as melt enthalpy were observed to increase indicating that the filler resulted in enhanced crystallinity.

The thermal degradation behavior of the BN composites is demonstrated in Figure 9.4. Table 9.2 also reports the temperatures for 10 and 50% weight loss, along with char yield at 700 °C. Pure PP exhibited the temperatures for 10 and 50% weight loss as 404 and 448 °C, which were enhanced as a function of BN content in the composites.

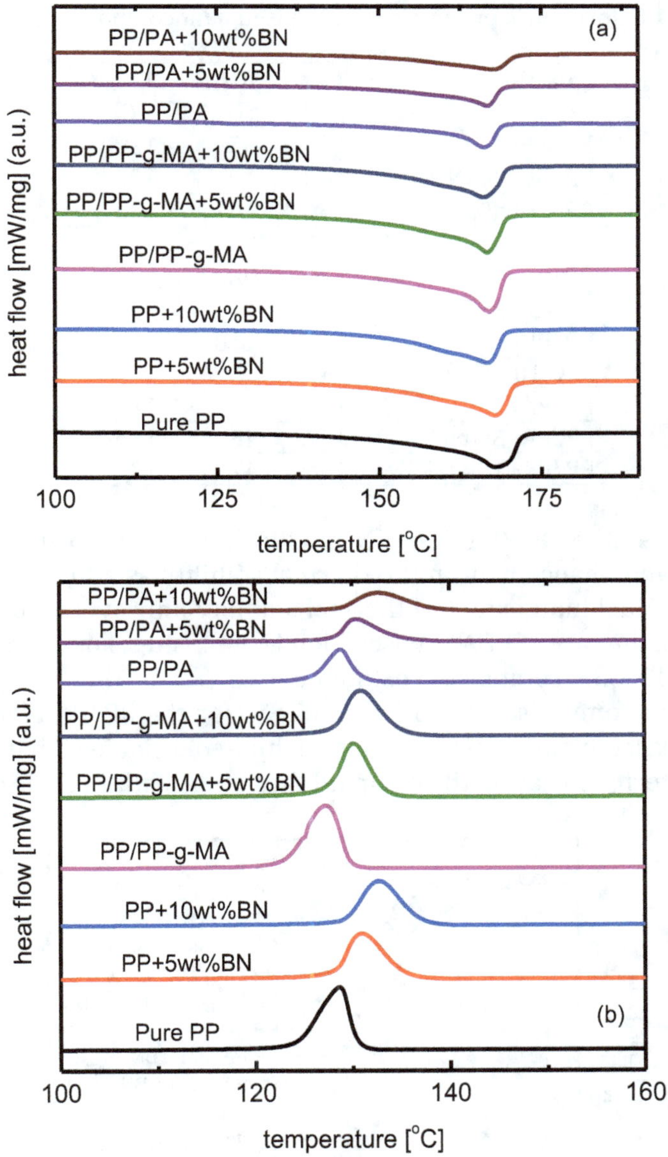

Figure 9.3 DSC (a) melting and (b) crystallization thermograms of BN composites.

The char yield also increased accordingly, with a value of 9.25% for the PP+10%BN composite as compared to 0.14% for pure PP. Addition of both PP-g-MA and PA to PP resulted in the enhancement in the initiation and peak degradation temperatures. Similar to the PP+BN

Table 9.1 Calorimetric properties of BN based nanocomposites

Sample	T_m (°C)	ΔH_m (J/g)	T_{crys} (°C)
Pure PP	168	95	129
PP+5wt%BN	168	101	130
PP+10wt%BN	166	102	132
PP/PP-g-MA	167	93	127
PP/PP-g-MA+5wt%BN	166	98	130
PP/PP-g-MA+10wt%BN	166	110	130
PP/PA	167	41	128
PP/PA+5wt%BN	167	45	130
PP/PA+10wt%BN	168	47	132

composites, the PP/PP-g-MA+BN and PP/PA+BN composites also exhibited an enhancement in the thermal stability as a function of BN content. Maximum extent of thermal stability improvement was observed in the case of PP/PA based blend composites. For instance, the PP/PA+10%BN composites exhibited $T_{10wt\%}$ and $T_{50wt\%}$ of 440 and 475 °C, in comparison to 410 and 459 °C for the PP/PA blend. The PP/PA based blend composites also exhibited a single step degradation pattern, indicating the compatibility of PA with PP. It is evident

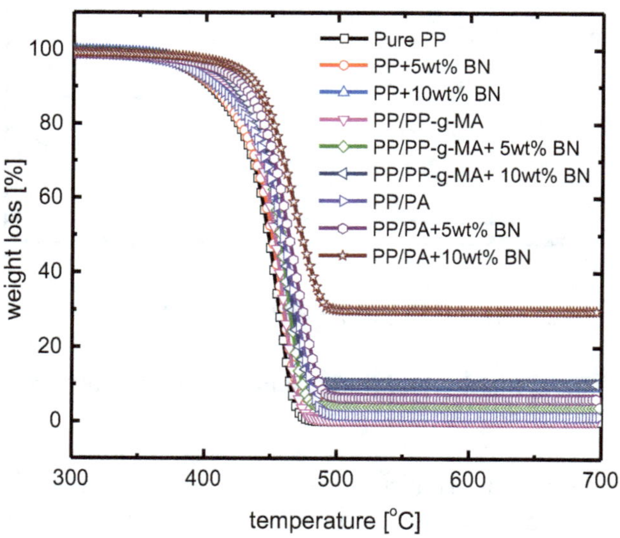

Figure 9.4 TGA thermograms of BN composites.

Table 9.2 Thermal stability and char yield of BN nanocomposites

Sample	$T_{10wt\%}$	$T_{50wt\%}$	Char yield @ 700 °C
Pure PP	404	448	0.14
PP+5wt%BN	406	452	4.80
PP+10wt%BN	424	458	9.25
PP/PP-g-MA	415	454	0.37
PP/PP-g-MA+5wt%BN	424	457	3.93
PP/PP-g-MA+10wt%BN	427	460	10.21
PP/PA	410	459	0.41
PP/PA+5wt%BN	430	465	6.16
PP/PA+10wt%BN	440	475	11.40

that BN phase in the composites formed a thermal barrier at the polymer-filler interface which delayed the release of thermal degradation products [12]. The weight loss as a function of time at a fixed temperature was also reduced in the composites, thus, confirming their superior thermal stability.

Figure 9.5 shows the X-ray diffractograms of the BN based composites. Pristine BN exhibited a sharp diffraction signal at 27°. The diffraction signal corresponding to BN was also observed in the PP+BN composites, indicating that the BN particles were not fully

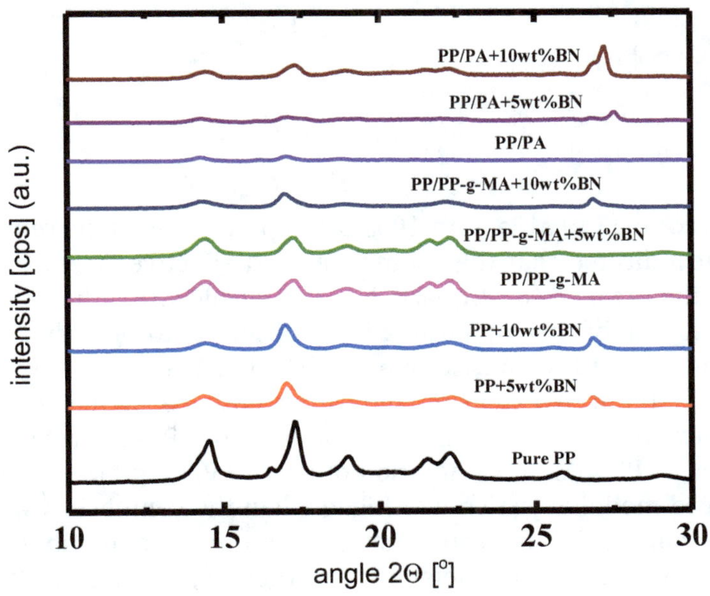

Figure 9.5 X-ray diffractograms of BN composites.

delaminated in the polymer. In the case of PP/PP-g-MA+BN composites, the BN signal was relatively reduced indicating higher degree of filler delamination due to the addition of the compatibilizer. The PP/PA based blend composites also exhibited a sharp diffraction peak corresponding to BN, the intensity of which was also dependent on the filler content. Overall, the crystalline phase of PP remained largely unaffected on the addition of BN, PP-g-MA and PA to PP. These observations indicated that enhancement of shear during the compounding process as well as increase in the compatibilizer content may be beneficial in achieving complete filler delamination.

The tensile properties of the BN based composites are reported in Table 9.3. PP exhibited a tensile modulus of 1044 MPa, which was

Table 9.3 Mechanical properties of BN based nanocomposites

Sample	UTS (MPa)	Modulus (MPa)	Extension at break (%)
Pure PP	33	1044	23
PP+5wt%BN	29	1051	12
PP+10wt%BN	21	1349	7
PP/PP-g-MA	32	956	7
PP/PP-g-MA+5wt%BN	29	1061	5
PP/PP-g-MA+10wt%BN	26	1423	5
PP/PA	36	1273	3
PP/PA+5wt%BN	25	1011	3
PP/PA+10wt%BN	24	853	3

enhanced to 1349 MPa with 10% BN, representing an increment of 30%. It indicated that the composites had effective stress transfer due to the presence of BN particles. The tensile strength and extension at break of the composites were observed to be reduced as a function of filler fraction. Addition of 5% PP-g-MA compatibilizer resulted in reduced extension indicating absence of plasticization. However, the tensile strength and modulus of PP were retained on addition of PP-g-MA. The addition of BN to PP/PP-g-MA resulted in enhanced modulus, which was higher than the composites without compatibilizer. This indicated that the addition of compatibilizer enhanced the extent of filler delamination, thereby, resulting in effective stress transfer. The observed changes in the polymer crystallinity on incorporating BN also resulted in corresponding effect on the tensile modulus. The reduction in tensile strength was also lesser in magnitude in the PP/PP-g-MA composites as compared to the composites without PP-g-MA. The PP/PA blend resulted in higher modulus as compared to PP due to the presence of stiffer PA. The PP/PA composite with 5% and 10% BN resulted in significant reduction in the modulus, probably due to the presence of filler aggregates. To further understand the factors leading to the observed behavior, it is needed to evaluate the morphology of the composites, especially phase distribution and filler localization.

Figure 9.6 presents the rheological performance of the BN composites as a function of angular frequency. As observed in Figure 9.6a, the materials exhibited an increase in the storage modulus with angular frequency. PP, PP/PP-g-MA and PP/PA exhibited similar trend for storage modulus as the tensile modulus, with PP/PA blend resulting in the highest storage modulus. The storage modulus of the composites was either same or lower than the corresponding polymer matrices. It indicated that the polymer melt was not restrained by the presence of low aspect ratio BN particles. Along with low aspect ratio, absence of strong interfacial interactions between the polymer and filler particles would have resulted in the observed phenomenon. PP+BN and PP/PP-g-MA+BN based composites also exhibited higher magnitude of loss modulus as compared to storage modulus, thus, indicating dominance of viscous response. As observed in Figure 9.6c, the PP/PA+BN composites exhibited highest value of viscosity. As expected, the viscosity of the materials decreased as a function of angular frequency.

The thermal conductivity performance of the composites is presented in Table 9.4 as a function of temperature. The increment in temperature resulted in a gradual decrease in the conductivity of the materials. At 25 °C, the thermal conductivity of PP was measured to be 0.225 W/mK, which was enhanced by more than 100% to a value of 0.545 W/mK on the addition of 5% BN. However, further increment in the BN content resulted in a decrease in the thermal conductivity of the composite, probably due to the presence of filler aggregates which resulted in a loss of conducting channels through the polymer. Addition of 5 wt% PP-g-MA further decreased the thermal conductivity to 0.173 W/mK, as compared to pure PP. However, the conductivity was observed to significantly increase with the addition of BN and the increment was also a function of BN content. For instance, the composite with 10% BN exhibited the thermal conductivity of 0.740 W/mK, which indicated an increment of >325%. It confirmed

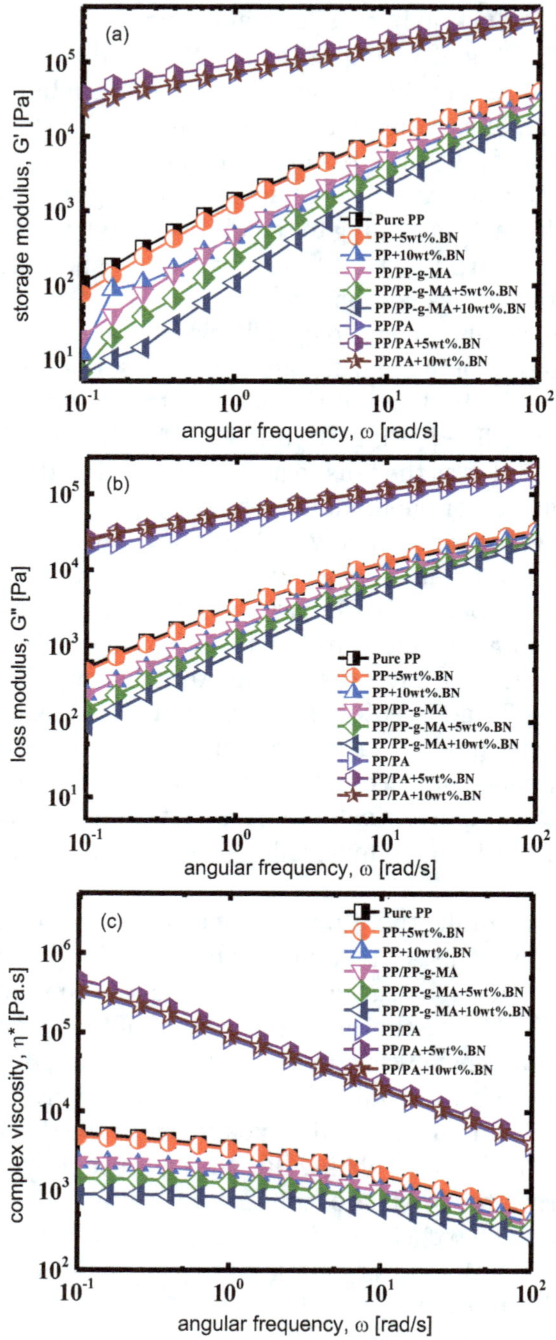

Figure 9.6 Storage modulus (a), loss modulus (b) and complex viscosity (c) of BN composites as a function of angular frequency.

that the addition of compatibilizer resulted in an effective filler dispersion, thereby leading to long-range thermally conducting paths. Blending PP with amorphous PA resulted in a significant reduction of thermal conductivity to 0.144 W/mK, as compared to pure PP. Similar to other cases, the addition of BN enhanced the thermal conductivity. The composite with 10% BN was observed to enhance the conductivity by >200%. Overall, the incorporation of BN was confirmed to be beneficial in significantly enhancing the thermal conductivity of the PP composites.

Table 9.4 Thermal conductivity of BN nanocomposites

Sample	Thermal conductivity (W/m K)			
	25 °C	50 °C	75 °C	100 °C
Pure PP	0.225	0.207	0.199	0.186
PP+5wt%BN	0.545	0.532	0.529	0.521
PP+10wt%BN	0.338	0.331	0.328	0.328
PP/PP-g-MA	0.173	0.168	0.168	0.165
PP/PP-g-MA+5wt%BN	0.688	0.678	0.672	0.669
PP/PP-g-MA+10wt%BN	0.740	0.738	0.729	0.721
PP/PA	0.144	0.144	0.138	0.128
PP/PA+5wt%BN	0.413	0.398	0.389	0.375
PP/PA+10wt%BN	0.435	0.431	0.428	0.422

The surface electrical resistivity of the BN based composites is presented in Table 9.5. Neat PP, PP/PP-g-MA and PP/PA displayed the surface resistivity of about 10^9 Ωm. The surface resistivity of PP remained unchanged by the incorporation of 5% and 10% BN. In the case of composites generated in the presence of PP-g-MA, the surface resistivity was observed to increase by an order of magnitude as compared to PP/PP-g-MA. As the surface resistivity is directly proportional to the insulating property of the composites, it indicated an enhancement of the electrical insulation of the materials. It is well known that the BN particle exhibit an electrical insulation property [13]. Thus, the addition of BN particles developed electrical insulating properties in the polymer matrix. Though the presence of compatibilizer led to enhancement in the surface resistivity due to enhanced filler dispersion, however, the resistivity was not affected by the filler fraction. Similarly, in the case of PP/PA composites, the resistivity was observed to increase by an order of magnitude as compared to the PP/PA matrix, but remained unchanged with the filler fraction.

Table 9.5 Electrical resistivity of BN nanocomposites

Sample	Electrical resistivity (Ωm)
Pure PP	2.0×10^{9}
PP+5wt%BN	2.0×10^{9}
PP+10wt%BN	2.1×10^{9}
PP/PP-g-MA	3.7×10^{9}
PP/PP-g-MA+5wt%BN	4.9×10^{10}
PP/PP-g-MA+10wt%BN	4.5×10^{10}
PP/PA	3.0×10^{9}
PP/PA+5wt%BN	3.0×10^{10}
PP/PA+10wt%BN	3.0×10^{10}

9.3.2 Graphene Composites

Table 9.6 demonstrates the calorimetric properties of graphene based composites. Figure 9.7 also shows the melting and crystallization thermograms of the nanocomposites. In the case of PP+G composites, no significant change in the peak melting and crystallization temperatures of PP was observed on the addition of graphene nanosheets. However, the melt enthalpy was observed to slightly increase as a function of graphene content, thus, indicating an increment in the polymer crystallinity [14]. Similarly, in the presence of the compatibilizer, the crystallinity of the nanocomposites was also observed to increase to a small extent, while the peak melting and crystallization

Table 9.6 Calorimetric properties of graphene based nanocomposites

Sample	T_m (°C)	ΔH_m (J/g)	T_{crys} (°C)
Pure PP	168	95	129
PP+3wt%G	167	95	128
PP+5wt%G	167	99	128
PP/PP-g-MA	167	93	127
PP/PP-g-MA+3wt%G	167	94	126
PP/PP-g-MA+5wt%G	167	95	126
PP/PA	167	41	128
PP/PA+3wt%G	168	52	127
PP/PA+5wt%G	168	59	128

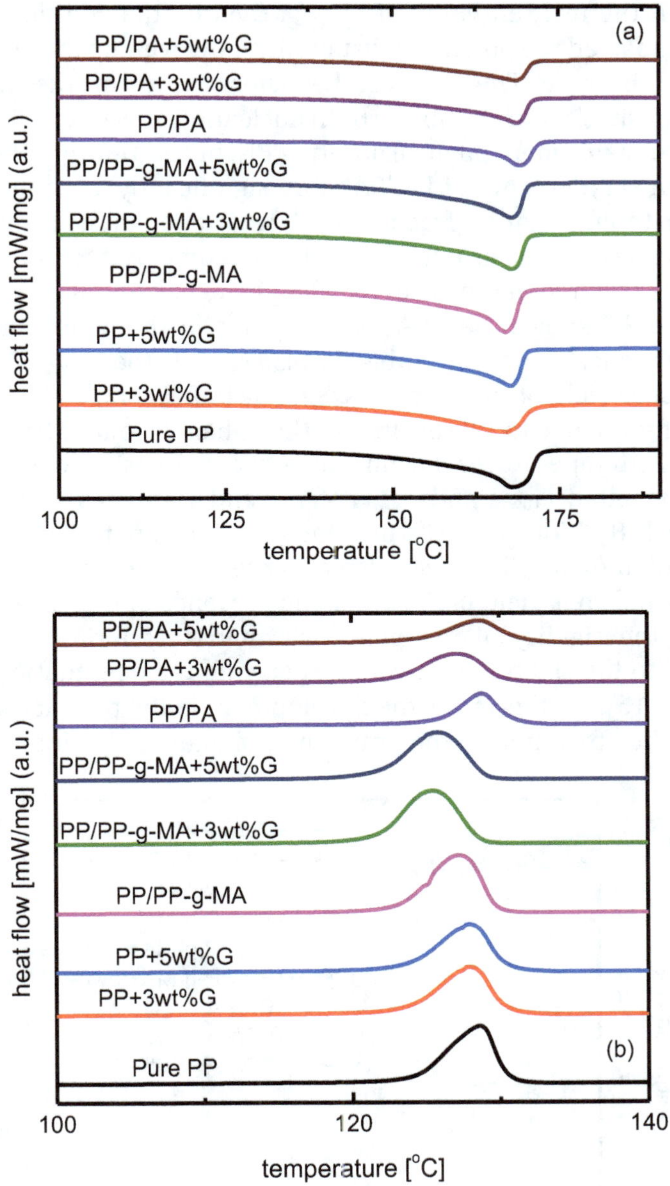

Figure 9.7 DSC (a) melting and (b) crystallization thermograms of graphene based composites.

temperatures remained constant. The composites with PP/PA also exhibited a significant increment in the melt enthalpy. For instance, as compared to 41 J/g for PP/PA, the composites with 5% graphene exhibited the melt enthalpy of 59 J/g. Overall, the addition of graphene impacted the polymer crystallinity, however, no nucleation effect was observed. The observed behavior was in contrast with the BN composites, which exhibited both nucleation effect as well as crystallinity enhancement on incorporating BN. In addition, graphene impacted the crystallinity of PP/PA more significantly as compared to BN, even though a larger fraction of BN was added in the composites. Larger aspect ratio of graphene platelets as compared to BN particles may have resulted in the observed composite behaviors.

Figure 9.8 shows the TGA thermograms of the nanocomposites with graphene. In addition, Table 9.7 also reports the $T_{10wt\%}$, $T_{50wt\%}$ as well as char yield of the materials. Similar to BN composites, the addition of graphene to PP resulted in the enhancement of both $T_{10wt\%}$ and $T_{50wt\%}$ temperatures. For instance, the composite with 5% graphene exhibited $T_{10wt\%}$ and $T_{50wt\%}$ of 440 and 470 °C, as compared to 404 and 448 °C for PP. In the presence of the compatibilizer and PA, the addition of graphene was also observed to increase the thermal stability and char yield of PP. Thus, the presence of the graphene increased the rigidity of the polymeric matrix resulting in the restriction to the chain mobility, thus, delaying the degradation of the polymer [15]. Furthermore, the development in the thermal stability can also be attributed to the graphene-polymer chains interactions,

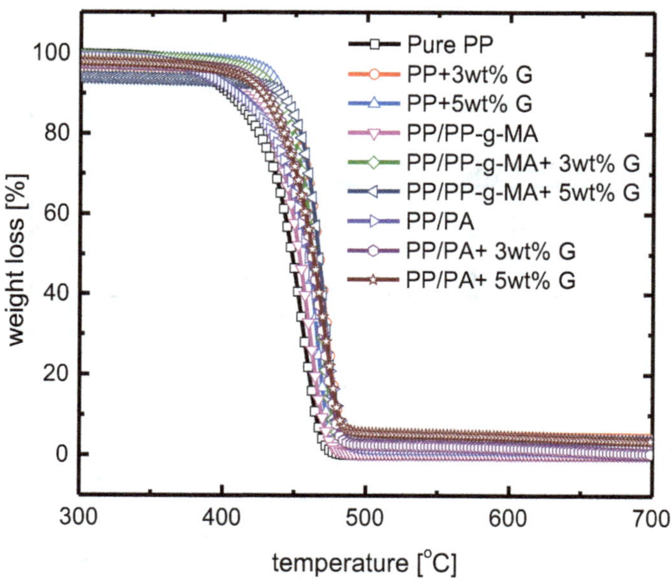

Figure 9.8 TGA thermograms of nanocomposites with graphene.

which lead to the formation of network-structured layer of graphene in the polymer matrix [14]. The composites also exhibited delayed degradation when tested under isothermal conditions at 190 °C, as a function of time. It, thus, confirmed that the composites have suitable thermal stability at the processing temperature used to generate the nanocomposites.

Table 9.7 Thermal stability and char yield of graphene nanocomposites

Sample	$T_{10wt\%}$	$T_{50wt\%}$	Char yield @ 700 °C
Pure PP	404	448	0.14
PP+3wt%G	437	466	4.16
PP+5wt%G	440	470	5.58
PP/PP-g-MA	415	454	0.37
PP/PP-g-MA+3wt%G	436	462	3.24
PP/PP-g-MA+5wt%G	440	466	3.32
PP/PA	410	459	0.41
PP/PA+3wt%G	428	461	1.40
PP/PA+5wt%G	430	462	3.53

The X-ray diffractograms of graphene nanocomposites are demonstrated in Figure 9.9. Graphene was observed to have the

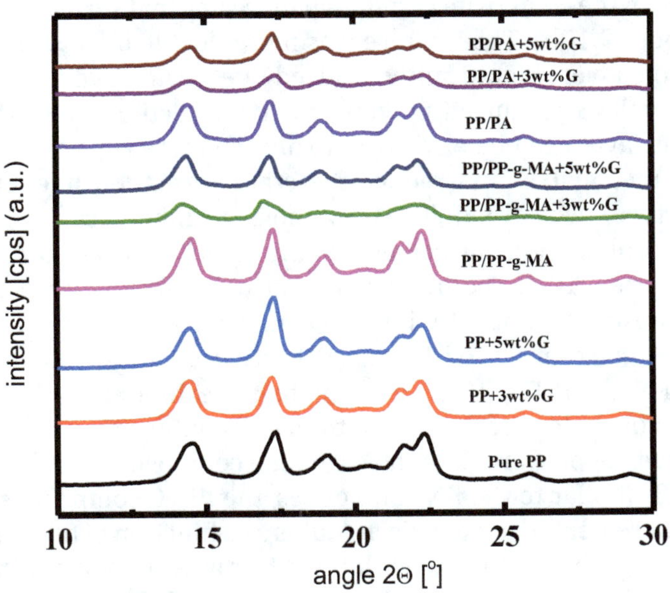

Figure 9.9 X-ray diffractograms of graphene nanocomposites.

characteristic diffraction peak at $2\theta \approx 26.5°$ (not shown). In the PP+G composites, the graphene characteristic peak was observed to be absent, thus, indicating a good degree of filler dispersion. PP exhibited crystalline diffraction peaks corresponding to (1 1 0), (0 4 0), (1 3 0), (1 1 1), (0 4 1), (0 6 0) and (2 2 0) planes (α monoclinic), which were observed to remain unchanged with the addition of graphene, thus, indicating no change in the PP crystalline structure. Similarly, the crystal structure of PP phase was also retained in both PP/PP-g-MA and PP/PA based composites. Diffraction peak corresponding to graphene was also absent in these composites due to effective filler delamination, even at 5 wt% filler content [16,17]. The observed behavior was in contrast with the BN composites, where the processing conditions employed for the generation of nanocomposites did not lead to effective filler delamination, and the filler characteristic diffraction peak was observed in the composite diffractograms.

Morphology of the fracture surface of the graphene composites was also studied with scanning electron microscopy, as demonstrated in Figure 9.10 for PP, PP/PP-g-MA and PP/PA composites with 5 wt% graphene. In the case of PP+G composite (Figure 9.10a), brittle failure was observed with significant extent of micro-voids present on the surface. Lack of adhesion between the polymer-filler phases may have resulted in such morphology. The addition of PP-g-MA to PP in the composite resulted in more optimal surface morphology (Figure 9.10b). The fracture was more ductile in nature, and the micro-voids were significantly reduced due to effective filler dispersion as well as enhanced polymer-filler interaction. The exfoliated graphene layers can be envisaged to absorb more rupture energy of the polymer matrix and reduce the crack formation phenomenon. The fracture surface of PP/PA+G composite exhibited completely different morphology as the corresponding PP and PP/PP-g-MA composites (Figure 9.10c). The fracture was also ductile in nature, though, the composite did not exhibit significant extension. Similar observations were also observed for the BN based composites, where the PP/PP-g-MA and PP/PA based composites exhibited more optimal surface morphology as compared to the PP composites.

The tensile properties of the graphene composites are detailed in Table 9.8. Similar to the BN composites, the PP+G composites exhibited an increment in the tensile modulus as a function of filler content. For instance, the composite with 5 wt% graphene resulted in an increase of 46% in the modulus, as compared to PP. In addition, the extent of modulus enhancement in the case of PP+G composites was

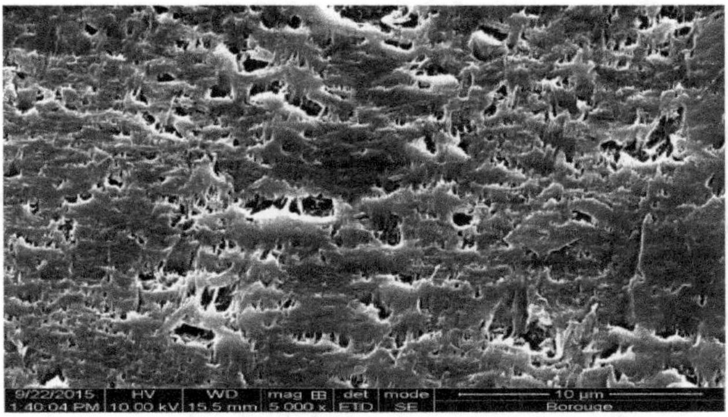

(a)

(b)

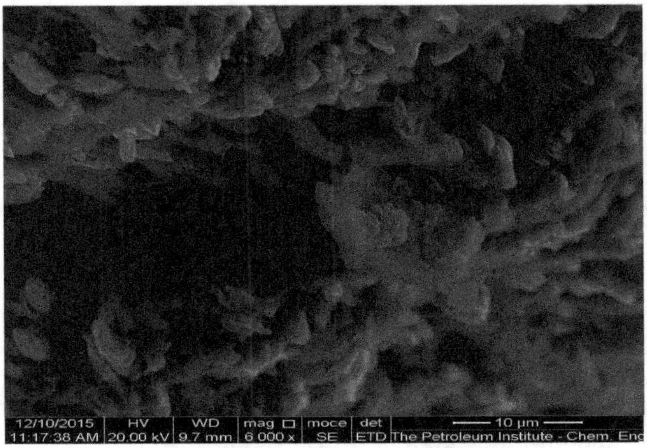

(c)

Figure 9.10 Scanning electron micrographs of the fracture surfaces of (a) PP+5wt%G, (b)PP/PP-g-MA+5wt%G, and (c) PP/PA+5wt%G nanocomposites.

Table 9.8 Mechanical properties of graphene based nanocomposites

Sample	UTS (MPa)	Modulus (MPa)	Extension at break (%)
Pure PP	33	1044	23
PP+3wt%G	29	1151	5
PP+5wt%G	22	1522	3
PP/PP-g-MA	32	956	7
PP/PP-g-MA+3wt%G	23	1488	3
PP/PP-g-MA+5wt%G	21	1903	2
PP/PA	36	1273	3
PP/PA+3wt%G	24	1852	2
PP/PA+5wt%G	23	2254	2

also significantly superior to the corresponding PP+BN composites. The tensile strength and extension of the PP+G composites were observed to be negatively impacted on the addition of graphene. The addition of the PP-g-MA to PP for the composite generation resulted in further increase in the modulus of the composites, as compared to PP+G composites. The PP/PP-g-MA composite with 5 wt% graphene exhibited a modulus of 1903 MPa, which indicated an increment of ~100%. Thus, the addition of compatibilizer resulted in enhanced polymer-filler interfacial interactions and higher extent of filler delamination, thereby, leading to effective stress transfer. Similarly, the PP/PA composites also had a significant enhancement in the tensile modulus on graphene incorporation. In comparison to 1273 MPa for PP/PA blend, the tensile modulus of the composite with 5 wt% graphene was measured to be 2254 MPa. Similar to the PP and PP/PP-g-MA composites, the strength and extension at break of the PP/PA composites were observed to reduce on graphene incorporation. Overall, the comparison among the graphene and BN composites revealed much higher extent of modulus enhancement in the case of graphene as reinforcing filler.

Figure 9.11 demonstrates the storage and loss modulus as well as complex viscosity of graphene composites as a function of angular frequency. As shown in Figure 9.11a, the addition of graphene to PP resulted in a significant enhancement in storage modulus, which was also a function of filler content. In addition, the storage modulus curves for the composites exhibited nearly frequency independent behavior. Similarly, the PP/PP-g-MA based composites also exhibited

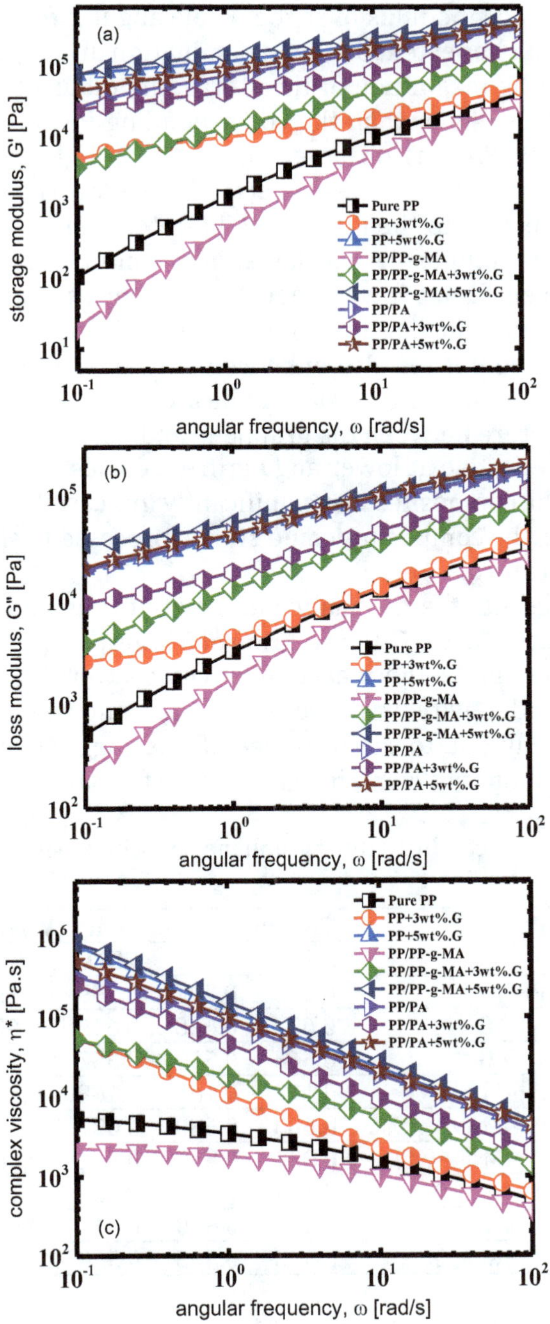

Figure 9.11 Storage modulus (a), loss modulus (b) and complex viscosity (c) of graphene nanocomposites as a function of angular frequency.

enhanced storage modulus over the whole angular frequency range. Owing to the higher extent of filler delamination due to polymer-filler interactions, the magnitude of the storage modulus of PP/PP-g-MA composites was higher than the corresponding PP composites. The modulus of PP/PA was observed to be much higher than PP and PP/PP-g-MA, though it was less affected by the incorporation of graphene as compared to the PP and PP/PP-g-MA composites. The loss modulus of the composites was lower in magnitude than the storage modulus, especially at lower angular frequencies, thus, indicating dominant elastic behavior. As shown in Figure 9.11c, the complex viscosity of the composites also enhanced with filler fraction. The observed increment in the viscosity can be attributed to the interlinked network structure between the graphene and polymer matrix, which is especially effective at lower angular frequencies. Even though the viscosity of the composites was significantly increased on incorporating graphene, the composites could still be processed using the same processing protocols.

Table 9.9 presents the thermal conductivity of the graphene based nanocomposites. At 25 °C, the addition of 3% graphene to PP resulted in an increment in the thermal conductivity by 180%. A slight reduction in the conductivity was observed in the composite with 5% graphene, probably due to the presence of filler aggregates in the absence of the compatibilizer which hinder the formation of conducting

Table 9.9 Thermal conductivity of graphene nanocomposites

Sample	Thermal conductivity (W/m K)			
	25 °C	50 °C	75 °C	100 °C
Pure PP	0.225	0.207	0.199	0.186
PP+3wt%G	0.632	0.628	0.619	0.620
PP+5wt%G	0.566	0.559	0.552	0.552
PP/PP-g-MA	0.173	0.168	0.168	0.165
PP/PP-g-MA+3wt%G	0.683	0.680	0.677	0.672
PP/PP-g-MA+5wt%G	0.821	0.818	0.811	0.807
PP/PA	0.144	0.144	0.138	0.128
PP/PA+3wt%G	0.525	0.520	0.517	0.515
PP/PA+5wt%G	0.568	0.563	0.558	0.553

channels in the polymer matrix. The effect of compatibilizer in enhancing the filler dispersion was evident from the higher magnitude of thermal conductivity in the PP/PP-g-MA composites, which was also a function of the filler content. For instance, the composite with 5% graphene was observed to have 380% increase in conductivity as compared to PP/PP-g-MA. Similar increase was observed in the case of PP/PA composites, where the composite with 5% graphene exhibited thermal conductivity of 0.568 W/mK as compared to 0.144 W/mK for PP/PA. Overall, the thermal conductivity of the composites was significantly improved on the addition of a small amount of graphene. In addition, the graphene composites also exhibited much higher extent of increase in the thermal conductivity as compared to the BN composites.

The surface resistivity of the graphene reinforced nanocomposites is presented in Table 9.10. Addition of graphene was observed to reduce the surface resistivity of PP by four orders of magnitude. As the surface resistivity is inversely proportional to the surface conductivity, it indicated a significant increase in the electrical conductivity.

Table 9.10 Electrical resistivity of graphene nanocomposites

Sample	Electrical resistivity (Ωm)
Pure PP	2.0×10^9
PP+3wt%G	3.0×10^5
PP+5wt%G	4.6×10^5
PP/PP-g-MA	3.7×10^9
PP/PP-g-MA+3wt%G	4.6×10^4
PP/PP-g-MA+5wt%G	1.4×10^5
PP/PA	3.0×10^9
PP/PA+3wt%G	4.8×10^4
PP/PA+5wt%G	6.4×10^6

Thus, the incorporation of graphene in the polymer increased the mobility of the conducting network and the density of the potential charge carriers [15]. PP/PP-g-MA and PP/PA composites with 3 wt% graphene also resulted in five orders of magnitude reduction in the surface resistivity. The composites with 5 wt% graphene had a small extent of increase in the surface resistivity, however, it was still 3-4 orders of magnitude lower than respective polymer matrices, thus, indicating enhanced electrical conductivity of the composites.

9.4 Conclusions

In the current study, mechanical, thermal, rheological and conductivity properties of PP, PP/PP-g-MA, PP/PA nanocomposites, prepared with graphene and boron nitride filler particles, were studied. The nanocomposites exhibited optimal filler dispersion in the presence of PP-g-MA, though the extent of dispersion of graphene was observed to be superior than BN particles. The filler particles enhanced the crystallinity as well as thermal stability of the polymer matrices. Furthermore, the tensile and elastic moduli of graphene composites were much higher in magnitude than the corresponding BN composites. Addition of 5 wt% graphene to the polymers resulted in significant increment in the thermal conductivity. 4-5 order of magnitude reduction in the surface resistivity of the composites was observed on the addition of 3 wt% graphene. On the other hand, the BN composites resulted in an enhancement of the surface resistivity of the composites. The developed nanocomposite materials represent advanced materials with high potential of application in the areas such as heat exchangers, electronic packaging, etc.

References

1. Pavlidoua, S., and Papaspyrides, C. D. (2008) A review on polymer–layered silicate nanocomposites. *Progress in Polymer Science*, **33**, 1119-1198.
2. Mittal, V. (2007) Polypropylene-layered silicate nanocomposites: Filler matrix interactions and mechanical properties. *Journal of Thermoplastic Composite Materials*, **20**, 575-599.
3. Mukhopadhyay, P., and Gupta, R. K. (2011) Trends and frontiers in graphene-based polymer nanocomposites. *Plastics Engineering*, **67**, 32-42.
4. Mittal, V. (2014) Functional polymer nanocomposites with graphene: A review. *Macromolecular Materials and Engineering*, **299**, 906-931.
5. Pawar, S., Mittal, V., and Matsko, N. (2017) PP/EVA-MA blend nanocomposites with graphene nano-sheets. In: *Functional Polymer Blends and Nanocomposites*, Mittal, V. (ed.), Central West Publishing. Australia, pp. 107-136.
6. Mittal, V., Pannirselvam, M., and Griffin, G. (2017) Polypropylene/polystyrene blend nanocomposites with graphene. In: *Functional Polymer Blends and Nanocomposites*, Mittal, V. (ed.), Central West Publishing. Australia, pp. 137-164.
7. Chen, H., Ginzburg, V. V., Yang, J., Yang, Y., Liu, W., Huanf, Y., Du, L., and Chen, B. (2016) Thermal conductivity of polymer-based composites: Fundamentals and applications. *Progress in Polymer Science*, **59**, 41-85.

8. Tavman, I. H., and Akinci, H. (2000) Transverse thermal conductivity of fiber reinforced polymer composites. *International Communications in Heat and Mass Transfer*, **27**(2), 253-261.
9. Kusunose, T., Yagi, T., Firoz, S. H., and Sekino, T. (2013) Fabrication of epoxy/silicon nitride nanowire composites and evaluation of their thermal conductivity. *Journal of Materials Chemistry A*, **1**(10), 3440-3445.
10. Yu, J., Chen, Y., Wuhrer, R., Liu, Z., and Ringer, S. P. (2005) In situ formation of BN nanotubes during nitriding reactions. *Chemistry of Materials*, **17**(20), 5172-5176.
11. Hill, R. F., and Supancic, P. H. (2002) Thermal conductivity of platelet-filled polymer composites. *Journal of the American Ceramic Society*, **85**(4), 851-857.
12. Li, T., and Hsu, S. L.-C. (2011) Preparation and properties of thermally conductive photosensitive polyimide/boron nitride nanocomposites. *Journal of Applied Polymer Science*, **121**, 916-922.
13. Hou, J., Li, G., Yang, N., Qin, L., Grami, M. E., Zhang, Q., Wang, N., and Qu, X. (2014) Preparation and characterization of surface modified boron nitride epoxy composites with enhanced thermal conductivity. *RSC Advances*, **4**, 44282-44290.
14. Zhao, S., Chen, F., Zhao, C., Huang, Y., Dong, J. Y., and Han, C. C. (2013) Interpenetrating network formation in isotactic polypropylene/graphene composites. *Polymer*, **54**(14), 3680-3690.
15. Fim, F. d. C., Basso, N. R. S., Graebin, A. P., Azambuja, D. S., and Galland, G. B. (2013) Thermal, electrical, and mechanical properties of polyethylene–graphene nanocomposites obtained by in situ polymerization. *Journal of Applied Polymer Science*, **128**(5), 2630-2637.
16. Qiu, F., Hao, Y., Li, X., Wang, B., and Wang, M. (2015) Functionalized graphene sheets filled isotactic polypropylene nanocomposites. *Composites, Part B: Engineering*, **71**, 175-183.
17. Milani, M. A., Gonzalez, D., Quijada, R., Basso, N. R. S., Cerrada, M. L., Azambuja, D. S., and Galland, G. B. (2013) Polypropylene/graphene nanosheet nanocomposites by in situ polymerization: Synthesis, characterization and fundamental properties. *Composites Science and Technology*, **84**, 1-7.

10

Polypropylene/Cyclohexylethylene-*co*-Ethylene Blends and Polypropylene/Cyclohexylethylene-*co*-Ethylene/Graphene Nanocomposites

10.1 Introduction

Polymer blending and composite generation are two of the widely acceptable techniques used to produce high performance polymeric materials. In comparison to the synthesis of new monomers or polymers, polymer blending is beneficial to upgrade the performance and processability of polymers in a simple, effective and economic way. Also, it presents benefits in terms of polymer recycling and waste management [1-4]. Polymer composites are engineering hybrid materials based on filler reinforced polymer matrices and have displayed significant potential to replace conventional materials in many applications areas owing to weight saving, strength, dimensional, thermal and chemical stability, flame retardancy, ease of processing and handling, cost effectiveness, etc. [5,6]. Polymer nanocomposites are an advanced class of promising composites, with one dimension of the filler phase lower than 100 nm. The nanomaterials like graphene, carbon nanotubes, nanoclay, etc., significantly improve the mechanical, thermal, rheological, flame retardation and gas barrier properties of polymers at very low loadings in comparison to other conventional polymer composites [7-10], thereby also retaining the optical clarity and low density. The dispersion state of these nanomaterials in the polymer matrices is the critical factor which decides the extent of interfacial interactions between the components and, thus, final properties [11].

On account of its superior mechanical, thermal and barrier properties, along with unique structural features, graphene has received extensive research attention for designing high performance polymer nanocomposites. Graphene is a two dimensional nanomaterial with a hexagonally organized sp^2 bonded sheet morphology and exhibits structural similarity to silicates as well as chemical similarity to carbon nanotubes [12-15]. Typically, graphene is successfully sourced

Anish M. Varghese and Vikas Mittal
Khalifa University of Science and Technology, Abu Dhabi, UAE

from graphite by means of four different routes, namely thermal reduction, chemical reduction, chemical vapor decomposition and micro-mechanical expansion [16-18]. Achievement of effective graphene dispersion in the polymer matrices, especially in the case of non-polar polymers, is a challenging task [19,20]. It can be overcome by reducing the polarity difference between the polymer and filler phases. In this context, the use of compatibilizers possessing both hydrophilic and hydrophobic components is one of the effective ways to improve the compatibility of graphene and its derivatives with polymer matrices in a non-covalent manner [21-23].

Polypropylene (PP) is one of the most commonly used commodity thermoplastic, which has attracted significant research interest because of its versatile properties, low cost and easy processability. The unique properties of PP drive its suitability for a wide range of applications such as automotive parts, household implements, building and construction materials, medical devices, electrical and electronic devices, other industrial materials, etc. [24-27]. Considering the rising importance of PP, a number of research studies have reported the incorporation of nanomaterials, especially graphene, to further enhance its properties.

In a recent study, Haghnegahdar *et al.* [28] studied the influence of graphene on the fracture toughness of ethylene propylene diene monomer (EPDM) toughened PP. The fracture toughness of the PP/EPDM blend was observed to improve after graphene addition, and the degree of improvement was higher for few-layer graphene (FLG) platelets in comparison with multi-layer graphene platelets due to the improved dispersion state. Also, the application of dynamic vulcanization was observed to be beneficial to further strengthen the fracture toughness of PP/EPDM/graphene nanocomposites. In another study, the authors reported enhanced electrical and thermal conductivity as well as thermal stability for graphene reinforced PP/EPDM blend nanocomposites [29]. The authors also made use of PP functionalized graphene and observed improved graphene dispersion in the PP/EPDM blend matrix. It helped to strengthen the interfacial bonding, which correspondingly exhibited enhanced tensile strength and modulus as well as reduced damping [30].

Parameswaranpillai *et al.* [31] employed styrene-(ethylene-*co*-butylene)-styrene triblock copolymer (SEBS) to obtain improved interfacial interactions between graphene and PP, which correspondingly resulted in improved toughness and thermal stability, attributed to the diminished graphene flocculation tendency. In another recent study, the authors reported the generation of PP/polystyrene/SEBS/graphene nanocomposites possessing superior toughness, stiffness and thermal stability for effective automotive applications [32]. Li *et al.* [33] also used PP-graft-maleic anhydride (PP-*g*-MA) as a compatibilizer for developing PP/graphene nanocomposites. Due to the influence of PP-*g*-MA, the graphene nanosheets were dispersed homogeneously in the PP matrix, which resulted in improved static and dynamic mechanical behavior, electrical properties and thermal stability.

In this study, cyclohexylethylene-*co*-ethylene (CHEE) copolymer was used to develop PP blends and FLG reinforced PP blend nanocomposites using direct melt blending technique. The effect of varying amounts of CHEE and graphene on the mechanical, thermal and morphological characteristics of PP was evaluated using universal testing (UTM) analysis, Izod impact test, differential scanning calorimetry (DSC), thermogravimetric analysis (TGA), laser flash analysis (LFA), wide angle X-ray diffraction (WAXD), Fourier-transform infrared (FTIR) spectroscopy, Raman spectroscopy and scanning electron microscopy (SEM). The ultimate objective of the study was to explore the benefit of graphene to simultaneously enhance tensile modulus, tensile strength, impact strength, thermal stability and thermal conductivity in the presence of CHEE copolymer.

10.2 Experimental

10.2.1 Materials

PP matrix was a homopolymer grade with a melt flow rate of 8 g/10 min at 230 °C/2.16 kg and a density in the range of 900-910 kg.m^{-3} from Abu Dhabi Polymers (Borouge), UAE. Cyclohexylethylene-*co*-ethylene with cyclohexylethylene content of 60%, average M_w of ~90 kDa, polydispersity index (PDI) of 1.12 and glass transition temperature (T_g) of 29 °C was procured from Signa Aldrich. FLG grade N002-PDR with carbon content of 96.3%, average thickness of 1.0-1.2 nm, average length of ≤10 μm and surface area of 400-800 m^2.g^{-1} was purchased from Angstron Materials, USA.

10.2.2 Preparation of PP/CHEE Blends

PP/CHEE blends with weight compositions of 95/5, 90/10, 85/15 and 80/20 were melt blended in a mini conical co-rotating twin screw extruder (HAAKE MiniLab, Thermo Scientific). The melt blending was performed at 180 °C for 15 minutes with a screw speed of 100 rpm. For the preparation of dumbbell and rectangular bar shaped specimens for respective tensile and Izod impact analysis, mini lab-scale injection molding machine (HAAKE MiniJet PRO, Thermo Scientific) was used. Cylinder temperature of 180 °C, mold temperature of 125 °C, injection pressure of 430 bar for 10 s and post pressure of 500 bar for 6 s were used.

10.2.3 Preparation of PP/CHEE/Graphene Nanocomposites

PP blend with 10 wt% CHEE (PP/10CHEE) was used as the matrix material for the fabrication of FLG nanocomposites. To study the effect of FLG content, PP/10CHEE blend nanocomposites with 0.5, 1, 3 and 5 wt% FLG were fabricated using mini conical co-rotating twin screw extruder (HAAKE MiniLab, Thermo Scientific) by following the same compounding parameters as used in the case of PP/CHEE blends. The resulting nanocomposites were processed to form dumbbell, rectangular bar and disc shaped specimens using mini lab-scale injection molding machine (HAAKE MiniJet PRO, Thermo Scientific), using the same conditions as mentioned above.

10.2.4 Characterization of Blends and Nanocomposites

Resil impactor (Ceast) equipped with a hammer energy of 4 J and a speed of 3.64 $m.s^{-1}$ was used to analyze the un-notched Izod impact strength of blends and nanocomposites as per ISO 180. The samples were analyzed at room temperature using 80 mm x 10 mm x 4 mm rectangular bar shaped specimens, and the reported Izod impact strength was the mean value of five test specimens.

Tensile testing was carried out on Instron 3345 UTM equipped with a load cell of 50 kN to analyze the mechanical properties (such as tensile modulus, tensile strength and elongation at break) of blends and nanocomposites samples according to ISO 527. The tensile properties of the samples were measured at room temperature using 75 mm x 5 mm x 2 mm dumbbell shaped specimens with a crosshead moving rate of 10 $mm.min^{-1}$ and a span length of 35 mm. The mean value of five test specimens was used to represent the tensile test results.

Differential scanning calorimetry (Discovery series of TA instruments) was used to measure peak melting temperature (T_m), peak crystallization temperature (T_c), melting enthalpy (ΔH_f), crystallization enthalpy (ΔH_c) and percentage crystallinity (X_c) of PP/CHEE blends and PP/10CHEE/FLG nanocomposites. The analysis was conducted in dry nitrogen atmosphere with a flow rate of 50 mL.min^{-1} using about 3-8 mg of the samples weighed in aluminum pans. The samples were subjected to two sets of heating and cooling cycles at a rate of 10 °C.min^{-1}, and the thermograms were generated using the second set of heating and cooling cycles, which included heating from -50 to 200 °C and cooling from 200 to -50 °C.

Thermogravimetric analysis (Discovery series of TA instruments) was used to examine the thermal stability and degradation behavior of the blends and nanocomposites samples. TGA thermograms were recorded by heating about 3-8 mg of the samples under dry nitrogen atmosphere at a heating rate of 10 °C.min^{-1} from 25 to 700 °C. The char yield of the nanocomposite samples was also measured using the thermograms.

Laser flash apparatus Netzsch LFA 447 was used to analyze the thermal conductivity of neat PP, PP/10CHEE blend and graphene nanocomposites as per ASTM E1461. The thermal conductivity of the samples at 25, 50, 75 and 100 °C was measured using disc shaped test specimens (12 mm diameter x 2 mm thickness, with Vespel as a standard) with the aid of Cowan method using the equation:

Thermal conductivity, $k = D\rho_d C_p$

where D is the density, ρ_d is the thermal diffusivity and C_p is the specific heat.

Fourier-transform infrared spectrometer Bruker Tensor II was used to analyze the chemical structure of neat PP, PP/10CHEE blend and FLG nanocomposites. FTIR spectra of the samples were generated using thin sections of the injection molded specimens by collecting 32 scans with a resolution of 4 cm^{-1} in the wavenumber range from 4000 to 400 cm^{-1} (transmission mode) with the aid of a diamond attenuated total reflectance (ATR) crystal.

Raman spectrometer Jobin Yvon Horiba LabRAM with a confocal microscope was used to collect the Raman spectra of the samples using a He-Ne laser with an excitation wavelength of 633 nm as the light source. The spectra were generated in a Raman shift range of 500-3500 cm^{-1} using thin injection molded blends and nanocomposites samples at room temperature with the application of a confocal objective at a magnification of 50x.

X'Pert PRO Panalytical powder diffractometer using Cu-Kα radiation (1.5406 Å wavelength) was used to study the wide angle X-ray diffraction behavior of the injection molded blends and nanocomposites samples at room temperature. The X-ray diffractograms of the samples was recorded in the 2θ span of 5-60° with a step rate of 0.017° s^{-1}.

Scanning electron microscope FEI Quanta FEG250 was used to examine fracture micro-morphology of PP/10CHEE blend and its graphene nanocomposites. For this purpose, impact fractured surface of the test specimens was sputter coated with gold.

10.3 Results and Discussion

In this study, PP/CHEE blends and PP/10CHEE/FLG nanocomposites were prepared using direct melt mixing. PP/CHEE blends containing 5, 10, 15 and 20 wt% of copolymer were developed in order to study the influence of copolymer content on the blend performance and to obtain optimal blend composition for the generation of high performance FLG nanocomposites. On the basis of preliminary mechanical characterization, PP/10CHEE blend was used to generate nanocomposites containing 0.5, 1, 3 and 5 wt% of FLG by following the same mixing conditions as used for blends.

Un-notched Izod impact strength of neat PP, PP/CHEE blends and PP/10CHEE/FLG nanocomposites is summarized in Table 10.1. The impact strength of PP was enhanced by ca. 3.5-fold after blending with 5 wt% of CHEE, followed by a decreasing tendency with further increase in CHEE content. These observations indicated the influence of the amount and resultant size of the dispersed CHEE copolymeric phase in the PP/CHEE blends. The enhanced impact strength can be attributed to the crack growth termination mechanism, where CHEE copolymer phase in the blend system serves as an effective stress concentrator and absorbs the impact load, thus, resulting in crazing and shear yielding in the PP matrix, subsequently leading to microvoids. Accordingly, the resultant matrix prevents the growth of cracks by undergoing plastic deformation, thus, exhibiting effective impact energy dissipation [34-36]. Also, the effect of 0.5-5 wt% of FLG on the impact strength of PP/10CHEE blend is summarized in Table 10.1. The impact strength of PP/10CHEE blend increased from 12 to 16 kJ.m^{-2} after the incorporation of 0.5 wt% of FLG, indicating ca. 33% improvement. The PP/10CHEE/FLG nanocomposites containing 1, 3, and 5 wt% FLG exhibited the impact strength values of 12,

Table 10.1 Impact and tensile properties of PP blends and nanocomposites

Sample	Impact strength (KJ/m²)	Tensile strength (MPa)	Tensile modulus (MPa)	Extension at break (%)
PP	7	39	1011	24
PP/5CHEE	25	40	1222	24
PP/10CHEE	12	40	1348	24
PP/15CHEE	8	39	1337	32
PP/20CHEE	6	39	1312	54
PP/10CHEE/0.5FLG	16	42	1390	13
PP/10CHEE/1FLG	12	44	1437	12
PP/10CHEE/3FLG	9	47	1565	10
PP/10CHEE/5FLG	8	45	1481	9

9, and 8 kJ.m^{-2} respectively, as compared to 7 kJ.m^{-2} for neat PP (Figure 10.1). At low FLG loading, i.e., 0.5 wt%, FLG served as effective crack inhibitor by disturbing crack propagation in the PP/10CHEE blend system with convoluted pathways, which correspondingly

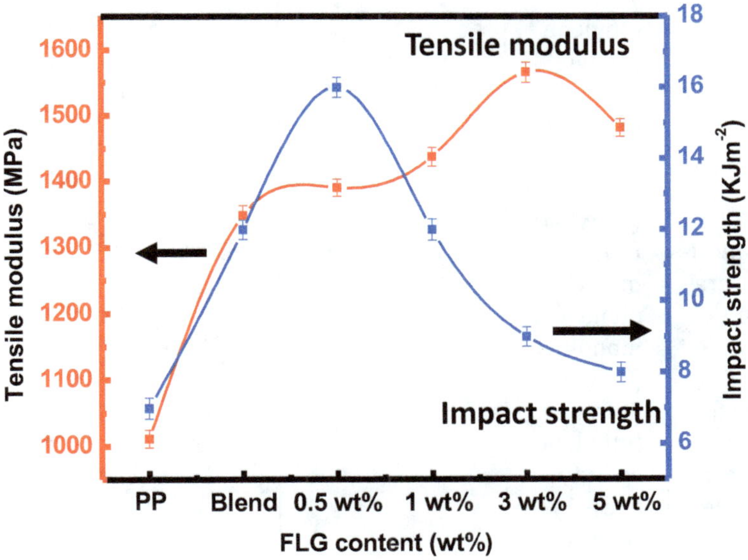

Figure 10.1 Tensile modulus and impact strength of PP, PP/10CHEE blend, and 0.5-5 wt% FLG filled PP/10CHEE/FLG nanocomposites.

enhanced the energy absorption as well as dissipation, thus, leading to enhanced impact performance [28,37,38]. Haghnegahdar *et al.*

[28] also reported an improvement of ca. 23.5% in impact strength for PP/20 wt% EPDM blend nanocomposite with 0.5 wt% FLG. In another study, Parameshwaranpillai *et al.* [31] observed enhanced impact strength values for PP/20 wt% SEBS blend nanocomposites with 0.05-1 wt% graphene nanoplatelets. Table 10.2 provides a comparison of the impact strength observed in this study with various compatibilized PP/graphene nanocomposite systems reported in literature studies.

Table 10.2 Comparison of the impact strength and tensile modulus observed in this study with the compatibilized PP/graphene nanocomposite systems reported in literature

PP nanocomposite system	Compatibilizer	inc. in impact str. (%)	opt. filler content (wt%)	inc. in tensile mod. (%)	opt. filler content (wt%)	Ref.
PP/few-layer graphene	Cyclohexylethylene-*co*-ethylene (10 wt%)	33.3	0.5	16.1	3	This work
PP/few-layer graphene	Ethylene propylene diene monomer (20 wt%)	23.5	0.5	17.2	0.5	[28]
PP/multi-layer graphene	Ethylene propylene diene monomer (20 wt%)	7.7	0.5	5.7	0.5	[28]
PP/exfoliated graphene nanoplatelets	Styrene-(ethylene-*co*-butylene)-styrene (20 wt%)	6	1	No effect	0.05-1	[31]
PP/functionalized graphene sheets	Maleic anhydride-*graft*-PP (4.5 wt%)	-	-	27	1.5	[33]

Table 10.1 also lists the tensile properties of neat PP, PP/CHEE blends and PP/CHEE/FLG nanocomposites. Incorporation of 10 wt% CHEE significantly enhanced the tensile modulus, while retaining the tensile strength of PP, followed by a decrease in the properties on further increase in the copolymer content. The observed performance is closely related to the capability of PP and CHEE in the blend system to undergo concurrent deformation during the application of tensile load, which is probably caused by the defect-free blend interface, resulting from the appreciable compatibility and interaction between the blend components [39,40]. Additionally, the elongation at break of neat PP was observed to enhance with CHEE copolymer amount. For instance, a value of 54% was observed for PP blend with 20 wt% CHEE. As expected, the addition of FLG enhanced the tensile modulus and strength of PP in the presence of 10 wt% CHEE. The properties were observed to increase with FLG content up to 3 wt%, followed by a decrease on further increasing the FLG content (Figure 10.1). Specifically, the tensile strength values for 0.5, 1, 3, and 5 wt% FLG containing PP/10CHEE nanocomposites were 42, 44, 47, and 45 MPa respectively in comparison with 39 MPa for neat PP (Figure 10.2). The enhanced tensile modulus and strength of nanocomposites indicated effective FLG dispersion in the PP matrix and strong interfacial interactions [28,41,42]. Table 10.2 also provides a comparison of the

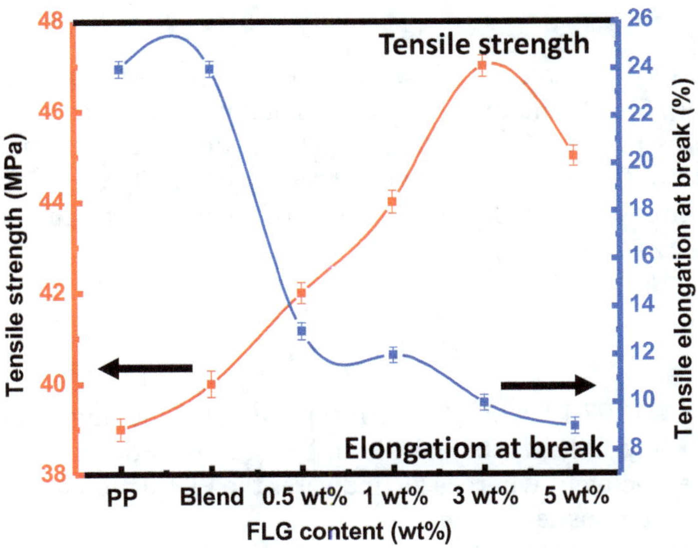

Figure 10.2 Tensile strength and elongation at break of PP, PP/10CHEE blend and 0.5-5 wt% FLG filled PP/10CHEE/FLG nanocomposites.

tensile modulus observed in this study with the compatibilized PP/graphene nanocomposite systems reported in literature. Furthermore, the mechanical performance of PP/10CHEE/FLG nanocomposites was in good agreement with the degree of crystallinity presented in Table 10.3. Ryu *et al.* [13] also reported similar trend for the mechanical performance of the PP/aminated GO nanocomposites.

Table 10.3 Calorimetric properties of neat PP, PP/CHEE blends and PP/10CHEE/FLG nanocomposites

Sample	T_c (°C)	ΔH_c (J/g)	T_m (°C)	ΔH_f (J/g)	X_c (%)	ΔX_c (%)
Neat PP	128	99	168	91	44	-
PP/5CHEE blend	126	96	167	86	44	0
PP/10CHEE blend	125	89	167	83	45	1
PP/15CHEE blend	125	85	167	79	45	1
PP/20CHEE blend	124	80	167	75	45	1
PP/10CHEE/0.5FLG nanocomposite	124	95	168	98	53	9
PP/10CHEE/1FLG nanocomposite	124	93	168	100	54	10
PP/10CHEE/3FLG nanocomposite	124	91	168	99	55	11
PP/10CHEE/5FLG nanocomposite	122	88	169	91	52	8

DSC second heating and cooling thermograms of neat PP, CHEE, PP/CHEE blends and PP/10CHEE/FLG nanocomposites are depicted in Figures 10.3 and 10.4. The calorimetric properties of the materials, such as T_m, T_c, ΔH_f, ΔH_c, X_c and percentage crystallinity variation (ΔX_c), are listed in Table 10.3. The following relationship was employed to calculate X_c of the samples [18,31]:

$$X_C = \frac{\Delta H}{\Delta H_0 w_{\mathrm{pp}}} \times 100$$

where ΔH_0 (207.1 J/g [31]) and ΔH represent the melting enthalpy of theoretically fully crystalline PP and PP samples developed in the study respectively, whereas w_{pp} indicates the PP fraction in the blend and nanocomposite samples.

The miscibility of CHEE copolymer with PP was supported by the presence of a single melting transition in the melting thermogram of PP/CHEE blends (Figure 10.3). From Table 10.3, it is clear that CHEE

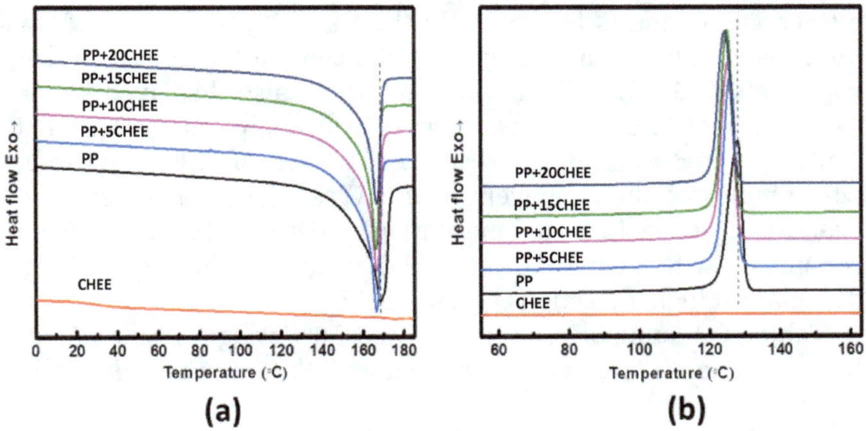

Figure 10.3 DSC (a) melting and (b) crystallization curves of neat PP, CHEE copolymer and 5-20 wt% CHEE containing PP/CHEE blends.

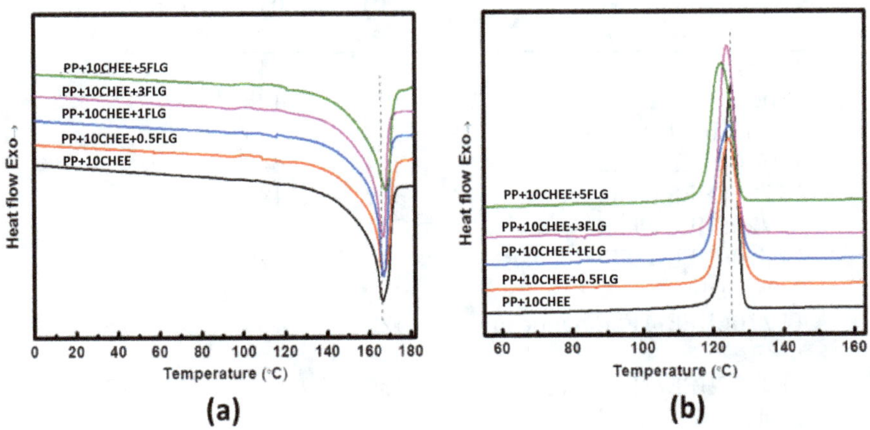

Figure 10.4 DSC (a) melting and (b) crystallization curves of PP/10CHEE blend and 0.5-5 wt% FLG containing PP/10CHEE/FLG nanocomposites.

phase had no significant influence on the T_m of PP (168 °C) [43]. Also, the T_c of the PP/CHEE blends exhibited a decrease of 2-4°C as compared to neat PP (128 °C). Both melting and crystallization enthalpies were observed to decrease with increase in CHEE content in the PP/CHEE blends. The X_c of neat PP and PP phase in PP/CHEE blends was almost consistent, thus, indicating that the presence of CHEE copolymer had no negative effect on the arrangement of PP crystallites [44]. The incorporation of FLG in PP/10CHEE blend had no notable effect on its T_m and T_c. Interestingly, the ΔH_f of PP/10CHEE blend was observed to increase markedly with FLG content till 3 wt%. The ΔH_c values of PP/10CHEE/FLG nanocomposites also displayed almost similar behavior. X_c of PP was observed to strongly enhance with FLG content up to 3 wt%, in the presence of 10 wt% of CHEE copolymer, which indicated the effectiveness of FLG as nucleating agent. However, the decrease in X_c and nucleating activity of FLG for the nanocomposites with 5 wt% FLG resulted due to the propensity of FLG to flocculate at higher fractions [11,32,45-47].

Table 10.4 summarizes the 10% and 50% weight loss temperatures ($T_{10\%}$ and $T_{50\%}$) and char yield at 700 °C for the blend and

Table 10.4 TGA data of neat PP, CHEE, PP/CHEE blends and PP/10CHEE/FLG nanocomposites ($T_{10\%}$: initial degradation temperature and $T_{50\%}$: maximum degradation temperature)

Sample	$T_{10\%}$ (°C)	$T_{50\%}$ (°C)	Char residue at 700 °C (%)
Neat PP	404	447	-
CHEE	403	438	-
PP/5CHEE blend	398	444	-
PP/10CHEE blend	397	443	-
PP/15CHEE blend	382	435	-
PP/20CHEE blend	377	434	-
PP/10CHEE/0.5FLG nanocomposite	423	450	1.03
PP/10CHEE/1FLG nanocomposite	428	455	2.15
PP/10CHEE/3FLG nanocomposite	435	462	5.12
PP/10CHEE/5FLG nanocomposite	440	466	6.58

nanocomposite materials. The TGA thermograms exhibited a single-step degradation behavior for all samples. In nitrogen environment, PP exhibited $T_{10\%}$ and $T_{50\%}$ values of 404 °C and 447°C respectively. The thermal stability of PP was observed to decrease after blending with CHEE copolymer. The incorporation of FLG enhanced the thermal stability of PP/10CHEE blend, as a function of FLG content. For example, the $T_{10\%}$ of PP/10CHEE blend was increased from 397 to 423, 428, 435, and 440 °C on the inclusion of 0.5, 1, 3 and 5 wt% of FLG respectively, owing to the FLG induced increase in the heat transfer efficiency caused by strong PP/10CHEE blend-FLG interactions [11,18,48,49]. The char yield of ca. 1.03, 2.15, 5.12 and 6.58% was observed for the nanocomposites with 0.5, 1, 3 and 5 wt% FLG. Achaby *et al.* [18] also reported single-step degradation behavior for the PP composites with graphene nanosheets. The authors confirmed the capability of graphene nanosheets to appreciably enhance the thermal stability of PP. An increment of ca. 40 °C in $T_{5\%}$ was observed for the composite with 3 wt% graphene nanosheets. In another study, Li *et al.* [48] highlighted the potential of graphene nanoplatelets to enhance the thermal stability of PP. Accordingly, the authors observed ca. 36 °C higher $T_{10\%}$ after the inclusion of 15 wt% graphene nanoplatelets in the PP matrix.

The thermal conductivity of the polymer nanocomposites depends on many factors, such as filler loading and dispersion state, polymer-filler interfacial interactions, etc. [50]. Figure 10.5 illustrates the thermal conduction behavior of neat PP, PP/10CHEE blend and PP/10CHEE/FLG nanocomposites. Neat PP displayed a thermal conductivity of ca. 0.203 W.(m.K)$^{-1}$, with PP/10CHEE blend also exhibiting a similar value. The addition of FLG enhanced the thermal conductivity as a function of FLG content. Table 10.5 also summarizes the thermal conductivity of the materials in the temperature range of 25-100 °C. On addition of 5 wt% FLG, the thermal conductivity of PP/10CHEE blend improved from 0.201 to 0.324 W.(m.K)$^{-1}$, owing to the increased phonon transfer diffusion attributed to the conducting network evolution in the polymer blend matrix [51,52]. These observations are in good agreement with other literature studies on PP/graphene nanocomposites. For example, Song *et al.* [50] reported improved thermal conductivity of ca. 0.396 W.(m.K)$^{-1}$ after the addition of 5 wt% PP grafted graphene in PP. In another study, Kalaitzidou *et al.* [53] observed a thermal conductivity of ca. 0.4 W.(m.K)$^{-1}$ for PP nanocomposite reinforced with 3 vol% of exfoliated graphite platelets.

Structural characterization of FLG, CHEE, PP, PP/10CHEE blend and PP/10CHEE/FLG nanocomposites was carried out using FTIR in transmission mode using thin sections of the injection molded specimens (Figure 10.6). The IR spectrum of FLG demonstrated two absorption bands at 1995 and 2114 cm^{-1} corresponding to C-H bending of aromatic groups and C≡C stretching of alkyne groups. The FTIR spectra of neat PP displayed absorption peaks corresponding to asymmetric stretching vibration of CH_3 groups at 2957 cm^{-1}, symmetric stretching vibration of CH_2 groups at 2916 cm^{-1}, symmetric stretching vibration of CH_3 groups at 2870 cm^{-1}, symmetric stretching vibration of CH_2 groups at 2837 cm^{-1}, C-H bending vibration at

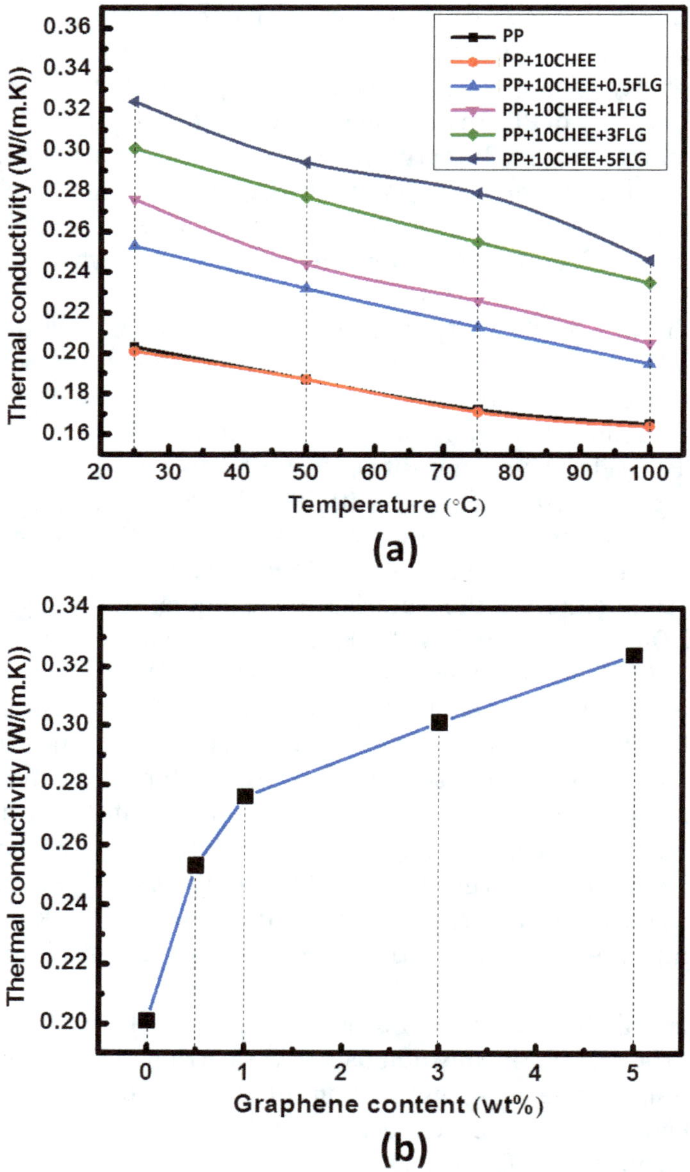

Figure 10.5 (a) Thermal conductivity variation of PP/10CE70/FLG nanocomposites with 0.5-5 wt% FLG in temperatures range from 25 to 100 °C and (b) thermal conductivity of PP/10CE70/FLG nanocomposites as a function of FLG content at 25 °C.

Table 10.5 Thermal conductivity of blends and nanocomposites

Sample	Thermal conductivity (W.(m.K)$^{-1}$)			
	at 25 °C	at 50 °C	at 75 °C	at 100 °C
Neat PP	0.203	0.187	0.172	0.165
PP/10CHEE blend	0.201	0.187	0.171	0.164
PP/10CHEE/0.5FLG	0.253	0.232	0.213	0.195
PP/10CHEE/1FLG	0.276	0.244	0.226	0.205
PP/10CHEE/3FLG	0.301	0.277	0.255	0.235
PP/10CHEE/5FLG	0.324	0.294	0.279	0.246

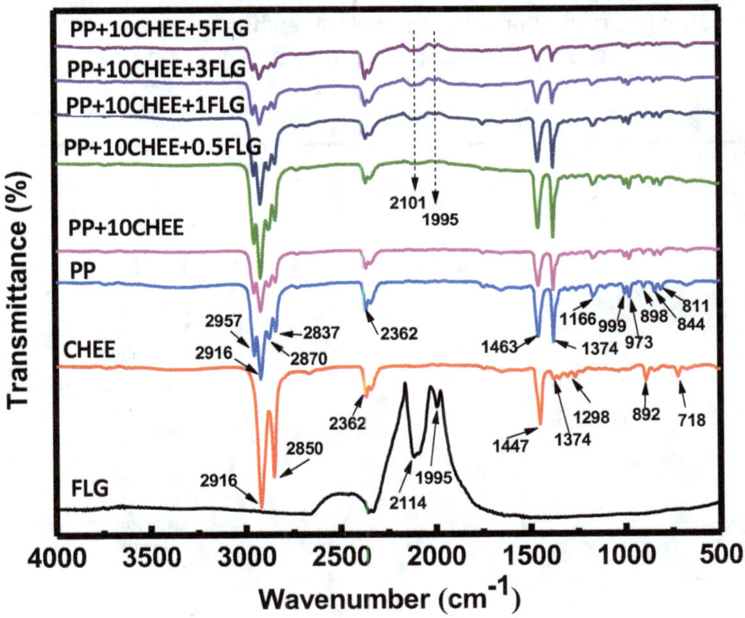

Figure 10.6 FTIR spectra of FLG, CHEE, neat PP, PP/10CHEE blend and PP/10CHEE/FLG nanocomposites.

2362 cm^{-1}, asymmetric bending vibration of CH_3 at 1463 cm^{-1}, symmetric bending vibration of CH_3 at 1374 cm^{-1}, tertiary methyl skeleton deformation at 999, 973, 844 and 811 cm^{-1} and perpendicular absorption in the stretch direction at 898 cm^{-1} [54]. In the case of CHEE, the characteristic IR peaks were observed for stretching vibration of CH_2 at 2916 and 2850 cm^{-1}, bending C-H vibration at 2362 cm^{-1}, bending vibration of CH_2 at 1447 cm^{-1}, CH_2 wagging at 1374 cm^{-1}, CH_2

twisting at 1298 cm^{-1} and CH_2 rocking at 892 and 718 cm^{-1} [4,54]. Figure 10.6 revealed the presence of two additional peaks at 2101 and 1995 cm^{-1} in PP/10CHEE/FLG nanocomposites due to FLG. In addition, the peak associated with the C≡C stretching vibration of alkyne at 2114 cm^{-1} on the FLG surface was shifted to 2101 cm^{-1} in the PP/CE-70/FLG nanocomposites, most probably due to its good interaction with the PP/10CHEE blend [55,56]. Zhou *et al.* [56] also reported a shifting of GO characteristic peak at 1703 cm^{-1} to 1690 cm^{-1} in the IR spectra of polypyrrole/GO nanocomposites. Table 10.6 also summarizes the FTIR assignments of FLG, neat PP, CHEE, PP/10CHEE blend and PP/CHEE/FLG nanocomposites.

Table 10.6 Absorption bands of samples in the IR region and their assignments (v: stretching; v_s: symmetric stretching; v_{as}: asymmetric stretching; δ: bending; δ_s: symmetric bending and δ_{as}: asymmetric bending)

Sample	Absorption band (cm^{-1})	Assignment
Neat PP	2957	v_{as} (CH_3)
	2916	v_s (CH_2)
	2870	v_s (CH_3)
	2837	v_s (CH_2)
	2362	δ (C-H)
	1463	δ_{as} (CH_3)
	1374	δ_s (CH_3)
	1166	$v_{C\text{-}C}$ of C-CH_3
	999	Tertiary methyl skeleton deformation
	973	Tertiary methyl skeleton deformation
	898	Perpendicular absorption in the stretch direction
	844	Tertiary methyl skeleton deformation
	811	Tertiary methyl skeleton deformation
Neat CHEE	2916	v (CH_2)
	2850	v (CH_2)
	2362	δ (C-H)
	1447	δ (CH_2)
	1374	CH_2 wagging
	1298	CH_2 twisting
	892	CH_2 rocking
	718	CH_2 rocking
PP/10CHEE blend	All absorption bands of neat PP	
FLG	2114	$v_{C\equiv C}$ of alkyne group
	1995	$\delta_{C\text{-}H}$ of aromatic group
PP/10CHEE/FLG nanocomposites	All absorption bands of neat PP, FLG band at 1995, and downshifted FLG band at 2101	

Figure 10.7 illustrates the Raman spectra of FLG, PP/10CHEE blend and PP/10CHEE/FLG composites. Raman spectrum of FLG

displayed characteristic D, G, 2D and D+G bands at ca. 1348, 1583, 2672 and 2946 cm^{-1} Raman shift regions respectively. In FLG, D band

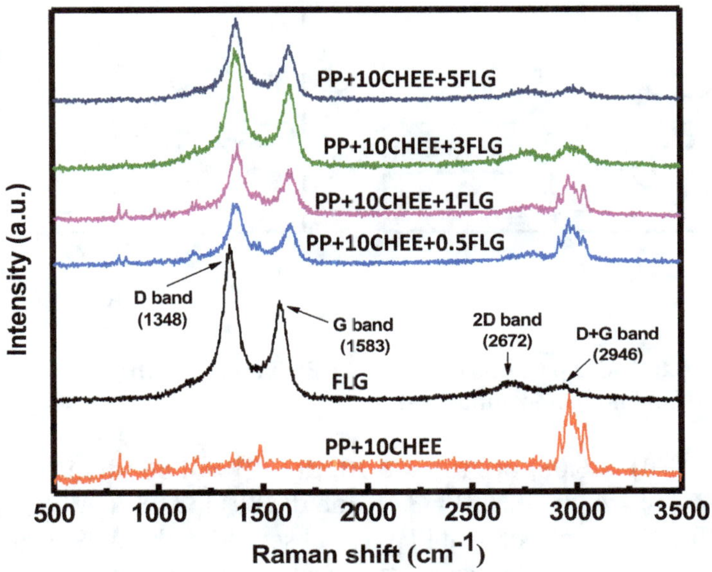

Figure 10.7 Raman spectra of PP/10CHEE blend, FLG and PP/10CHEE/FLG nanocomposites.

associated with the Raman mode inspired C-C ring structural disorders, while G band signified the stretching vibrations of C-C bond. 2D band, also known as second order D band, indicated graphene sheet stacking in the direction of the c-axis [46,57,58]. The Raman spectra of PP/10CHEE/FLG nanocomposites exhibited shifts in the characteristic D, G and 2D bands of FLG as well as merged D+G band of PP/CHEE blend. In addition, the intensity of the characteristic D, G and 2D bands of FLG increased with FLG content up to 3 wt%. However, the intensity of D+G band of PP/10CHEE blend in the 2878-3089 cm^{-1} Raman shift region was observed to decrease with increasing FLG content.

Wide angle X-ray diffraction patterns of PP/CHEE blends and PP/CHEE/FLG nanocomposites are depicted in Figure 10.8. For neat PP, characteristic diffraction peaks were observed at 14.55°, 17.26°, 19.03°, 21.47°, 22.18°, 25.76° and 29.01° 2θ corresponding to α crystalline planes (110), (040), (130), (111), combinatorial (131) and (041), (160) and (220) respectively [33,46,48,59]. The inter-planar distances of the diffraction peaks associated with the neat polymer

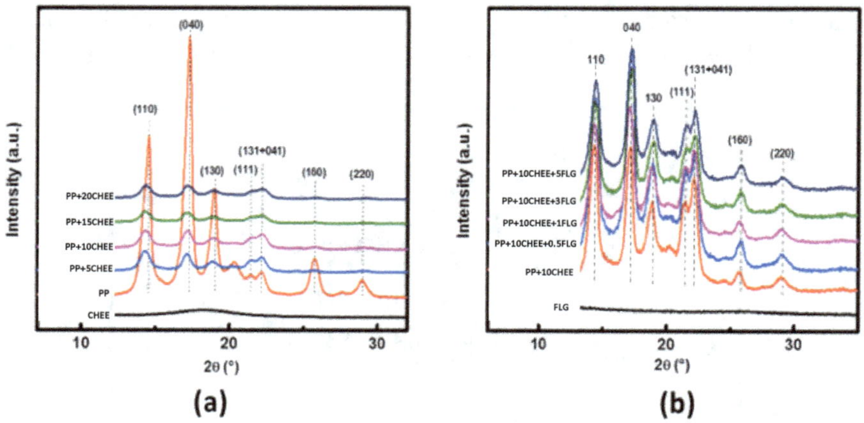

Figure 10.8 WAXD patterns of (a) PP/CHEE blends and (b) PP/10CHEE/FLG nanocomposites.

were d(110) = 6.08, d(040) = 5.13, d(130) = 4.66, d(111) = 4.14, d(131+041) = 4.00, d(160) = 3.46 and d(220) = 3.08 Å. All α crystalline diffraction peaks of neat PP were observed in the WAXD patterns of PP/CHEE blends, however, the peak position of the diffraction peaks corresponding to (110), (040) and (130) planes was shifted to 14.30°, 17.17° and 18.89° 2θ angles, which correspondingly enhanced the inter-planar distances of these peaks to 6.19, 5.16, and 4.69 Å respectively (Table 10.7). The observed changes indicated a

Table 10.7 WAXD data of samples (2θ: diffraction angle and *d*: inter-planar distance)

Sample	Crystalline planes					
	110		040		130	
	2θ (°)	*d* (Å)	2θ (°)	*d* (Å)	2θ (°)	*d* (Å)
Neat PP	14.55	6.08	17.26	5.13	19.03	4.66
PP/CHEE blends	14.30	6.19	17.17	5.16	18.89	4.69
PP/CHEE/FLG nanocomposites	14.30	6.19	17.17	5.16	18.89	4.69

good level of compatibility between the blend constituents. The WAXD pattern of FLG did not exhibit any diffraction peak indicating exfoliated morphology and the formation of isolated randomly distributed layered structure [33,60]. In the case of PP/10CHEE/FLG nanocomposites, the X-ray diffraction patterns did not depict any peak associated with FLG, thus, confirming the absence of filler re-stacking during the melt mixing process [28,61]. Also, the diffraction peak positions as well as inter-planar distances in the nanocomposites were consistent with PP/10CHEE blend.

SEM images of FLG, as presented in Figure 10.9, confirmed the layered surface morphology. In addition, the micro-morphology of the impact fractured surfaces of neat PP, PP/10CHEE blend and PP/10CHEE/FLG nanocomposites was studied with SEM, as shown in

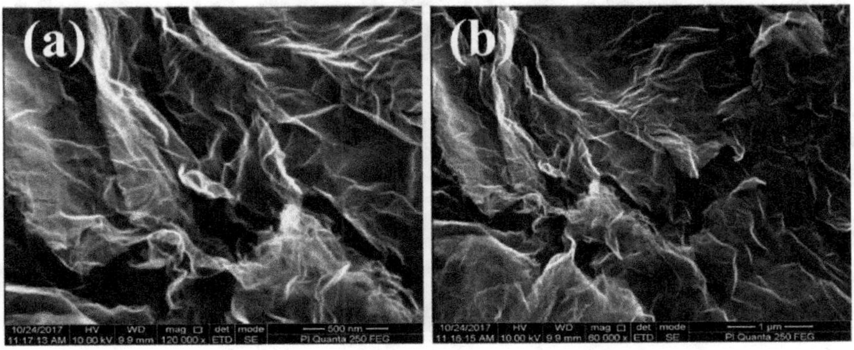

Figure 10.9 SEM images of FLG at (a) high and (b) low magnification.

in Figures 10.10 and 10.11. Higher extent of CHEE dispersed phase pull-outs or microvoids (highlighted circles in SEM images) were observed for PP/10CHEE blend (Figure 10.10b) and its nanocomposites with 0.5 and 1 wt% FLG (Figure 10.10c & 10.10d), in comparison with PP/10CHEE/3FLG (Figure 10.10e) and PP/10CHEE/5FLG (Figure 10.10f) nanocomposites. The observed pull-out of the CHEE dispersed phase from the continuous PP matrix indicated debonding or cavitation upon application of impact force. According to the crack termination mechanism, the CHEE copolymer phase can function as successful stress absorber and generates crazing as well as shear yielding in the continuous PP matrix, thereby, forming microvoids through the debonding or cavitation at the interface [28]. The morphological observations from the un-notched Izod impact fractured samples were, thus, in good agreement with the observed impact strength behavior. The fractograph of PP/10CHEE blend in Figure 10.10b revealed effective distribution of CHEE phase in the PP phase, accordingly the impact strength of PP was improved. It could be seen from Figure 10.10c that the incorporation of 0.5 wt% of FLG in PP/10CHEE was beneficial to improve the interface between the PP and CHEE copolymer (in terms of void filling) phases, resulting from the effective FLG dispersion. The strengthened interfaces would have

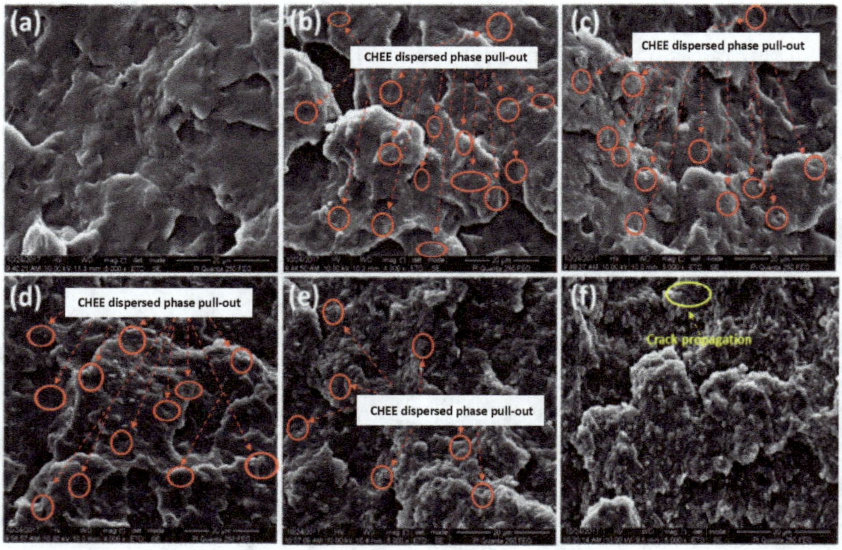

Figure 10.10 SEM micrographs showing the micro-morphology of the impact fractured surfaces of (a) neat PP, (b) PP/10CHEE blend and PP/10CHEE/FLG nanocomposites with FLG content of (c) 0.5, (d) 1, (e) 3 and (f) 5 wt%.

acted as effective crack inhibitors, thus, offering improved impact performance [31,62]. In the case of PP/10CHEE/1FLG nanocomposite (Figure 10.10d), the void filling capability of 1 wt% FLG was not as effective as observed for the 0.5 wt% nanocomposite, which resulted in reduced impact strength. With increase in the FLG content in the PP/10CHEE/FLG nanocomposites, the impact strength was observed to decrease due to the flocculation tendency of FLG [31]. The fractograph of PP/10CHEE/FLG nanocomposite containing 5 wt% FLG also revealed the crack propagation caused by the stress liberation through the FLG flocculates. Additionally, the surface roughness of the PP/10CHEE blend was observed to increase with FLG content due to the flocculation tendency of FLG (Figure 10.11).

10.4 Conclusions

In this study, the successful generation of PP/CHEE blends and PP/10CHEE/graphene nanocomposites using direct melt mixing has

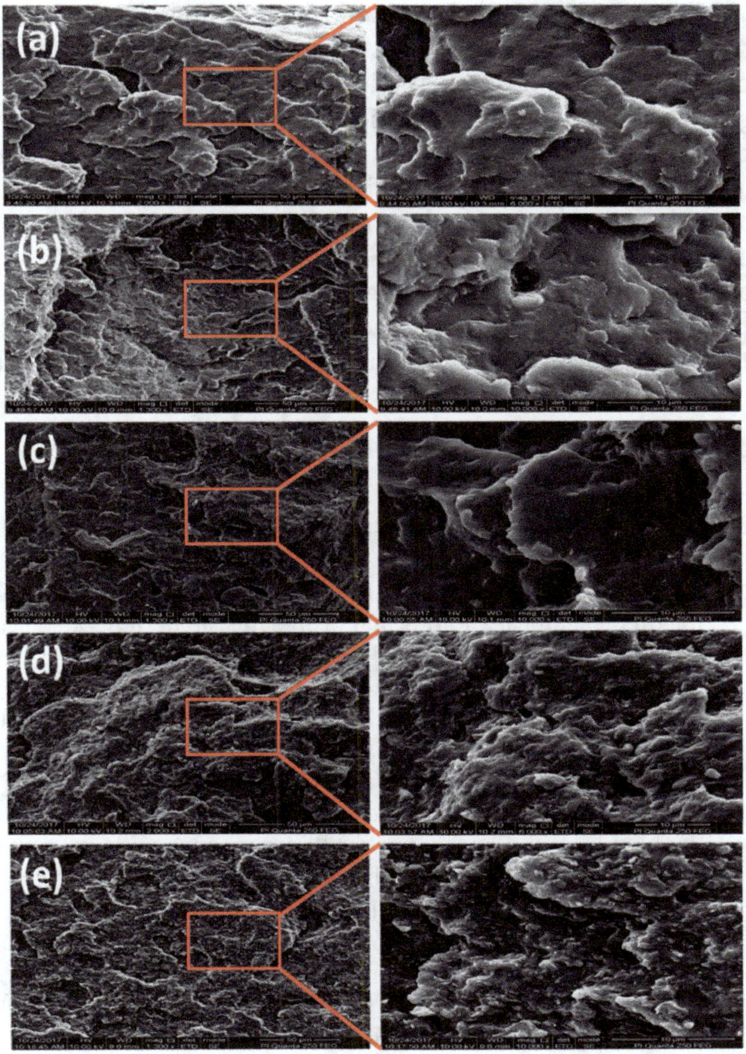

Figure 10.11 SEM images showing the micro-morphology of the impact fractured surfaces of (a) PP/10CHEE blend and PP/10CHEE/FLG nanocomposites with FLG content of (b) 0.5, (c) 1, (d) 3 and (e) 5 wt%.

been demonstrated. Among the various PP/CHEE blend compositions, the blend with 10 wt% CHEE exhibited optimal impact and tensile properties. The incorporation of a low amount of FLG (0.5 wt%) in the nanocomposite formulation was advantageous to enhance the impact performance by ca. 129% due to its capability to act as effective crack inhibitor. Tensile results revealed an optimal enhancement of ca. 55% and 20% in tensile modulus and strength respectively for the nanocomposite with 3 wt% FLG, which was further supported by the observations from DSC. With increasing FLG content in the nanocomposites, the heat transfer efficiency was observed to increase, which resulted in enhanced thermal stability. The resultant PP nanocomposites also exhibited increasing thermal conductivity with FLG content, and the extent of enhancement was ca. 62% in the case of PP/10CHEE nanocomposite with 5 wt% FLG. The observed findings support the potential use of the developed materials for a wide range of applications in the field of automotive materials, electronic and electrical devices, building and construction materials, among others.

References

1. *Functional Polymer Blends: Synthesis, Properties, and Performance*, Mittal, V. (ed.), CRC Press, USA (2012).
2. Ibrahim, B. A., and Kadum, K. M. (2010) Influence of polymer blending on mechanical and thermal properties. *Modern Applied Science*, **4**, 157-161.
3. Parameswaranpillai, J., Thomas, S., and Grohens, Y. (2014) Polymer blends: state of the art, new challenges, and opportunities. In: *Characterization of Polymer Blends*, Thomas, S., Grohens, Y., and Jyotishkumar, P. (eds.), Wiley-VCH, Germany, pp. 1-6.
4. Mittal, V., Luckachan, G. E., and Matsko, N. B. (2014) PE/chlorinated-PE blends and PE/chlorinated-PE/graphene oxide nanocomposites: Morphology, phase miscibility, and interfacial interactions. *Macromolecular Chemistry and Physics*, **215**, 255-268.
5. Hong, C. H., Lee, Y. B., Bae, J. W., Jho, J. Y., Nam, B. U., and Hwang, T. W. (2005) Preparation and mechanical properties of polypropylene/clay nanocomposites for automotive parts application. *Journal of Applied Polymer Science*, **98**, 427-433.
6. Ramos Filho, F. G., Mélo, T. J. A., Rabello, M. S., and Silva, S. M. (2005) Thermal stability of nanocomposites based on polypropylene and bentonite. *Polymer Degradation and Stability*, **89**, 383-392.
7. Fu, X., Yao, C., and Yang, G. (2015) Recent advances in graphene/polyamide 6 composites: a review. *RSC Advances*, **5**, 61688-61702.
8. Müller, K., Bugnicourt, E., Latorre, M., Jorda, M., Echegoyen Sanz, Y., Lagaron, J. M., Miesbauer, O., Bianchin, A., Hankin, S., and Bölz, U. (2017) Review on the processing and properties of polymer nanocomposites and nanocoatings and their applications in the packaging, automotive and solar energy fields. *Nanomaterials*, **7**, 74.
9. Varghese, A. M., Rangaraj, V. M., Mun, S. K., Macosko, C. W., and Mittal, V. (2018) Effect of graphene on polypropylene/maleic

anhydride-graft-ethylene-vinyl acetate (PP/EVA-g-MA) blend: Mechanical, thermal, morphological, and rheological properties. *Industrial & Engineering Chemistry Research*, **57**, 7834-7845.

10. Alexandre, M., and Dubois, P. (2000) Polymer-layered silicate nanocomposites: preparation, properties and uses of a new class of materials. *Materials Science and Engineering, R: Reports*, **28**, 1-63.
11. Mittal, V., and Chaudhry, A. U. (2016) Polyethylene-thermally reduced graphene nanocomposites: comparison of masterbatch and direct melt mixing approaches on mechanical, thermal, rheological, and morphological properties. *Colloid and Polymer Science*, **294**, 1659-1670.
12. Vallés, C., Abdelkader, A. M., Young, R. J., and Kinloch, I. A. (2014) Few layer graphene-polypropylene nanocomposites: the role of flake diameter. *Faraday Discussions*, **173**, 379-390.
13. Ryu, S. H., and Shanmugharaj, A. (2014) Influence of hexamethylene diamine functionalized graphene oxide on the melt crystallization and properties of polypropylene nanocomposites. *Materials Chemistry and Physics*, **146**, 478-486.
14. Mukhopadhyay, P., and Gupta, R. K. (2011) Trends and frontiers in graphene-based polymer nanocomposites. *Plastics Engineering*, **67**, 32-42.
15. Mittal, V., and Chaudhry, A. U. (2015) Polymer–graphene nanocomposites: effect of polymer matrix and filler amount on properties. *Macromolecular Materials and Engineering*, **300**, 510-521.
16. Park, S., and Ruoff, R. S. (2015) Synthesis and characterization of chemically modified graphenes. *Current Opinion in Colloid & Interface Science*, **20**, 322-328.
17. Kuilla, T., Bhadra, S., Yao, D., Kim, N. H., Bose, S., and Lee, J. H. (2010) Recent advances in graphene based polymer composites. *Progress in Polymer Science*, **35**, 1350-1375.
18. El Achaby, M., Arrakhiz, F. E., Vaudreuil, S., el Kacem Qaiss, A., Bousmina, M., and Fassi-Fehri, O. (2012) Mechanical, thermal, and rheological properties of graphene-based polypropylene nanocomposites prepared by melt mixing. *Polymer Composites*, **33**, 733-744.
19. Mittal, V., Chaudhry, A. U., and Luckachan, G. E. (2014) Biopolymer-thermally reduced graphene nanocomposites: Structural characterization and properties. *Materials Chemistry and Physics*, **147**, 319-332.
20. *Graphite, Graphene, and their Polymer Nanocomposites*, Mukhopadhyay, P., and Gupta, R. K. (eds.), CRC Press, USA (2012).
21. Mittal, V., and Chaudhry, A. U. (2015) Effect of amphiphilic compatibilizers on the filler dispersion and properties of polyethylene - thermally reduced graphene nanocomposites. *Journal of Applied Polymer Science*, **132**, doi: 10.1002/app.42484.
22. Chaudhry, A. U., and Mittal, V. (2015) Masterbatch approach to gen-

erate HDPE/CPE/graphene nanocomposites. In: *Synthesis Techniques for Polymer Nanocomposites*, Mittal, V. (ed.), Wiley-VCH, Germany, pp. 31-50.

23. Wang, Y., Wang, J., and Chen, S. (2014) Role of surfactant molecular weight on morphology and properties of functionalized graphite oxide filled polypropylene nanocomposites. *Express Polymer Letters*, **8**, 881-894.
24. Bikiaris, D., Vassiliou, A., Chrissafis, K., Paraskevopoulos, K., Jannakoudakis, A., and Docoslis, A. (2008) Effect of acid treated multi-walled carbon nanotubes on the mechanical, permeability, thermal properties and thermo-oxidative stability of isotactic polypropylene. *Polymer Degradation and Stability*, **93**, 952-967.
25. Dai Lam, T., Hoang, T. V., Quang, D. T., and Kim, J. S. (2009) Effect of nanosized and surface-modified precipitated calcium carbonate on properties of CaCO 3/polypropylene nanocomposites. *Materials Science and Engineering A*, **501**, 87-93.
26. Lonkar, S. P., Morlat-Therias, S., Caperaa, N., Leroux, F., Gardette, J.-L., and Singh, R. (2009) Preparation and nonisothermal crystallization behavior of polypropylene/layered double hydroxide nanocomposites. *Polymer*, **50**, 1505-1515.
27. Wang, Y., and Tsai, H. B. (2012) Thermal, dynamic-mechanical, and dielectric properties of surfactant intercalated graphite oxide filled maleated polypropylene nanocomposites. *Journal of Applied Polymer Science*, **123**, 3154-3163.
28. Haghnegahdar, M., Naderi, G., and Ghoreishy, M. (2017) Fracture toughness and deformation mechanism of un-vulcanized and dynamically vulcanized polypropylene/ethylene propylene diene monomer/graphene nanocomposites. *Composites Science And Technology*, **141**, 83-98.
29. Haghnegahdar, M., Naderi, G., and Ghoreishy, M. (2017) Electrical and thermal properties of a thermoplastic elastomer nanocomposite based on polypropylene/ethylene propylene diene monomer/graphene. *Soft Materials*, **15**, 82-94.
30. Haghnegahdar, M., Naderi, G., and Ghoreishy, M. (2017) Microstructure and mechanical properties of nanocomposite based on polypropylene/ethylene propylene diene monomer/graphene. *International Polymer Processing*, **32**, 72-83.
31. Parameswaranpillai, J., Joseph, G., Shinu, K., Sreejesh, P., Jose, S., Salim, N. V., and Hameed, N. (2015) The role of SEBS in tailoring the interface between the polymer matrix and exfoliated graphene nanoplatelets in hybrid composites. *Materials Chemistry and Physics*, **163**, 182-189.
32. Parameswaranpillai, J., Joseph, G., Shinu, K., Jose, S., Salim, N. V., and Hameed, N. (2015) Development of hybrid composites for automotive applications: effect of addition of SEBS on the morphology,

mechanical, viscoelastic, crystallization and thermal degradation properties of PP/PS–x GnP composites. *RSC Advances*, **5**, 25634-25641.

33. Li, C.-Q., Zha, J.-W., Long, H.-Q., Wang, S.-J., Zhang, D.-L., and Dang, Z.-M. (2017) Mechanical and dielectric properties of graphene incorporated polypropylene nanocomposites using polypropylene-graft-maleic anhydride as a compatibilizer. *Composites Science and Technology*, **153**, 111-118.
34. Liu, G., and Qiu, G. (2013) Study on the mechanical and morphological properties of toughened polypropylene blends for automobile bumpers. *Polymer Bulletin*, **70**, 849-857.
35. Panda, B. P., Mohanty, S., and Nayak, S. K. (2015) Mechanism of toughening in rubber toughened polyolefin - A review. *Polymer-Plastics Technology and Engineering*, **54**, 462-473.
36. Bu, H., Qiu, W., Tan, Z., Li, Q., and Rong, J. (2015) Study on toughening of poly (4-methyl-1-pentene) with various thermoplastic elastomers. *Journal of Thermoplastic Composite Materials*, **28**, 1334-1342.
37. Fu, S., Li, N., Wang, K., Zhang, Q., and Fu, Q. (2015) Reduction of graphene oxide with the presence of polypropylene micro-latex for facile preparation of polypropylene/graphene nanosheet composites. *Colloid and Polymer Science*, **293**, 1495-1503.
38. Shokrieh, M., and Joneidi, V. (2015) Characterization and simulation of impact behavior of graphene/polypropylene nanocomposites using a novel strain rate–dependent micromechanics model. *Journal of Composite Materials*, **49**, 2317-2328.
39. Zhu, L., Fan, H.-N., Yang, Z.-Q., and Xu, X.-H. (2010) Evaluation of phase morphology, rheological, and mechanical properties based on polypropylene toughened with poly (ethylene-co-octene) *Polymer-Plastics Technology and Engineering*, **49**, 208-217.
40. Contreras, V., Cafiero, M., Da Silva, S., Rosales, C., Perera, R., and Matos, M. (2006) Characterization and tensile properties of ternary blends with PA-6 nanocomposites. *Polymer Engineering & Science*, **46**, 1111-1120.
41. Chammingkwan, P., Matsushita, K., Taniike, T., and Terano, M. (2016) Enhancement in mechanical and electrical properties of polypropylene using graphene oxide grafted with end-functionalized polypropylene. *Materials*, **9**, 240.
42. Yuan, B., Bao, C., Song, L., Hong, N., Liew, K. M., and Hu, Y. (2014) Preparation of functionalized graphene oxide/polypropylene nanocomposite with significantly improved thermal stability and studies on the crystallization behavior and mechanical properties. *Chemical Engineering Journal*, **237**, 411-420.
43. Ferrer, G. G., Sánchez, M. S., Sánchez, E. V., Colomer, F. R., and Ribelles, J. L. G. (2000) Blends of styrene–butadiene–styrene triblock

copolymer and isotactic polypropylene: morphology and thermomechanical properties. *Polymer International*, **49**, 853-859.

44. Liao, C. Z., and Tjong, S. C. (2011) Effects of carbon nanofibers on the fracture, mechanical, and thermal properties of PP/SEBS-g-MA blends. *Polymer Engineering & Science*, **51**, 948-958.
45. Milani, M. A., González, D., Quijada, R., Basso, N. R., Cerrada, M. L., Azambuja, D.S., and Galland, G.B. (2013) Polypropylene/graphene nanosheet nanocomposites by in situ polymerization: synthesis, characterization and fundamental properties. *Composites Science and Technology*, **84**, 1-7.
46. Zhao, S., Chen, F., Zhao, C., Huang, Y., Dong, J.-Y., and Han, C. C. (2013) Interpenetrating network formation in isotactic polypropylene/graphene composites. *Polymer*, **54**, 3680-3690.
47. Triantou, M. I., Stathi, K. I., and Tarantili, P. A. (2014) Rheological and thermomechanical properties of graphene/ABS/PP nanocomposites. World Academy of Science, Engineering and Technology, *International Journal of Chemical, Molecular, Nuclear, Materials and Metallurgical Engineering*, **8**, 967-972.
48. Li, Y., Zhu, J., Wei, S., Ryu, J., Sun, L., and Guo, Z. (2011) Poly (propylene)/graphene nanoplatelet nanocomposites: melt rheological behavior and thermal, electrical, and electronic properties. *Macromolecular Chemistry and Physics*, **212**, 1951-1959.
49. Chatterjee, A., and Deopura, B. (2006) Thermal stability of polypropylene/carbon nanofiber composite. *Journal of Applied Polymer Science*, **100**, 3574-3578.
50. Song, P., Cao, Z., Cai, Y., Zhao, L., Fang, Z., and Fu, S. (2011) Fabrication of exfoliated graphene-based polypropylene nanocomposites with enhanced mechanical and thermal properties. *Polymer*, **52**, 4001-4010.
51. Inuwa, I., Hassan, A., and Shamsudin, S. (2014) Thermal properties, structure and morphology of graphene reinforced polyethylene terephthalate/polypropylene nano composites. *The Malaysian Journal of Analytical Sciences*, **18**, 466-477.
52. Huang, X., Zhi, C., and Jiang, P. (2012) Toward effective synergetic effects from graphene nanoplatelets and carbon nanotubes on thermal conductivity of ultrahigh volume fraction nanocarbon epoxy composites. *The Journal of Physical Chemistry C*, **116**, 23812-23820.
53. Kalaitzidou, K., Fukushima, H., and Drzal, L. T. (2007) Multifunctional polypropylene composites produced by incorporation of exfoliated graphite nanoplatelets. *Carbon*, **45**, 1446-1452.
54. Deshmane, C., Yuan, Q., Perkins, R., and Misra, R. (2007) On striking variation in impact toughness of polyethylene-clay and polypropylene–clay nanocomposite systems: the effect of clay-polymer interaction. *Materials Science and Engineering: A*, **458**, 150-157.
55. Wen, X., Wang, Y., Gong, J., Liu, J., Tian, N., Wang, Y., Jiang, Z., Qiu, J.,

and Tang, T. (2012) Thermal and flammability properties of polypropylene/carbon black nanocomposites. *Polymer Degradation and Stability*, **97**, 793-801.

56. Zhou, H., Han, G., Xiao, Y., Chang, Y., and Zhai, H.-J. (2014) Facile preparation of polypyrrole/graphene oxide nanocomposites with large areal capacitance using electrochemical codeposition for supercapacitors. *Journal of Power Sources*, **263**, 259-267.
57. Ryu, S. H., and Shanmugharaj, A. (2014) Influence of long-chain alkylamine-modified graphene oxide on the crystallization, mechanical and electrical properties of isotactic polypropylene nanocomposites. *Chemical Engineering Journal*, **244**, 552-560.
58. Pimenta, M., Dresselhaus, G., Dresselhaus, M. S., Cancado, L., Jorio, A., and Saito, R. (2007) Studying disorder in graphite-based systems by Raman spectroscopy. *Physical Chemistry Chemical Physics*, **9**, 1276-1290.
59. Gopakumar, T., and Page, D. (2004) Polypropylene/graphite nanocomposites by thermo-kinetic mixing. *Polymer Engineering & Science*, **44**, 1162-1169.
60. Kim, H., and Macosko, C. W. (2008) Morphology and properties of polyester/exfoliated graphite nanocomposites. *Macromolecules*, **41**, 3317-3327.
61. Liang, J., Huang, Y., Zhang, L., Wang, Y., Ma, Y., Guo, T., and Chen, Y. (2009) Molecular-level dispersion of graphene into poly (vinyl alcohol) and effective reinforcement of their nanocomposites. *Advanced Functional Materials*, **19**, 2297-2302.
62. He, S., Petkovich, N. D., Liu, K., Qian, Y., Macosko, C. W., and Stein, A. (2017) Unsaturated polyester resin toughening with very low loadings of GO derivatives. *Polymer*, **110**, 149-157.

11

Simultaneous Toughening and Reinforcement of Polypropylene

11.1 Introduction

Polypropylene finds a wide range of uses in consumer products, automotive parts, packaging materials, medical and industrial applications, etc., due to its superior physical and mechanical properties, easy processability and recyclability [1-4]. However, the engineering applications of PP are limited by its poor impact strength, especially at low temperatures [5-8]. Thus, further efforts are needed to overcome these limitations, thereby improving the standing of PP as a useful engineering plastic.

Blending of PP with different polymers or copolymers is one of the effective and economic techniques for improving its toughness [5,6,9-11]. Among these, elastomers or thermoplastic elastomers are the most widely accepted materials [3,10,12,13]. Till date, elastomers such as ethylene-propylene-diene terpolymer (EPDM) [14,15], ethylene-propylene rubber (EPR) [5,16,17], styrene butadiene rubber (SBR) [18], styrene-butadiene-styrene block copolymer (SBS) [19,20], styrene-ethylene-butylene-styrene triblock copolymer (SEBS) [21-23], ethylene-octene copolymer (POE) [7,24,25], ethylene-1-hexene copolymer (EHR) [26], etc., have been employed for improving the toughness and load bearing ability of PP. These studies suggested that the elastomers improved the toughness of PP via termination of crack growth mechanism. The addition of elastomers in the PP matrix initiates crazing and causes the shear yielding of the surrounding matrix, thereby, forming microvoids which subsequently terminate the crack growth [12,27]. However, these studies also reported reduced stiffness or elastic modulus of the polymer [2,27-29].

Rigid particles have also been reported to improve the toughness of thermoplastics [30-38]. These include calcium carbonate ($CaCO_3$) [3,360], calcium stearate [31], silica nanoparticles [32], montmorillonite (MMT) [33], nano-$CaCO_3$ [34], glass particles [35], nano-TiO_2

Anish Varghese, M. R. Vengatesan and Vikas Mittal
Khalifa University of Science and Technology, Abu Dhabi, UAE

[37], ultra-high molecular weight polyethylene particles [38], etc. However, in many cases, the rigid particles have been observed to improve the stiffness or modulus of PP at the expense of toughness [2,27,29]. The toughening mechanism of rigid particles is observed to be almost similar to elastomers [39]. Overall, as compared to elastomers, the rigid particles have the potential to generally improve both toughness and modulus of the material. However, the main drawback of the system includes the reduction in the processability due to the increase in the viscosity and difficulty in the dispersion of the rigid particulate filler [31].

To resolve this issue, several studies have reported a combined system of rigid particles and elastomers, named as ternary or hybrid composites [40]. For example, the different combinations in matrix polymers include: nano-silica and SEBS [41], montmorillonite and SEBS [10], organo-montmorillonite (org-MMT) and polyethylene octane copolymer (POE) [42], modified $CaCO_3$ and EPDM [43], glass fibers and SEBS [40], organo-clay and SEBS [44], talc and EPR and/or EBR [13], organophilic layered silicates (organo-clay) and POE [45], etc. Though most of these studies exhibited a good toughness-stiffness balance in PP using appropriate composition of the components, however, the requirement of more than one component presents a major drawback of the system.

The current study presents further efforts in the direction of achieving good toughness-stiffness balance of PP. In this respect, PP has been incorporated with different weight fractions of a variety of copolymers. The effect of the copolymers on the simultaneous toughening and reinforcement of PP, among other properties, has been subsequently investigated.

11.2 Experimental

11.2.1 Materials

Polypropylene homopolymer (HD915CF), with melt flow rate of 8 g/10 min (230 °C/2.16 kg) and density of 900-910 kg/m^3, was supplied by Abu Dhabi Polymers, UAE. The impact modifiers such as ethylene vinyl acetate-graft-maleic anhydride (EVA-MA, Fusabond C250, melt flow rate of 1.4 g/10 min (190 °C/2.16 kg) and density of 0.962 g/cm^3), ethylene butyl acrylate copolymer (EBA, Elvaloy AC 34035, butyl acrylate content of 35% by weight, melt flow rate of 40 g/10 min (190 °C/2.16 kg) and density of 0.93 g/cm^3) were supplied by

DuPont. Ethylene-α-olefin copolymer (EAO, Affinity EG 8200G, melt flow rate of 5.0 g/10 min (121 °C) and density of 0.870 g/cm^3) was purchased from Dow.

11.2.2 Preparation of Polypropylene/Copolymer Blends

Different weight ratios (5, 10, 15, 20 and 30%) of each copolymer were melt blended with PP using mini twin screw conical extruder (HAAKE MiniLab, Thermo scientific) at a temperature of 180 °C at 100 rpm for 15 min. The test specimens from the blended samples were prepared using mini injection molding machine (HAAKE MiniJet PRO, Thermo scientific) at 180 °C and 400 bar pressure for 10 s.

11.2.3 Characterization

Differential scanning calorimetric (DSC) analysis of PP/copolymer blends was carried out on a Discovery Series DSC from TA instruments under nitrogen atmosphere. The blended samples (3-7 mg weight) were heated to 200 °C using a heating rate of 10 °C/min. The samples were maintained at 200 °C for 1 min. Subsequently, the samples were cooled to -60 °C using a cooling rate of 10 °C/min, followed by isothermal conditions at -60 °C for 1 min. Afterwards, in the second cycle, the samples were heated to 200 °C and finally cooled to 25 °C using the same heating and cooling rates. Thermal degradation behavior of the blend samples (3-7 mg weight) was studied using Discovery Series thermogravimetric analyzer (TGA) from TA instruments under nitrogen atmosphere. The samples were analyzed from 25 °C to 700 °C using a heating rate of 10 °C/min.

Tensile properties such as tensile strength, modulus and extension at break of the blend samples were evaluated at ambient temperature using universal testing machine (Instron 3345), according to ASTM D638, using a crosshead speed of 10 mm/min. The dumbbell shaped test specimens were used for the tensile analysis. Un-notched impact strength of the blend samples was studied using Resil Impactor of Ceast with 4 J hammer energy and 3.46 m/s speed, according to ASTM D256. The rectangular bar shaped specimens were used to conduct the impact test at room temperature. An average of five test values was recorded in all cases.

The melt rheological characteristics of PP and its blends were carried out in shear mode using AR 2000ex Rheometer from TA Instruments, according to ASTM D4440. Parallel plate configuration was used with the plate diameter of 25 mm and geometric gap of 1.4 mm. Injection molded disc shaped samples of 25 mm diameter and 1.5 mm thickness were used. The samples were scanned in the dynamic oscillatory mode from 0.1 to 100 rad/s at 180 °C with a strain of 2%.

Analytical powder diffractometer (X'Pert PRO) (45 kV, 40 mA) with Cu-Kα radiation (wavelength 1.5406 A°) was utilized for X-ray diffraction analysis of the blend samples. The samples were measured at a rate of 0.0170°/s from 5° to 60° 2θ. The microstructure of the surface of the un-notched impact fractured PP/copolymer blends was analyzed using scanning electron microscope (SEM) (FEI Quanta, FEG250, USA) at accelerating voltages of 10-20 kV. The fractured surface of the samples was sputter-coated with 3 nm thick gold layer.

11.3 Results and Discussion

Figures 11.1 and 11.2 present the DSC thermograms of the copolymer

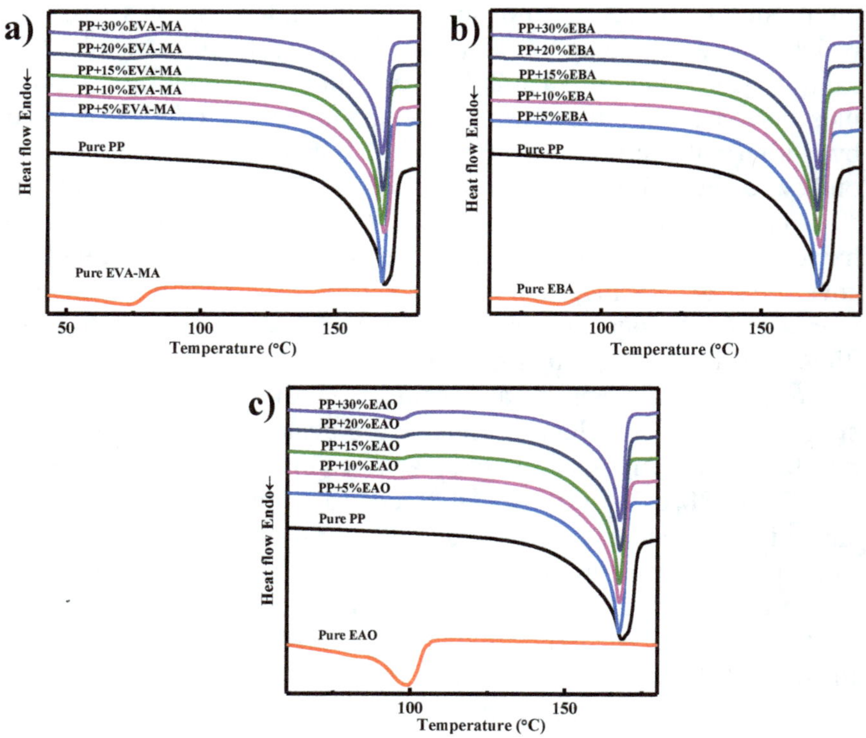

Figure 11.1 Melting endotherms of a) PP/EVA-MA blends, b) PP/EBA blends and c) PP/EAO blends.

blends with PP. The values of peak melting and crystallization temperature, melting and crystallization enthalpy and percentage crystallinity of PP and PP/copolymer blends are also shown in Table 11.1.

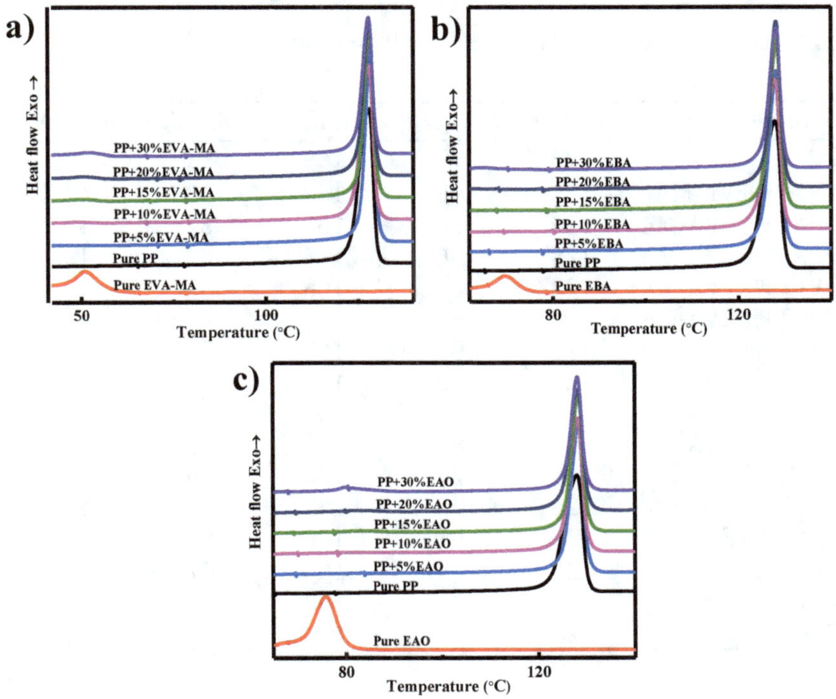

Figure 11.2 Crystallization exotherms of a) PP/EVA-MA blends, b) PP/EBA blends and c) PP/EAO blends.

From the melting endotherms of PP/EVA-MA, PP/EBA and PP/EAO blends in Figure 11.1 and Table 11.1, no significant change in the peak melting temperature was observed as compared to pure PP. This indicated that the crystal structure of PP remained unaffected on the addition of copolymers owing to good phase compatibility. The observation was also in good agreement with the reported studies [44,46]. The peak crystallization temperatures of the blend samples also exhibited a similar value as pure PP (128 °C) (Figure 11.2 and Table 11.1). This indicated that the soft segments in the copolymers did not negatively impact the arrangement of PP chains [5,47]. The blends exhibited a single endothermic peak up to a particular composition. For example, PP/EVA-MA blends had a single endotherm up to a PP:EVA-MA composition of 80:20. Increasing the EVA-MA content beyond 20 wt% resulted in a minor peak near the peak melting temperature of EVA-MA. This indicated a good miscibility between the PP and EVA-MA phases up to 20 wt% copolymer content. The same effect was also observed in other blends.

Table 11.1 DSC data of PP and PP/copolymer blends

Sample	T_c (°C)	ΔH_c (J/g)	T_m (°C)	ΔH_f (J/g)	X_c (%)
PP	128	99	169	91	44
PP+5% EVA-MA	128	95	168	92	47
PP+10% EVA-MA	128	90	168	87	47
PP+15% EVA-MA	128	86	168	85	48
PP+20% EVA-MA	128	81	168	81	49
PP+30% EVA-MA	128	70	167	72	50
PP+5% EBA	128	96	168	92	47
PP+10% EBA	128	90	168	83	45
PP+15% EBA	128	88	168	85	48
PP+20% EBA	128	81	168	79	44
PP+30% EBA	128	71	168	69	47
PP+5% EAO	128	96	168	97	49
PP+10% EAO	128	91	168	88	47
PP+15% EAO	128	88	168	87	49
PP+20% EAO	128	83	168	79	47
PP+30% EAO	128	70	168	65	45

The percentage crystallinity (X_c) of the polymer in the blends was calculated by using the equation [28,47]:

$$X_c = \frac{\Delta H}{\Delta H_0 w_{pp}} \times 100$$

where ΔH_0 is the enthalpy of fusion for 100% crystalline isotactic PP (207.1 J/g [28]), ΔH is the enthalpy of fusion measured from DSC and w_{pp} is the fraction of PP in the blend. PP/copolymer blends exhibited a small increment in PP crystallinity as compared to pure PP. Specifically, PP/EVA-MA blends had a uniform increment in crystallinity with copolymer content, even though nucleation effect did not seem to exist. The observed results also agreed well with the reported SEBS-MA/PP blends [4], where 17% increase in crystallinity was observed for the blends with 15 and 30 wt% SEBS-MA. The observed change in the extent of crystallinity represents an important opportunity to affect the mechanical performance of the blends.

The thermal stability and degradation behavior of pure PP, copolymers and PP/copolymer blends was investigated by TGA. Figure 11.3 presents TGA thermograms of materials. In addition, the temperatures at 5% and 50% weight loss ($T_{5\%}$ and $T_{50\%}$) for the materials

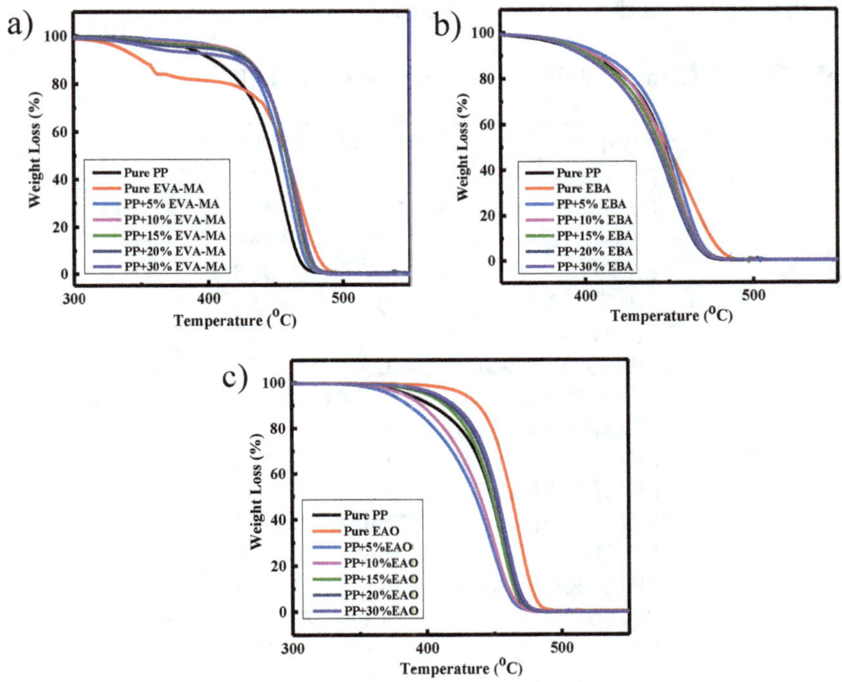

Figure 11.3 TGA thermograms of a) PP/EVA-MA blends, b) PP/EBA blends and c) PP/EAO blends.

are also listed in Table 11.2. $T_{5\%}$ for PP/EVA-MA blends increased with EVA-MA content up to 10 wt%, followed by a decrease beyond 10 wt% EVA-MA. The blend with 30% EVA-MA exhibited even lower thermal stability ($T_{5\%}$) as compared to pure PP (Table 11.2). In addition, at higher EVA-MA content, the blend exhibited two-step degradation, whereas the blends with lower content of EVA-MA (5 and 10 wt%) depicted a single step degradation. Vinyl acetate moiety in the EVA-MA structure can undergo swift chain scission at elevated temperatures, therefore, together with phase miscibility, a higher content of EVA-MA in the PP blends is expected to result in the observed behavior. PP/EBA blends exhibited a one-step degradation behavior, though $T_{5\%}$ and $T_{50\%}$ values were observed to decrease with increase in the EBA content. For PP/EAO blends, $T_{5\%}$ and $T_{50\%}$ values increased with increase in the EAO content. PP/EAO blends also exhibited single step degradation behavior. Overall, improved thermal stability of PP/copolymer blends, especially at lower copolymer contents, can be attributed to the phase miscibility and probable formation of β crystal phase of PP in the presence of copolymers.

Table 11.2 Thermal stability of PP, copolymers and PP/copolymer blends

Sample	$T_{5\%}$ (°C)	$T_{50\%}$ (°C)
PP	386	448
EVA-MA	335	458
EBA	389	449
EAO	427	463
PP+5% EVA-MA	409	455
PP+10% EVA-MA	413	458
PP+15% EVA-MA	407	458
PP+20% EVA-MA	396	459
PP+30% EVA-MA	359	459
PP+5% EBA	395	450
PP+10% EBA	390	447
PP+15% EBA	389	445
PP+20% EBA	386	443
PP+30% EBA	387	443
PP+5% EAO	373	436
PP+10% EAO	382	439
PP+15% EAO	400	448
PP+20% EAO	403	451
PP+30% EAO	406	453

The tensile properties of pure PP and PP blends are presented in Table 11.3. The tensile strength of PP/copolymer blends was observed to decrease with increase in the copolymer content. For instance, the PP blend with 30 wt% EVA-MA content had a 33% reduction in strength as compared to pure PP. Similarly, 30 wt% of both EBA and EAO decreased the tensile strength of PP by 41% and 31% respectively. This was in good agreement with the reported literature studies [19,28]. The tensile modulus of the PP blends was affected depending on the nature and structure of the copolymers. The tensile modulus values gradually increased with increase in EVA-MA content up to 20 wt%, followed by a significant drop for the blend with 30 wt% EVA-MA. In the case of PP/EBA and PP/EAO blends, the tensile modulus showed higher values up to 20 wt% of copolymer content as compared to PP, though a gradual decrease in the modulus was observed with the copolymer content. The higher tensile modulus of PP/copolymers blends, especially at lower copolymer contents, might be due to the presence of optimal phase compatibility and strong interfacial adhesion between the PP matrix and copolymers, which led to the simultaneous deformation in the PP/copolymer blends, thereby, preventing the growth of voids and defects at the interface during stretching. The enhancement in crystallinity would have also contributed to the observed improvement in modulus. In literature, the tensile modulus of the polymer/copolymer blends is generally reported to deteriorate due to the presence of the copolymer phase [28,42,44,45], however, a contrary trend was observed for the systems analyzed in the current study.

Table 11.3 Mechanical properties of PP/copolymer blends

Sample	Impact strength (kJ/m^2)	Tensile strength (MPa)	Tensile modulus (MPa)	Elongation at break (%)
PP	7	39	1011	24
PP+5% EVA-MA	20	34	1042	31
PP+10% EVA-MA	26	33	1095	38
PP+15% EVA-MA	40	31	1114	42
PP+20% EVA-MA	42	30	1143	43
PP+30% EVA-MA	52	26	984	52
PP+5% EBA	48	35	1126	30
PP+10% EBA	54	32	1115	44
PP+15% EBA	65	31	1105	136
PP+20% EBA	74	26	1036	299
PP+30% EBA	103	23	873	187
PP+5% EAO	9	36	1284	27
PP+10% EAO	19	35	1244	28
PP+15% EAO	83	33	1196	179
PP+20% EAO	96	30	1084	263
PP+30% EAO	103	27	999	326

Elongation at break is an important parameter providing insights about the toughness of the materials. UTM results indicated that the copolymers significantly improved the elongation at break of PP. PP/EVA-MA blends showed a steady increase in the elongation at break with copolymer content, i.e. 7% increment for 5 wt% EVA-MA to 28% increment for 30 wt% EVA-MA in the PP matrix. In the case of PP/EBA blends, a significant increment in the elongation at break up to 20 wt% EBA content was observed, followed by a small decrease for 30 wt% EBA. For PP/EAO blends, a continuous increment up to 302% was noticed. The observed results also agreed well with other literature reports [19,28,42]. The significant increase of elongation at break of PP/copolymer blends indicated corresponding increase in the toughness and impact strength of PP. Thus, the observed tensile performance confirmed the simultaneous toughening and reinforcement of the PP matrix.

Impact strength is one of the important parameters to express the toughness of a material. It is the ability of a material to absorb energy and undergo plastic deformation without fracture. Table 11.3 presents the Izod un-notched impact strength of PP and blends. The results indicated that the incorporation of copolymers enhanced the impact strength of PP as a function of the copolymer content. For example, the lower content of (5 wt%) of EVA-MA enhanced the impact strength of PP form 7 kJ/m^2 to 20 kJ/m^2. The addition of higher EVA-MA content significantly enhanced the impact strength of PP to 52 kJ/m^2, which was 7 times higher than pure PP. In PP/EBA blends, the magnitude of enhancement was even higher as compared to EVA-MA. For instance, the Izod impact strength of PP/EBA blends ranged from 48 kJ/m^2 for 5 wt% EBA to 103 kJ/m^2 for 30 wt% EBA. PP/EAO blends also exhibited similar response. The improved impact strength of the blends can be attributed to the good phase compatibility, as also observed in other studies [12,24,28,42,47-49]. In the present study, the fracture and toughening behavior of PP/copolymer blends might have followed the crazing and de-bonding mechanism. The copolymer in the PP matrix absorbs most of the sudden force and initiates crazing and shear yielding of the surrounding matrix, thereby forming microvoids. This leads to the generation of substantial plastic deformation of the matrix, which subsequently terminates the crack growth. In addition, the copolymers may de-bond or cavitate from the PP blend on impact, which releases the stress at the distinct area of the crack tip, thus, absorbing more energy and producing plastic deformation. A variation in the impact strength of PP/copolymer blend also indicated the effect of the copolymer structure, polarity, molecular weight and composition [28].

The SEM fractographs of the un-notched impact fractured surfaces of PP/EVA-MA, PP/EBA and PP/EAO blends (15 and 30 wt% copolymer content) are shown in Figure 11.4. The morphology of the fractured surface of each copolymer blend was different, and the blends exhibited a ductile fracture on impact. Copolymers were observed to be uniformly distributed in the PP matrix. In the case of 15 wt% copolymer blends, EAO system exhibited an optimum impact strength of 83 kJ/m^2 due to the presence of very small domains, which provided a large number of craze initiation sites, thereby, dissipating a greater amount of impact energy. In the case of blends with 30 wt% copolymer, both EBA and EAO systems exhibited high impact strength of 103 kJ/m^2 due to more craze initiation sites as compared to EVA-MA. For the EVA-MA blend, the copolymer domains were roughly spherical in shape and resulted in relatively lower impact strength of 52 kJ/m^2. From the SEM images, the domain size of the copolymer phase was observed to increase on enhancing its content in the blends.

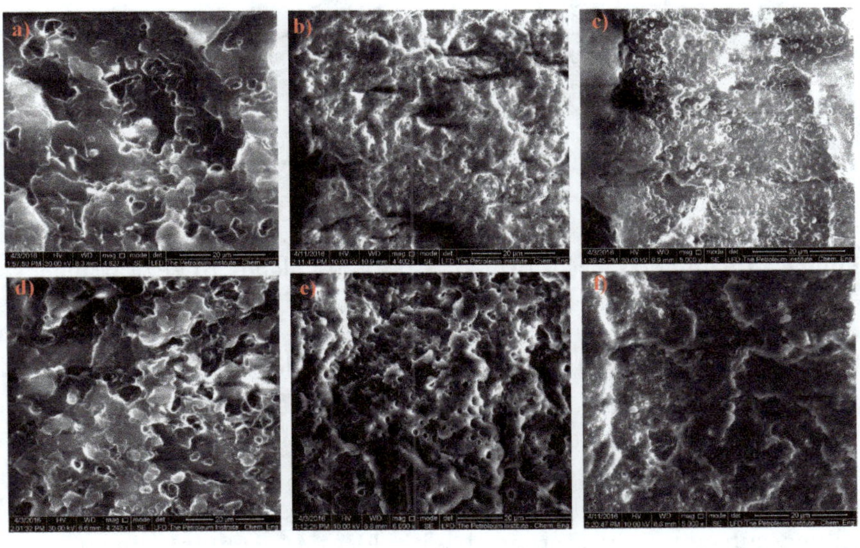

Figure 11.4 Images of blends with a) 15 wt% EVA-MA, b) 15wt% EBA, c) 15 wt% EAO, d) 30 wt% EVA-MA, e) 30 wt% EBA and f) 30 wt% EAO.

Melt rheological properties of PP/copolymer blends were studied by shear rheometer at 180 °C. Figure 11.5-11.7 present the frequency sweep curves of complex viscosity, storage modulus (G') and loss modulus (G") of the blends. The complex viscosity of PP and PP blends decreased with increase in ω, indicating a shear thinning behavior due to the pseudo-plastic nature [7]. This suggested the presence of finely dispersed copolymer phase in the PP matrix. In the blends, the magnitude of η* increased with an increase in content for

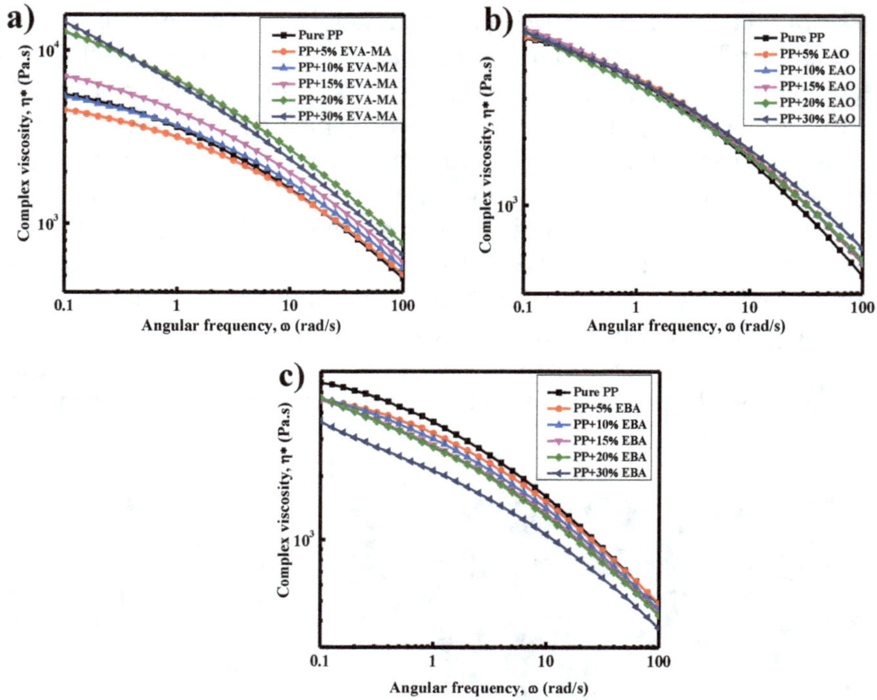

Figure 11.5 Complex viscosity of a) PP/EVA-MA blends, b) PP/EAO blends and c) PP/EBA blends.

both EVA-MA and EAO, whereas the magnitude of η* decreased with an increase in the EBA content. At lower ω, the blends with 5 and 10 wt% EVA-MA had lower η* as compared to pure PP, due to the low surface friction between the PP matrix and dispersed copolymer phase, which resulted in interlayer slipperiness in the blend. The η* values subsequently increased with EVA-MA content, due to the higher interaction between the PP and EVA-MA phases. Among the PP/copolymer blends, the blend with 30 wt% EVA-MA showed an optimum magnitude of η* (14545 Pa.s) at lower ω.

As observed from Figures 11.6 and 11.7, G' and G'' of PP and PP blends increased with ω, however, the rate of increase was higher for G' as compared to G''. Further, the G' of PP blends increased with increase in the copolymer content. Also the G' curves almost merged with each other at higher angular frequency values. The increased G' of the blends could be related to the relaxation of the dispersed phase in the PP blends. Overall, among the PP/copolymer blends, the blend with 30 wt% of EVA-MA exhibited an optimum G'.

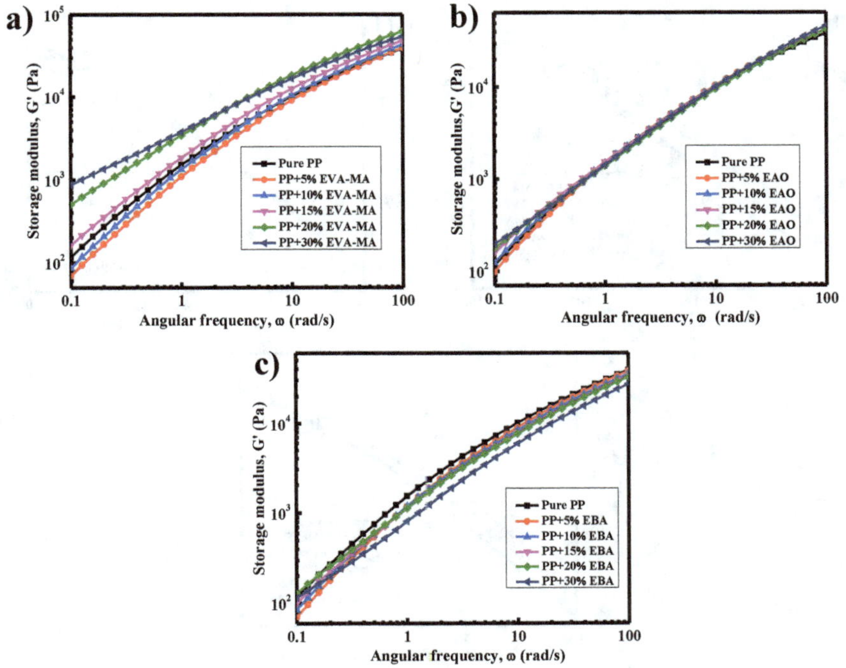

Figure 11.6 Storage modulus of a) PP/EVA-MA blends, b) PP/EAO blends and c) PP/EBA blends.

The G" plots of the PP/copolymer blends also exhibited a similar behavior as G'. Also, in general, G" values were observed to be higher that G' values for the materials, thus, indicating the dominance of the viscous character.

Wide angle XRD spectra of pure PP, pure copolymers and PP/copolymer blends are presented in Figure 11.8. Pure PP exhibited seven characteristic diffraction peaks of α crystal form at 2θ values of 14.36°, 17.19°, 18.91°, 21.50°, 22.19°, 25.82° and 29.03° corresponding to 110, 040, 130, 111, 041, 160 and 020 lattice planes. The blends of PP with EVA-MA, EBA and EAO also exhibited the same diffraction peaks for pure PP. In addition, the blends exhibited an additional peak at 16.53° corresponding to 300 lattice plane due to the formation of β form of crystal structure. This indicated that the copolymers acted as β nucleating agent (β- NA) for the formation of β crystals in PP. It also agreed with the increased extent of crystallinity observed from DSC, though no change in the peak crystallization temperature was observed. The formation of β crystals enhanced the toughness and thermal performance of the PP/copolymer blends,

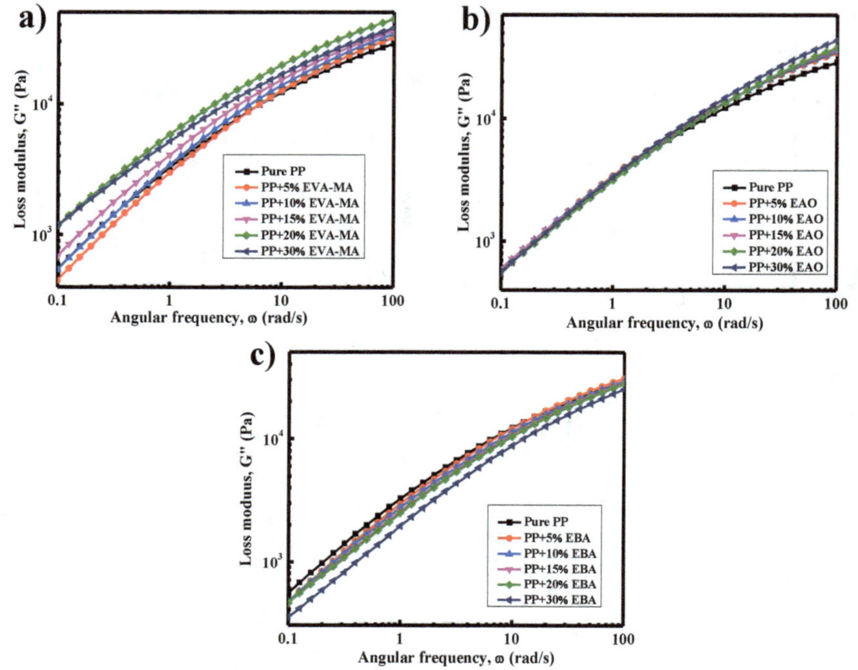

Figure 11.7 Loss modulus of a) PP/EVA-MA blends, b) PP/EAO blends and c) PP/EBA blends.

which also supported the results from mechanical and thermal analysis [50]. In PP/copolymer blends, α and β crystal formation corresponded to the monoclinic and trigonal crystal lattices respectively.

11.4 Conclusions

In the present work, polypropylene (PP) was melt blended with various weight fractions (5, 10, 15, 20 and 30 wt%) of three different copolymers such as maleic anhydride grafted ethylene vinyl acetate (EVA-MA), ethylene butyl acrylate (EBA) and ethylene alpha olefin (EAO). A significant improvement in the tensile modulus, extension at break and impact strength was observed for the PP/copolymer blends. Overall, a well-balanced toughness-stiffness behavior was observed in the PP/copolymer blends with the copolymer content of 20 wt%. Specifically, the blend with 20 wt% EAO had the most optimal toughness-stiffness properties. DSC results confirmed that the copolymers possessed a good compatibility with the polymer up to particular content and a higher content resulted in a low extent of phase

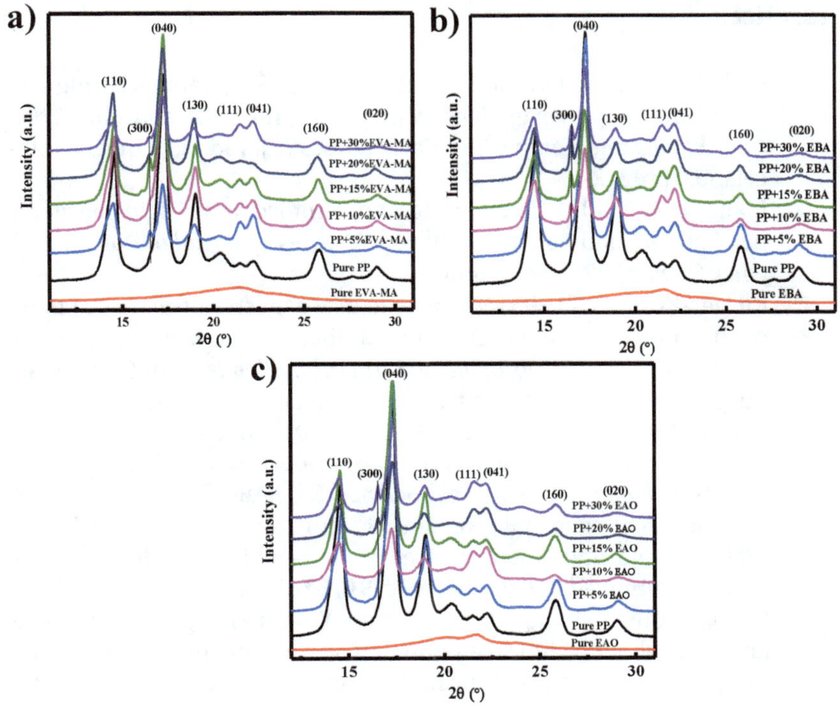

Figure 11.8 X-ray diffractograms of a) PP/EVA-MA blends, b) PP/EBA blends and c) PP/EAO blends.

separation. PP/EVA-MA blends exhibited enhanced thermal stability as compared to other PP/copolymer blends. XRD analysis confirmed the formation of β crystals in the blends, which resulted in improved mechanical and thermal properties. Microscopy analysis indicated a uniform dispersion of the copolymers in the polymer matrix. Rheological analysis also confirmed the specific effect of the copolymers based on their nature and amount. This work provides a facile, efficient and economical approach for the simultaneous improvement of the toughness and stiffness of PP, among other properties. The PP/copolymer blends developed in the study exhibit high potential for many engineering applications.

Acknowledgement

The authors sincerely thank the Petroleum Institute Research Centre, The Petroleum Institute (now part of Khalifa University of Science and Technology), UAE, for the financial support.

References

1. Galli, P., Haylock, J. C., and Simonazzi T. (1995) Manufacturing and properties of polypropylene copolymers. In: *Polypropylene Structure, Blends and Composites*, Karger-Kocsis, J. (ed.), Springer, Netherlands, pp. 1-24.
2. Liang, J. Z. (2002) Toughening and reinforcing in rigid inorganic particulate filled poly (propylene): A review. *Journal of Applied Polymer Science*, **83**, 1547-1555.
3. Karger-Kocsis, J. (1995) Microstructural aspects of fracture in polypropylene and in its filled, chopped fiber and fiber mat reinforced composites. In: *Polypropylene Structure, Blends and Composites*, Karger-Kocsis, J. (ed.), Springer, Netherlands, pp. 142-201.
4. Liao, C. Z., and Tjong, S. C. (2011) Effects of carbon nanofibers on the fracture, mechanical, and thermal properties of PP/SEBS-g-MA blends. *Polymer Engineering & Science*, **51**, 948-958.
5. Zebarjad, S., Bagheri, R., Reihani, S., and Lazzeri, A. (2003) Deformation, yield and fracture of elastomer-modified polypropylene. *Journal of Applied Polymer Science*, **90**, 3767-3779.
6. Chen, F., Qiu, B., Wang, B., Shangguan, Y., and Zheng, Q. (2015) Balanced toughening and strengthening of ethylene–propylene rubber toughened isotactic polypropylene using a poly (styrene-b-ethylene-propylene) diblock copolymer. *RSC Advances*, **5**, 20831-20837.
7. Zhu, L., Fan, H.-N., Yang, Z.-Q., and Xu, X.-H. (2010) Evaluation of phase morphology, rheological, and mechanical properties based on polypropylene toughened with poly (ethylene-co-octene). *Polymer-Plastics Technology and Engineering*, **49**, 208-217.
8. Zhang, M., Liu, Y., Zhang, X., Gao, J., Huang, F., Song, Z., Wei, G., and Qiao, J. (2002) The effect of elastomeric nano-particles on the mechanical properties and crystallization behavior of polypropylene. *Polymer*, **43**, 5133-5138.
9. Dikobe, D., and Luyt, A. (2009) Morphology and properties of polypropylene/ethylene vinyl acetate copolymer/wood powder blend composites. *Express Polymer Letters*, **3**, 190-199.
10. Bao, S., and Tjong, S. (2007) Impact essential work of fracture of polypropylene/montmorillonite nanocomposites toughened with SEBS-g-MA elastomer. *Composites, Part A: Applied Science and Manufacturing*, **38**, 378-387.
11. Wang, Z., Hao, W., Zhang, Z., and Cui, Z. (2008) Toughening of propylene-ethylene copolymer/calcium carbonate composites with high-density polyethylene. *Journal of Macromolecular Science, Part B: Physics*, **47**, 576-588.
12. Su, J.-j., Geng, C.-z., Wang, K., Yang, G.-h., and Fu, Q. (2014) Comparison of the toughening behavior for poly (ethylene terephthalate)

with spherulitic or ellipsoid elastomer-particles. *Journal of Polymer Research*, **21**, 1-12.

13. Yokoyama, Y., and Ricco, T. (1998) Toughening of polypropylene by different elastomeric systems. *Polymer*, **39**, 3675-3681.
14. Choudhary, V., Varma, H., and Varma, I. (1991) Polyolefin blends: effect of EPDM rubber on crystallization, morphology and mechanical properties of polypropylene/EPDM blends. *Polymer*, **32**, 2534-2540.
15. Da Costa, H. M., Ramos, V. D., and Rocha, M. C. (2006) Analysis of thermal properties and impact strength of PP/SRT, PP/EPDM and PP/SRT/EPDM mixtures in single screw extruder. *Polymer Testing*, **25**, 498-503.
16. Zebarjad, S., Lazzeri, A., Bagheri, R., Reihani, S. S., and Frounchi, M. (2003) Fracture mechanism under dynamic loading of elastomer-modified polypropylene. *Materials Letters*, **57**, 2733-2741.
17. Kim, M. H., Alamo, R. G., and Lin, J. (1999) The cocrystallization behavior of binary blends of isotactic polypropylene and propylene-ethylene random copolymers. *Polymer Engineering & Science*, **39**, 2117-2131.
18. Cook, R. F., Koester, K. J., Macosko, C. W., and Ajbani, M. (2005) Rheological and mechanical behavior of blends of styrene-butadiene rubber with polypropylene. *Polymer Engineering & Science*, **45**, 1487-1497.
19. Das, V., Gautam, S. S., and Pandey, A. K. (2011) Effect of SBS content on low temperature impact strength, morphology and rheology of PP-cp/SBS blends. *Polymer-Plastics Technology and Engineering*, **50**, 825-832.
20. Hu, J., Zhang, Y., and Zhou, X. (2011) Study on modified recycling PP by thermoplastic elastomer SBS. *China Plastics Industry*, **39**, 37-40.
21. Mae, H., Omiya, M., and Kishimoto, K. (2008) Material ductility and toughening mechanism of polypropylene blended with bimodal distributed particle size of styrene-ethylene-butadiene-styrene triblock copolymer at high strain rate. *Journal of Applied Polymer Science*, **110**, 3941-3953.
22. Bassani, A., and Pessan, L. (2002) Toughening of polypropylene with styrene/ethylene-butylene/styrene tri-block copolymer: Effects of reactive and nonreactive compatibilization. *Journal of Applied Polymer Science*, **86**, 3466-3479.
23. Balkan, O., Demirer, H., and Kayalı, E. S. (2011) Effects of deformation rates on mechanical properties of PP/SEBS blends. *Journal of Achievements in Materials and Manufacturing Engineering*, **47**, 26-33.
24. Liu, G., and Qiu, G. (2013) Study on the mechanical and morphological properties of toughened polypropylene blends for automobile bumpers. *Polymer Bulletin*, **70**, 849-857.

25. Yang, J., Zhang, Y., and Zhang, Y. (2003) Brittle–ductile transition of PP/POE blends in both impact and high speed tensile tests. *Polymer*, **44**, 5047-5052.
26. Yamaguchi, M., Nitta, K. H., Miyata, H., and Masuda, T. (1997) Rheological properties for binary blends of i-PP and ethylene-1-hexene copolymer. *Journal of Applied Polymer Science*, **63**, 467-474.
27. Liang, J., and Li, R. (2000) Rubber toughening in polypropylene: A review. *Journal of Applied Polymer Science*, **77**, 409-417.
28. Keskin, R., and Adanur, S. (2011) Improving toughness of polypropylene with thermoplastic elastomers in injection molding. *Polymer-Plastics Technology and Engineering*, **50**, 20-28.
29. Dasari, A., Zhang, Q.-X., Yu, Z.-Z., and Mai, Y.-W. (2010) Toughening polypropylene and its nanocomposites with submicrometer voids. *Macromolecules*, **43**, 5734-5739.
30. Zuiderduin, W., Westzaan, C., Huetink, J., and Gaymans, R. (2003) Toughening of polypropylene with calcium carbonate particles. *Polymer*, **44**, 261-275.
31. Yordem, O. S., Simanke, A. G., and Lesser, A. J. (2011) Reinforcing and toughening of polypropylene with self-assembled low molar mass additives. *Polymer Engineering & Science*, **51**, 550-558.
32. Wu, C. L., Zhang, M. Q., Rong, M. Z., and Friedrich, K. (2005) Silica nanoparticles filled polypropylene: effects of particle surface treatment, matrix ductility and particle species on mechanical performance of the composites. *Composites Science and Technology*, **65**, 635-645.
33. Zehetmeyer, G., Soares, R. M., Brandelli, A., Mauler, R. S., and Oliveira, R. V. (2012) Evaluation of polypropylene/montmorillonite nanocomposites as food packaging material. *Polymer Bulletin*, **68**, 2199-2217.
34. Lazzeri, A., Zebarjad, S. M., Pracella, M., Cavalier, K., and Rosa, R. (2005) Filler toughening of plastics. Part 1-the effect of surface interactions on physico-mechanical properties and rheological behaviour of ultrafine $CaCO_3$/HDPE nanocomposites. *Polymer*, **46**, 827-844.
35. Thio, Y., Argon, A., and Cohen, R. (2004) Role of interfacial adhesion strength on toughening polypropylene with rigid particles. *Polymer*, **45**, 3139-3147.
36. Zuiderduin, W., Huetink, J., and Gaymans, R. (2006) Rigid particle toughening of aliphatic polyketone. *Polymer*, **47**, 5880-5887.
37. Zhang, H., Zhang, Z., Yang, J.-L., and Friedrich, K. (2006) Temperature dependence of crack initiation fracture toughness of various nanoparticles filled polyamide 66. *Polymer*, **47**, 679-689.
38. Ranade, R. A., Wunder, S. L., and Baran, G. R. (2006) Toughening of dimethacrylate resins by addition of ultrahigh molecular weight polyethylene (UHMWPE) particles. *Polymer*, **47**, 4318-4327.

39. Thio, Y. S. (2003) *Toughening Mechanisms in Polypropylene Filled with Rigid Particles*, PhD Thesis, Massachusetts Institute of Technology, USA.
40. Tjong, S. C., Xu, S.-A., Li, R. K.-Y., and Mai, Y.-W. (2002) Mechanical behavior and fracture toughness evaluation of maleic anhydride compatibilized short glass fiber/SEBS/polypropylene hybrid composites. *Composites Science and Technology*, **62**, 831-840.
41. Panaitescu, D. M., Vuluga, Z., Radovici, C., and Nicolae, C. (2012) Morphological investigation of PP/nanosilica composites containing SEBS. *Polymer Testing*, **31**, 355-365.
42. Lim, J., Hassan, A., Rahmat, A., and Wahit, M. (2006) Rubber-toughened polypropylene nanocomposite: Effect of polyethylene octene copolymer on mechanical properties and phase morphology. *Journal of Applied Polymer Science*, **99**, 3441-3450.
43. Gu, J., Jia, D., and Cheng, R. (2008) Polypropylene composite toughened by a novel modified nano-$CaCO_3$. *Polymer-Plastics Technology and Engineering*, **47**, 583-589.
44. Vuluga, Z., Panaitescu, D. M., Radovici, C., Nicolae, C., and Iorga, M. D. (2012) Effect of SEBS on morphology, thermal, and mechanical properties of PP/organoclay nanocomposites. *Polymer Bulletin*, **69**, 1073-1091.
45. Lim, J. W., Hassan, A., Rahmat, A. R., and Wahit, M. U. (2006) Morphology, thermal and mechanical behavior of polypropylene nanocomposites toughened with poly (ethylene-co-octene). *Polymer International*, **55**, 204-215.
46. Ferrer, G. G., Sánchez, M. S., Sánchez, E. V., Colomer, F. R., and Ribelles, J. L. G. (2000) Blends of styrene–butadiene–styrene triblock copolymer and isotactic polypropylene: morphology and thermomechanical properties. *Polymer International*, 49, 853-859.
47. Bu, H., Qiu, W., Tan, Z., Li, Q., and Rong, J. (2015) Study on toughening of poly (4-methyl-1-pentene) with various thermoplastic elastomers. *Journal of Thermoplastic Composite Materials*, **28**, 1334-1342.
48. Jingyun, Z., Jianqiang, D., and Heying, Z. (2007) Study on properties of pp modified with POE. *China Synthetic Resins and Plastics*, **24**, 14-22.
49. Panda, B. P., Mohanty, S., and Nayak, S. K. (2015) Mechanism of toughening in rubber toughened polyolefin - A review. *Polymer-Plastics Technology and Engineering*, **54**, 462-473.
50. Bao, R.-Y., Cao, J., Liu, Z.-Y., Yang, W., Xie, B.-H., and Yang, M.-B. (2014) Towards balanced strength and toughness improvement of isotactic polypropylene nanocomposites by surface functionalized graphene oxide. *Journal of Materials Chemistry A*, **2**, 3190-3199.

12

Polypropylene/GO Nanocomposites with Amine Functionalized Polypropylene Compatibilizer

12.1 Introduction

Polypropylene (PP), a thermoplastic semi-crystalline material, is one of the most commonly used high volume commodity polymers because of its interesting price/performance ratio, low density, excellent processing, chemical resistance, corrosion resistance, recyclability, hardness and transparency. The commercial applications of PP include packaging, household goods, automobile parts and textiles, among others [1-3]. On the other hand, the application of PP as an engineering plastic is not satisfactory due to the poor toughness and impact resistance, specifically under extreme conditions like high strain rates or low temperature [4]. The crystallinity of PP makes it strong, but lowers the toughness and also enhances its chances for environmental stress cracking [5-8]. Another disadvantage of PP is its poor interfacial bonding with polar filler materials, owing to its non-polar nature.

The addition of soft rubbery phase elastomers like poly(ethylene-co-1-octene), ethylene-propylene random copolymer (EPR) and ethylene-propylene-diene terpolymer (EPDM) to PP, as impact modifiers, helps to enhance the toughness [9,10], however, does not specially help to achieve interfacial interactions with fillers. Decreasing the polarity of the fillers by surface treatment improves filler dispersion to some extent, but the elasticity and toughness are negatively impacted. On the other hand, increasing the polarity of the PP matrix, by using grafted PP or amphiphilic compatibilizers, along with an appropriate filler, may be a viable alternative to resolve the issues associated with PP [11-13]. Amphiphilic compatibilizers, having polar and non-polar groups, result in much improved dispersion of the filler in the polymer matrix, thereby also enhancing the polymer toughness.

Graphene is a two-dimensional nanomaterial with unique properties [14]. It has one atom thick planar sheets consisting of sp^2-bonded

Haleema Saleem, M. R. Vengatesan and Vikas Mittal
Khalifa University of Science and Technology, Abu Dhabi, UAE

carbon atoms which are densely packed in a honeycomb crystal lattice structure. The properties of graphene include good fracture strength (~130 GPa), excellent Young's modulus (~1 TPa), high surface area (~2600 m^2/g), superior electrical (~6000 S/cm) and thermal conductivity (~5000 W/mK), etc., which make graphene suitable for various applications [15]. With respect to polymer composites, the physical properties of polymers can be improved by the incorporation of graphene sheets at very low loading, and the surface of graphene can be tuned for achieving the compatibility with the polymer matrices. Thus, various studies have been carried out on the preparation as well as analysis of the structure-property relationships of graphene/polymer nanocomposites in the past several years [16,17]. In other studies, Ahmad *et al.* [18] examined the effect of graphene loading on the PP nanocomposites and reported enhanced elastic modulus. Song *et al.* [19] reported an enhancement of 75% in yield strength and 74% in stiffness by adding 0.42 vol% of graphene. Also, a remarkable improvement in the thermal oxidative stability of PP was achieved with graphene addition. However, graphene generally suffers from poor dispersibility in the polymer matrices [20]. As a precursor of graphene, graphene oxide (GO) is readily obtainable by the chemical oxidation of graphite [21-24]. It is more preferred than graphene in reinforcing polar polymers due to the presence of oxygenated functional groups on the surface. In the case of non-polar polymers like PP, the use of GO together with an amphiphilic compatibilizer can result in better dispersion as well as *in-situ* reduction due to the chemical interaction with the compatibilizer phase, leading to improved properties, especially thermal and electrical conductivity [25].

Shin *et al.* [26] generated GO/PP nanocomposites with octyl-triethoxysilane (OTES) as compatibilizer. The presence of OTES enhanced the interaction between GO and PP, thus, improving the strength. The elastic modulus increased by 127% and the fracture toughness was enhanced by 117% by the addition of 1.0 wt% GO filler. Ryu and Shanmugharaj [27] studied the effect of hexamethylene diamine (HMDA) functionalized GO on the properties of PP nanocomposites. The composites were prepared by melt blending using PP-graft-maleic anhydride (PP-g-MA) compatibilizer. Due to the strong interfacial bonding between the modified filler and polymer, improvement in the mechanical (tensile strength and Young's modulus enhanced by 14% and 35% respectively), thermal and electrical properties was observed, as compared to pristine GO filled PP composites. Bao *et al.* [28] generated isotactic PP (iPP) nanocomposites reinforced with surface functionalized GO. An increment of about 100% in toughness and 30% in the tensile strength were observed, with a loading of 0.1 wt% GO-N (β-nucleating agent on the surface of octadecylamine functionalized GO). In another study, Fuad *et al.* [29] reported the effect of PP-g-MA compatibilizer on the PP/nano-precipitated calcium carbonate (NPCC) nanocomposites. The introduction of PP-g-MA improved the impact strength, tensile strength and modulus of the PP nanocomposites. Page and Gopakumar [30] also observed that the compatibilization effect helped the graphite dispersion on a nanoscale, which improved the impact strength of the PP composites. Hemmati *et al.* [31] analyzed the properties of PP reinforced with single-walled nanotubes (SWNTs), compatibilized using PP-g-MA. Tensile modulus, Izod impact strength as well as tensile strength of the materials were observed to improve significantly due to the combined effect of the filler and compatibilizer phases.

Several studies have confirmed that the amine-functionalized PP exhibited better adhesion effects with the PP matrix when compared to the maleated PP compatibilizer. However, Cui *et al.* [32] observed no advantage of amine functionalized PP over PP-g-MA. In another study, Kobayashi *et al.* [33] examined the use of amino-functionalized polyethylene for enhancing the bonding between polyolefin and polyurethanes, in comparison with PE-g-MA. Amine functionalized polymer resulted in higher adhesion strength than PE-g-MA.

In the current study, an amine functionalized PP compatibilizer (PP-g-DA6) has been explored with GO nanofiller to improve the morphological, mechanical, thermal and rheological properties of PP. The amphiphilic compatibilizer, consisting of both polar and non-polar groups, can be expected to act as a bridge between the polar filler and non-polar host polymer. During the nanocomposite generation, the amine groups in the compatibilizer can react with the epoxy and carboxyl groups of GO, thus, forming covalent bonds. As a result, the chemical reduction of GO leads to the formation of reduced GO (r-GO). Figure 12.1 demonstrates the schematic of the composite generation.

12.2 Experimental

12.2.1 Materials

The isotactic PP, with a melt flow rate of 8 g/10 min, was provided by

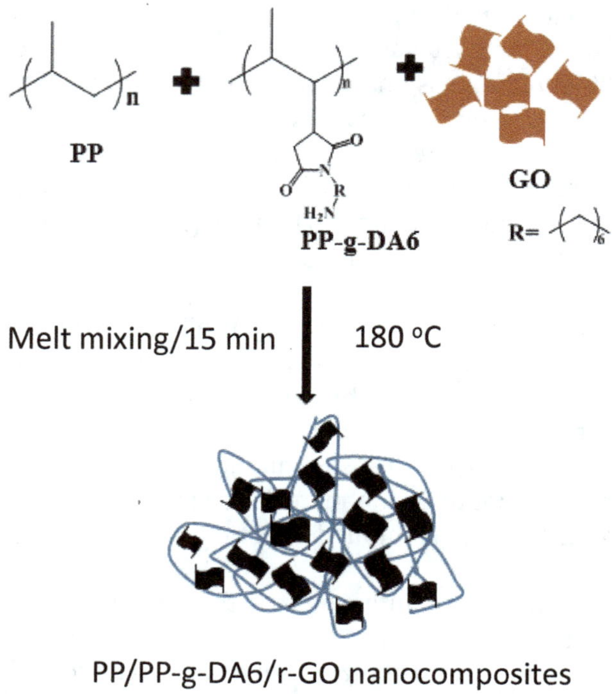

Figure 12.1 Generation of PP/r-GO nanocomposites.

Abu Dhabi Polymers. Polypropylene-graft-maleic anhydride and hexamethylenediamine (HMDA), used for the compatibilizer preparation, were supplied by Sigma Aldrich, Germany, and were used without any purification. PP-g-MA contained 8-10 wt% maleic anhydride and had an weight average molecular weight of about 9100 g/mol, viscosity of 4.0 poise and peak melting point of 156 °C. HMDA had a melt density of 0.89 g/mL at 25 °C and melting point range of 42-45 °C (lit.). GO was prepared using modified Hummer's method as reported before [34], by oxidation of graphite.

12.2.2 Compatibilizer Preparation

To prepare the compatibilizer PP-g-DA6, 10 g PP-g-MA and 20 g HMDA were weighed and added together in a round bottomed flask. The flask was immersed in a heating oil bath and kept for melt mixing for 30 min at 170°C under reflux conditions. The flask was kept under inert atmosphere during heating. The mixture was allowed to cool to room temperature after completion of the reaction. Afterwards, 200 mL distilled water was added to the flask, and the mixture was stirred overnight. Subsequently, the mixture was filtered using a Buchner funnel, followed by washing of the residue with water, ethanol and acetone. The extracted sample was then dried in an air oven, at a temperature of 50 °C for 24 h. The chemical reaction for the formation of PP-g-DA6 is shown in Figure 12.2.

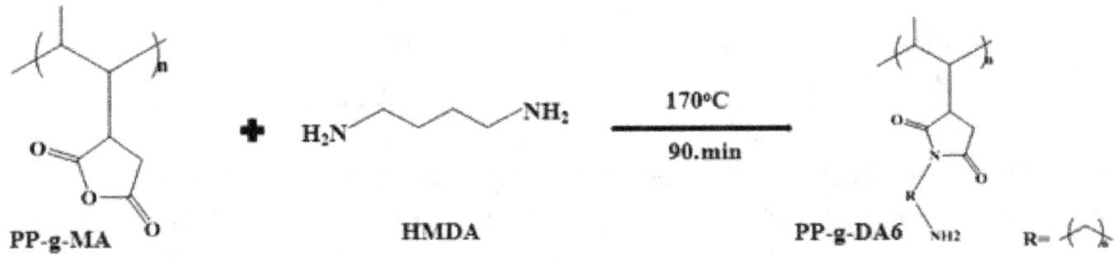

Figure 12.2 Preparation of PP-g-DA6.

12.2.3 Compatibilizer/Filler Hybrid Preparation

In order to identify the reaction between the amine functionalized PP compatibilizer and GO, hybrid of the two components was generated. In brief, 0.2 g of GO was added to 100 mL DMF solvent in a beaker, followed by ultra-sonication until the uniform dispersion of GO was attained. Afterwards, 0.2 g of PP-g-DA6 was added to the dispersion and stirred at 110 °C for 5 h. The mixture was then centrifuged, and the solvent was separated. The obtained hybrid PP-g-DA6/GO was dried in vacuum oven and analyzed using Fourier transform infra-red (FTIR) spectroscopy to examine the chemical linkage between the components.

12.2.4 Generation of Nanocomposites

PP nanocomposites were prepared by melt mixing PP, GO filler and PP-g-DA6 compatibilizer, using a twin conical screen extruder (Thermo Scientific Haake Minilab II , Germany), operated at 180 °C and 100 rpm in co-rotating mode. PP, GO and PP-g-DA6 were added to the extruder together after prior physical mixing and melt mixed for 15 min. A series of PP/PP-g-DA6/GO nanocomposites was generated by varying the content of GO as 0.5, 1.0, 3.0 and 5.0 wt%, as detailed in Table 12.1. The content of compatibilizer was fixed at 10 wt% in all nanocomposite formulations.

Table 12.1 Compositions of the different composites (in wt%)

Polymer/ Nanocomposites	PP (%)	PP-g-DA6 (%)	GO (%)
Pure PP	100	0	0.0
PP/PP-g-DA6	90.0	10	0.0
PP/PP-g-DA6/0.5% GO	89.5	10	0.5
PP/PP-g-DA6/1.0% GO	89.0	10	1.0
PP/PP-g-DA6/3.0% GO	87.0	10	3.0
PP/PP-g-DA6/5.0% GO	85.0	10	5.0

Test specimens were generated by injection molding of the extruded samples using an injection molding machine (Thermo Scientific Haake Minijet Pro, Germany). The cylinder temperature and mold temperature were set as 190 °C and 135 °C, respectively. Further, the injection pressure and post pressure were 430 bars and 500 bars, respectively, for 10 s. Specimens of different shapes, like dumbbell, disc and rectangular bar, were prepared for each formulation.

12.2.5 Characterization

FTIR analysis of the samples was carried out using a Nicolet iS10 spectrometer, equipped with SmartiTR diamond ATR accessory (45° angle of incidence), DTGS KBr detector as well as KBr beam splitter. The diamond ATR crystal had an index of refraction of 2.4 and a depth permeation of about 2 μm at 1000 cm^{-1} for the specimens having refractive index of 1.5. The characterization was performed in the transmission mode. The sample spectra were recorded utilizing OMNIC software in the IR range from 4000 cm^{-1} to 600 cm^{-1}, with a resolution of 4 cm^{-1} from 32 scans.

Wide angle X-ray diffraction (WAXRD) analyses of pure PP, GO and nanocomposites were performed on a powder diffractometer Panalytical X'Pert Pro using Cu Kα irradiation in the reflection mode. The filament current and accelerated voltages were 40 mA and 40 kV, respectively. To remove the noise, a zero-background holder was utilized. The analysis was carried out in the 2θ range from 5° to 60° at room temperature.

The thermal stability of the materials was analyzed by thermogravimetric analysis (TGA) using a TA Discovery thermogravimetric analyzer. The analysis was performed in an inert nitrogen atmosphere, and the temperature was ramped at 10 °C/min rate from 35 °C to 700 °C. The melting and crystallization behavior of the materials was analyzed using differential scanning calorimetry (DSC) under nitrogen atmosphere, using a TA Discovery DSC. Extruded samples of about 5-7 mg weight were used for the analysis. Two heating and cooling cycles were performed. In the first cycle, the samples were heated from room temperature (35 °C) to 200 °C, and subsequently cooled to -60 °C at a rate of 10 °C/min. In the second cycle, the samples were again heated to 200 °C to record the melt temperature and then cooled to 25 °C at a rate of 10 °C/min. The melting temperature (T_m), crystallization temperature (T_C), enthalpy of melting (ΔH_m) and enthalpy of crystallization (ΔH_C) were recorded.

The thermal conductivity of the samples was measured using laser flash analysis (LFA). This method is adequate for performing non-contact measurements of thermal conductivity, heat capacity and thermal diffusivity of various materials. The thermal conductivity (λ) was determined by using the relation $\lambda = \alpha.\rho.C_p$, where α, ρ and C_p are the thermal diffusivity, density and specific heat of the specimens, respectively, determined using the LFA 447 equipment from Netzsch, Germany (operated with Nanoflash software).

The melt rheological behavior of the specimens (storage modulus (G'), complex viscosity (η'), loss modulus (G") and mechanical loss factor (tan δ)) was determined using AR2000 813901 TA Instruments rheometer, having 25 mm parallel-plate geometry. Disc shaped samples of 25 mm diameter and 1.8 mm thickness were used. The tests were performed at 180 °C using the strain amplitude of about 2% in the angular frequency range 0.1-500 rad/sec. A gap of 1.7 mm was applied between the parallel plates.

The tensile analysis of the samples was carried out using universal testing machine, Instron 5567 UTM, in accordance with ASTM D 638. The tests were performed at room temperature, and a loading rate of 10 mm/min was employed. Mechanical properties like tensile modulus, elongation at break and tensile strength were recorded using the Bluehill software. Five replicated dumbbell shaped specimens were examined for each formulation and the mean value was recorded. Unnotched impact strength of the rectangular bar shaped samples was tested on a Resil 50 B Impactor from Ceast. The tests were carried out at room temperature in accordance with ASTM D256. Hammer energy of about 4 J and a speed of 3.46 m/s were used to impact the unnotched samples. The average value from five test samples was reported as the impact strength.

Raman analysis of the samples was carried out using a confocal microRaman spectrometer (Horiba LabRAM HR), equipped with a

633 nm laser source. The dispersion of GO nanofiller in the PP matrix was analyzed using transmission electron microscope (TEM, FEI, TECNAI) at 200 kV without staining. Here, ultra-thin sections of samples were microtomed (60 nm) at room temperature using PowerTome, equipped with a diamond knife. The sliced sections were collected on 400 mesh plain copper electron microscopy grids. Digital Micrograph software (Gatan) was used for the TEM image processing.

12.3 Results and Discussion

To examine the nature of interactions between PP, PP-g-DA6 compatibilizer and GO filler, FTIR analysis was carried out. Figure 12.3 shows the FTIR spectra of pristine PP, GO filler, PP-g-DA6 compatibilizer, PP-g-DA6/GO hybrid and PP nanocomposites. Pristine PP exhibited

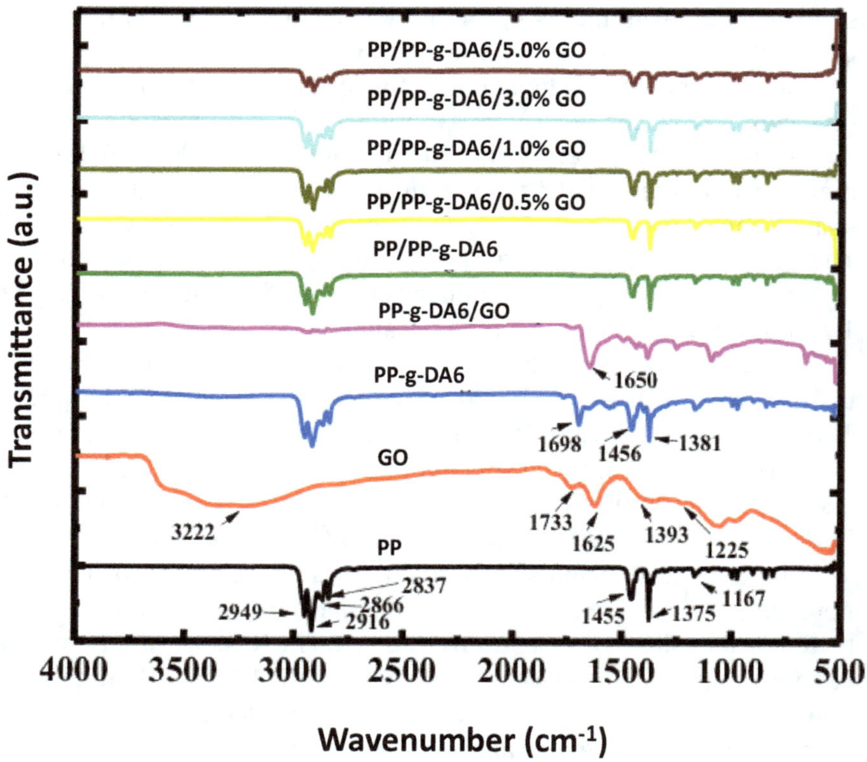

Figure 12.3 FTIR Spectra of pure PP, GO, PP-g-DA6, PP-g-DA6/GO hybrid and PP/PP-g-DA6/GO nanocomposites.

characteristic peaks at 2949 cm^{-1} (-CH_3 group's C-H asymmetric stretching), 2916 cm^{-1} (-CH_2 group's C-H asymmetric stretching), 2866 cm^{-1} (-CH_3 group's C-H symmetric stretching), 2837 cm^{-1} (-CH_2 group's C-H symmetric stretching), 1455 cm^{-1} (-CH_3 and -CH_2 groups' asymmetric C-H bending vibration), 1375 cm^{-1} (-CH_3 and -CH_2 groups' symmetric C-H bending vibration) and 1167 cm^{-1} (C-CH_3 group's C-C stretching). In the GO spectrum, a broad peak at 3222 cm^{-1} (O-H stretching vibration) was present. In addition, several other peaks at 1733 cm^{-1} (C=O stretching vibration), 1625 cm^{-1} (C=C stretching vibration), 1393 cm^{-1} (OH deformation vibration) and 1225 cm^{-1} (C-O-C stretching vibration) were also noted [35]. In the case of pure PP-g-DA6 compatibilizer, characteristic peaks at 1698 cm^{-1} (primary amide C=O stretching), 1456 cm^{-1} (NH secondary amine stretching) and 1381 cm^{-1} (amide carbonyl stretching) were observed. The FTIR spectrum of PP-g-DA6/GO hybrid indicated the disappearance of the peaks associated with the epoxide groups, thus, signaling the reduction of GO [36]. Further, a shift of peak from 1625 cm^{-1} (aromatic C=C stretching vibration) to 1650 cm^{-1} was observed due to the formation of r-GO. Thus, these observations confirmed the successful bond formation between GO and the amine functionalized PP compatibilizer.

Raman spectroscopic analysis was carried out to determine the interfacial interaction between the PP-g-DA6 compatibilizer and GO, as shown in Figure 12.4. Raman spectra provide insights about the ratio of the ordered crystalline and disordered sp^2 carbons present in the samples. The primary features of the Raman spectrum of GO are prominent D and G peaks appearing at approximately 1352 and 1577 cm^{-1}, respectively [37]. The D peak (defect band) indicates the extent of graphene disorder due to the structural edge effects [38], like the covalent bonding of the epoxides to the basal plane. The G peak corresponds to the graphite carbon structure [39], the planar configuration of the sp^2 bonded carbon which forms graphene. The sp^2 structure of graphene is disturbed by the oxidation process, hence, GO is not electrically conductive. However, the conductivity can be rapidly revived by reducing its oxygen functionalities. Several studies have shown that the electrical conductivity enhanced instantly during the epoxide reduction, and slowly during the reduction of other oxygen functionalities [40]. In Figure 12.4, the Raman spectra of pure GO and PP-g-DA6/GO hybrid, examined in the range 800-2000 cm^{-1}, indicated that the D band of the samples remained unchanged. However, the G band of the PP-g-DA6/GO hybrid shifted from 1577 cm^{-1} to 1595 cm^{-1} due to the covalent bonding between GO and the amine functionalized PP compatibilizer, to form r-GO. This type of blue shift in G band confirmed the stronger interaction between the compatibilizer and r-GO filler. The Raman I_D/I_G ratio (I_D: D peak Raman intensity and I_G: G peak Raman intensity) can be used to evaluate the quality of graphene as well as graphene based materials [41]. The intensity ratio (I_D/I_G) of GO and PP-g-DA6/GO hybrid are detailed in Table 12.2, which were calculated to be 1.00 and 1.07, respectively. It indicated that the I_D/I_G ratio of the hybrid increased by 7%, as compared to pure GO. The elimination of the oxygen containing groups in GO resulted in an increase in the extent of aromatic domain. On the other hand, it reduced the average size of the sp^2 domain, which enhanced the I_D/I_G ratio. Overall, the increase in the intensity ratio revealed a higher degree of reduction.

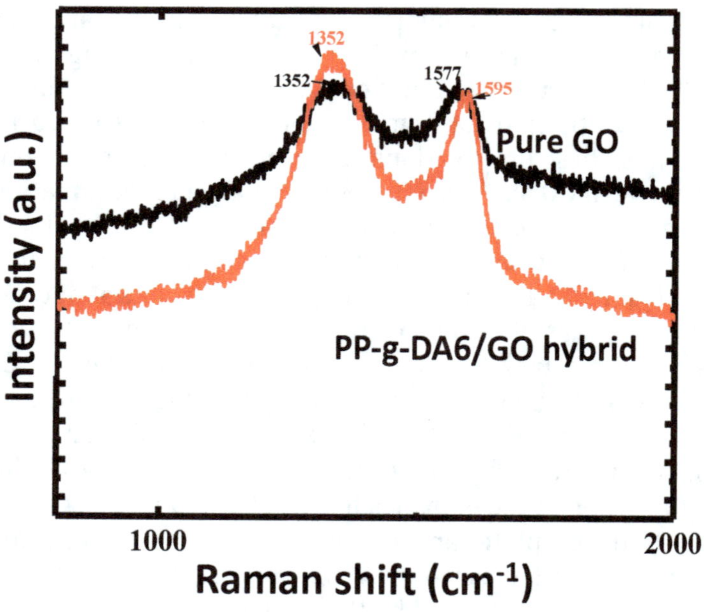

Figure 12.4 Raman spectra of pure GO and PP-g-DA6/GO hybrid.

Table 12.2 Raman intensity ratio (I_D/I_G) of GO and PP-g-DA6/GO hybrid

Material	D band (cm^{-1})	G band (cm^{-1})	Raman Intensity ratio (I_D/I_G)
Pure GO	1352	1577	1.00
PP-g-DA6/GO hybrid	1352	1595	1.07

The mechanical properties of the polymer composites have been reported to depend on the percentage crystallinity, interaction between the polymer matrix and nanofiller and filler dispersion state [42-45]. Table 12.3 presents the tensile modulus, tensile strength and elongation at break of pristine PP, PP/PP-g-DA6 and PP/PP-g-DA6/GO nanocomposites. As compared to the pristine PP, the tensile

Table 12.3 Tensile properties of pure PP, PP/PP-g-DA6 and PP/PP-g-DA6/GO nanocomposites

Material	Elongation at break (%)	Tensile stress at yield (MPa)	Tensile modulus (MPa)
Pure PP	7.4	38	879
PP/PP-g-DA6	6.9	37	893
PP/PP-g-DA6/0.5% GO	7.6	37	920
PP/PP-g-DA6/1.0% GO	7.2	37	916
PP/PP-g-DA6/3.0% GO	7.0	38	917
PP/PP-g-DA6/5.0% GO	7.7	37	920

modulus of the PP/PP-g-DA6/GO nanocomposites demonstrated a slight increase. For example, by the addition of 3.0 wt% and 5.0 wt% GO, along with 10.0 wt% PP-g-DA6 compatibilizer, the tensile modulus of the nanocomposites increased to 917 MPa and 920 MPa respectively from 879 MPa for pure PP. This indicated that the presence of the PP-g-DA6 compatibilizer enhanced the interfacial interaction between PP and GO, thus, resulting in the uniform distribution of the filler in the polymer matrix [46,47]. Owing to this, the elongation at break as well as yield stress values of the composites were also retained, irrespective of the amount of filler. In literature, a large number of studies have generally reported reduction in elongation and strength as a function of filler content owing to the presence of filler aggregates, however, no such reduction was observed in the current study.

Fracture toughness indicates the resistance of a material to fracture or crack propagation and is dependent on the energy absorbed during fracturing. Generally, toughness is characterized by the Charpy or Izod impact energy [48]. Polymers filled with micro or nanoparticles may exhibit different toughening mechanisms such as multiple crazing [48], crack-pinning, micro-cracking, shear yielding and particle bridging. Toughness increase of the semi-crystalline PP due to the incorporation of the nanoscale particles is mainly caused by the micro-cracking mechanism [49-51]. Further, the shear yielding mechanism is also significant. The fracture process in PP matrix begins with the plastic deformation in advance of the initial crack formation. Debonding of particles (forming holes), which is the initial damage mechanism, occurs, and the plastic flow takes place between the cavities. These zones are stretched and finally ruptured by tearing. Izod impact analysis of the specimens developed in the current study was carried out at room temperature, and the results are illustrated in Table 12.4. The pristine PP had impact strength of 7.96 kJ/m^2, while the blend of PP/PP-g-DA6 exhibited decreased impact strength of 6.30 kJ/m^2. Upon incorporating 0.5 wt%, 1.0 wt% and 3.0 wt% GO into PP, along with 10% compatibilizer PP-g-DA6, the impact strength of the samples increased up to 9.28 kJ/m^2 (16.6% increase), 19.45 kJ/m^2 (144.3% increase) and 19.60 kJ/m^2 (146.2% increase) respectively. The stronger interaction between the filler and PP matrix, due to the presence of PP-g-DA6 compatibilizer, played a significant role in inhibiting the motility of PP matrix and enhancing the energy need for crack initiation and propagation [52,53]. On the other hand, further increase in the amount of GO (5 wt% loading) caused a reduction in the crack growth initiation energy, probably due to the stacking of graphene sheets in the polymer matrix [54].

Table 12.4 Impact strength of pure PP, PP/PP-g-DA6 and PP/PP-g-DA6/GO nanocomposites

Material	Un-notched Izod impact strength (kJ/m^2)
Pure PP	7.96
PP/PP-g-DA6	6.30
PP/PP-g-DA6/0.5% GO	9.28
PP/PP-g-DA6/1.0% GO	19.45
PP/PP-g-DA6/3.0% GO	19.60
PP/PP-g-DA6/5.0% GO	14.31

To determine the processing properties and internal network structure of PP-g-DA6 compatibilized PP/GO nanocomposites, rheological analysis was carried out. The rheological behavior of the nanocomposites also serves as a measure of the filler dispersion in the polymer matrix [55]. The G' (Figure 12.5), G" (Figure 12.6), tan δ (Figure 12.7) and η^* (Figure 12.8) of PP, PP/PP-g-DA6 blend and PP/PP-g-DA6/GO nanocomposites were plotted as a function of angular frequency (ω). G' represents a property which is very sensitive to the

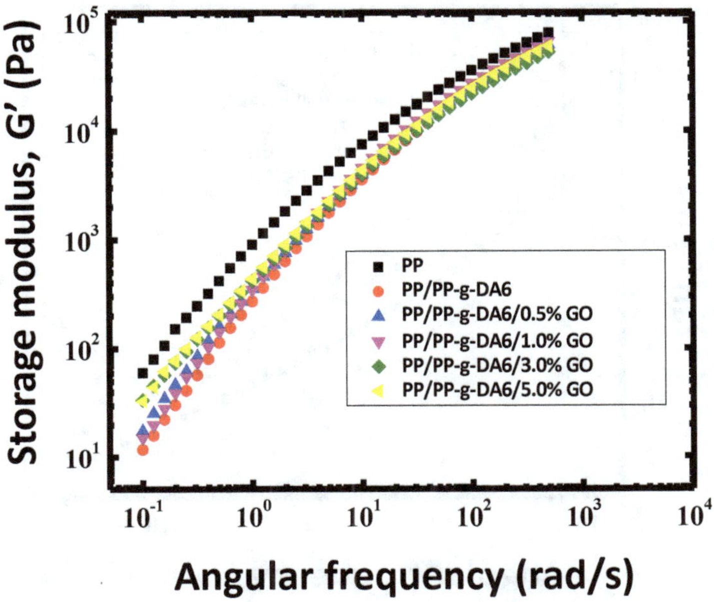

Figure 12.5 Storage modulus vs. angular frequency for pure PP, PP/PP-g-DA6 blend and PP/PP-g-DA6/GO nanocomposites.

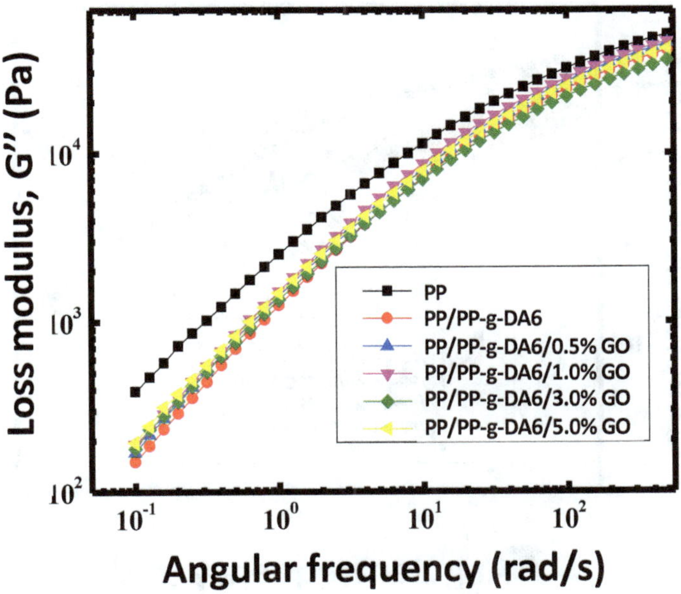

Figure 12.6 Loss modulus vs. angular frequency for pure PP, PP/PP-g-DA6 blend and PP/PP-g-DA6/GO nanocomposites.

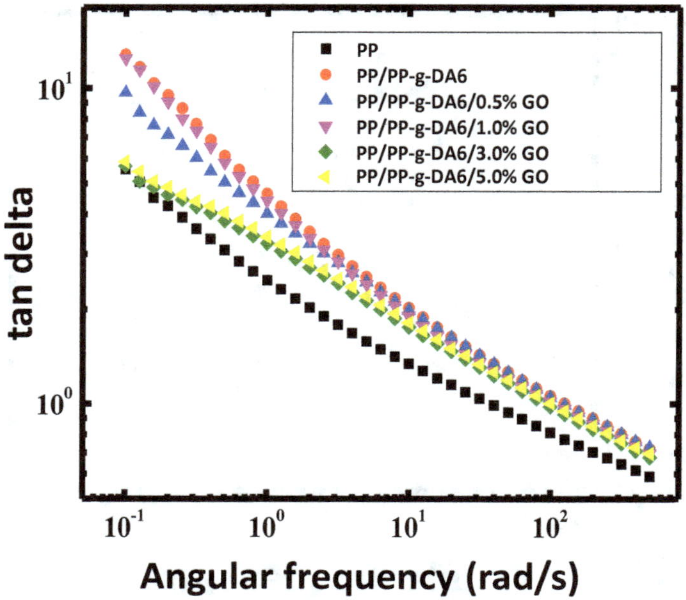

Figure 12.7 Tan delta vs. angular frequency for pure PP, PP/PP-g-DA6 blend and PP/PP-g-DA6/GO nanocomposites.

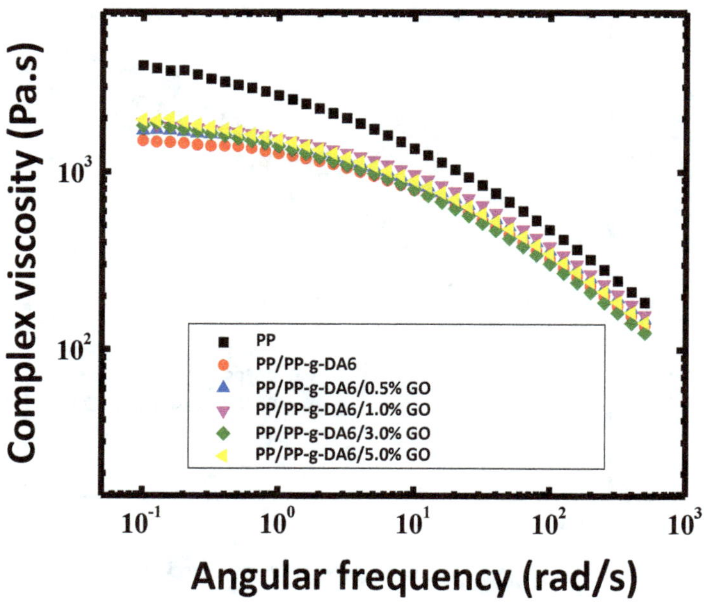

Figure 12.8 Complex viscosity (η^*) vs. angular frequency for pure PP, PP/PP-g-DA6 blend and PP/PP-g-DA6/GO nanocomposites.

formation of network structure in the nanocomposites. G' of the nanocomposites was observed to be lower that pure PP, due to matrix plasticization [56]. This decrease in G' can also be attributed to the PP/filler interlayer slipperiness, thus, the nanocomposite melt performed like a viscous PP liquid [57]. At higher frequencies, the G' curves were seen to be converging with each other, hence revealing the concentration independence [56]. Comparing the storage and loss modulus, G" of the materials was higher than G' at lower angular frequency, thus, revealing the dominance of viscous part. At higher frequencies, a transition was observed where the G' > G", which indicated the strong elastic behavior with lower relaxation time [34]. Similar to G' curves, the G" curves of the materials also converged with each other at higher angular frequency.

Tan δ is defined as the ratio of the loss modulus to the storage modulus [58-60]. In viscoelastic state, it is relatively higher as compared to the rubbery as well as glassy states. The tan δ of the nanocomposites was observed to be higher as compared to pure PP. Also, pure PP and nanocomposites exhibited viscoelastic state near 0.2 rad/sec, which was in good agreement with the literature studies [59]. For pristine PP, PP/PP-g-DA6 and PP/PP-g-DA6/GO nanocomposites, η* decreased with increased ω, indicating the shear thinning behavior [61,62]. The shear thinning nature at higher frequencies can be attributed to the dis-entanglement process. Also, η* of PP nanocomposites was observed to be lower than the pure PP, for the same reasons mentioned earlier for storage modulus.

Table 12.5 presents the calorimetric properties of PP and PP/PP-g-DA6/GO nanocomposites. The melting and cooling thermograms of

Table 12.5 Calorimetric behavior of PP and PP/ PP-g-DA6/GO nanocomposites

	ΔH_m (J/g)	T_m (°C)	ΔH_c (J/g)	T_c (°C)	X_c (%)
Pure PP	86.9	167	95.1	128	42
PP/PP-g-DA6/0.5% GO	94.3	166	93.1	127	45
PP/PP-g-DA6/1.0% GO	93.6	166	94.0	127	45
PP/PP-g-DA6/3.0% GO	92.2	167	91.3	127	45
PP/PP-g-DA6/5.0% GO	91.1	167	93.0	127	46

the materials are also presented in Figure 12.9. From the second heating curve, the degree of crystallinity was determined using the equation:

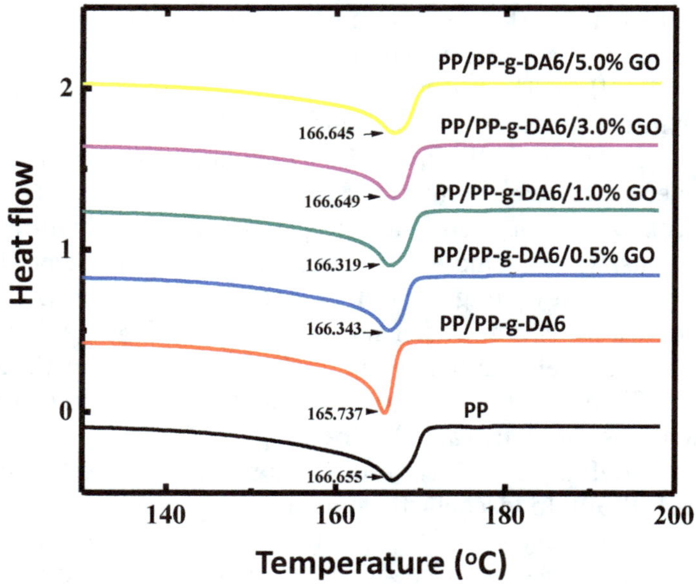

(a)

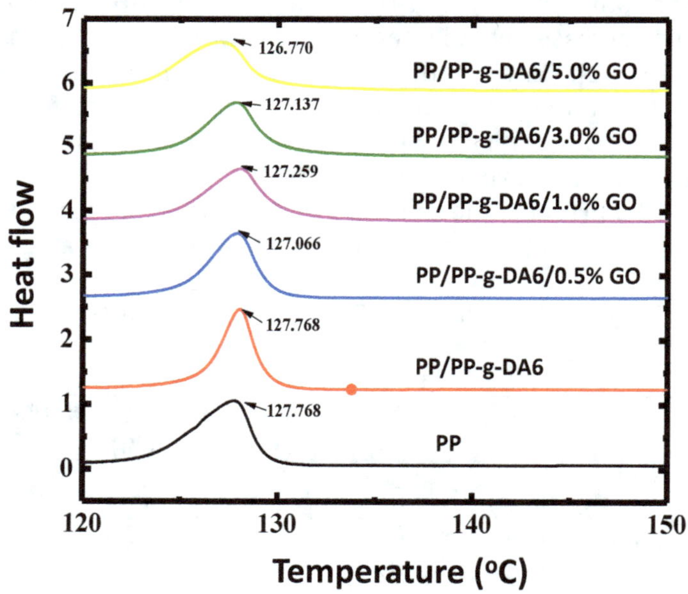

(b)

Figure 12.9 (a) Melting and (b) crystallization thermograms of pure PP, PP/PP-g-DA6 and PP/PP-g-DA6/GO nanocomposites.

$$X_c = [\,\Delta H_m / [(1\text{-}p)\,\Delta H_{100\%cr}]]\times 100$$

where X_c is the degree of crystallinity, ΔH_m is the melt enthalpy of PP in J/g (estimated from DSC), p is the filler fraction and $\Delta H_{100\%cr}$ is the enthalpy of fusion of 100% crystalline PP [63].

For pure PP and PP nanocomposites, a single endothermic peak appeared at a temperature above 160 °C. This confirmed the presence of α-crystalline structure without the presence of β-crystals [28]. Also, the addition of PP-g-DA6 compatibilizer and GO to the PP matrix increased the polymer crystallinity. For example, the crystallinity of PP in PP/PP-g-DA6/5.0% GO increased by 10%, compared to pure PP. The observed increase in crystallinity could be associated with the heterogeneous nucleation effect in PP. Further, no significant variation in T_m and T_c of PP was observed after the addition of nanofiller and compatibilizer. Previous reports acknowledged that grafting of long chain alkylamines like octadecylamine, hexadecylamine and decylamine on the 2-dimensional nanosheets resulted in the formation of highly oriented crystalline arrangement of alkylamines [64,65]. Thus, the addition of amine functionalized PP compatibilizer to PP would have provided the heterogeneous nucleation sites, thus, enhancing the crystal growth of PP [43].

Figure 12.10 illustrates the TGA weight loss curves of pure PP, PP/PP-g-DA6 and PP nanocomposites. The initial decomposition temperature (T_{ini} - the temperature at which 5% weight loss occurs), $T_{40\%}$ (the temperature at which 40% weight loss occurs) and char yield (the mass obtained at 700 °C) are also summarized in Table 12.6. For PP nanocomposites, the degradation temperatures for 5% and 40% weight loss enhanced with the filler loading till 1.0 wt %. It indicated enhanced thermal stability of the nanocomposites, when compared to the pure PP matrix [66]. For instance, PP/PP-g-DA6/5.0% GO sample exhibited an increase of 9.7% and 10.1% in T_{ini} and $T_{40\%}$ respectively as compared to pristine PP. Further, the char yield of the PP/PP-g-DA6/GO nanocomposites was observed to increase with GO loading. The enhanced thermal stability of the nanocomposites could be attributed to the improved interactions between the PP matrix and nanofiller. Hence, the heat transfer at the interface could be improved, thereby increasing the thermal degradation activation energy of the PP nanocomposites [67,68]. The dynamic TGA analysis (measurement of the weight loss as a function of time at a particular temperature) also exhibited delayed weight loss in the nanocomposites as compared to pure PP.

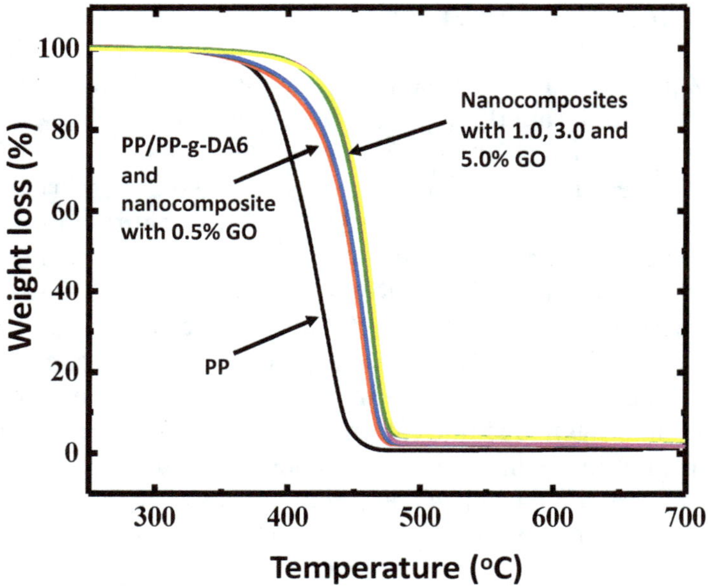

Figure 12.10 TGA thermograms of pure PP, PP/PP-g-DA6 and PP/PP-g-DA6/GO nanocomposites.

Table 12.6 Thermal degradation temperatures and char yield of pure PP, PP/PP-g-DA6 blend and PP/PP-g-DA6/GO nanocomposites

Composition	$T_{ini\%}$ (°C)	$T_{40\%}$ (°C)	Char yield at 700°C (wt%)
Pure PP	375	414	1.3
PP/PP-g-DA6	377	442	1.7
PP/PP-g-DA6/0.5% GO	383	444	1.8
PP/PP-g-DA6/1.0% GO	410	452	1.8
PP/PP-g-DA6/3.0% GO	410	452	3.0
PP/PP-g-DA6/5.0% GO	411	456	3.1

The thermal conductivity of suspended graphene is about 5000 W/mK [69], while the pristine PP has a very low thermal conductivity of approximately 0.2 W/mK [70]. Kalaitzidou *et al.* [70] reported that the addition of 3 wt% exfoliated graphene nanoplatelets to PP enhanced the thermal conductivity to 0.4 W/mK. The thermal conductivity analysis of the materials developed in the current study is depicted in Figure 12.11. The presence of nanofiller was observed to enhance the thermal conductivity of PP remarkably, as a function of filler loading. In the case of PP/PP-g-DA6/5.0% GO, the thermal conductivity reached 0.376 W/mK, which indicated an increase of 88%, as compared to pure PP.

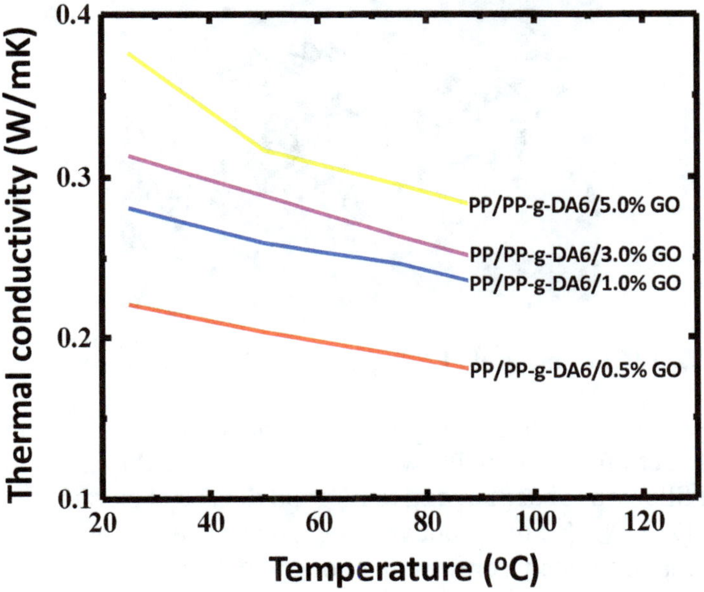

Figure 12.11 Thermal conductivity of PP/PP-g-DA6/GO nanocomposites.

The competency of the nanofiller to reinforce the polymer matrix is mainly decided by the degree of dispersion in the matrix. The strong interaction between the polymer and filler phases is mainly responsible for the physico-chemical properties of polymer nanocomposites [71]. Figure 12.12 presents the TEM micrograph of the PP/PP-g-MA/5.0% GO composite. The filler was observed to be uniformly distributed in the polymer matrix. No extensive filer aggregation was observed in the composite morphology. Thus, the presence of compatibilizer led to a superior degree of filler exfoliation in the polymer, which correspondingly resulted in the enhanced composite properties.

Figure 12.13 illustrates the WAXRD patterns of pure PP, GO, PP-g-DA6, PP/PP-g-DA6 blend and PP/PP-g-DA6/GO nanocomposites. For pure GO, a sharp reflection was observed at $2\theta = 9.6°$, while the diffraction peaks of pristine PP were observed at 14.5°, 17.2°, 18.8° and 22.1°, which corresponded to (110), (040), (130) and (111) crystal

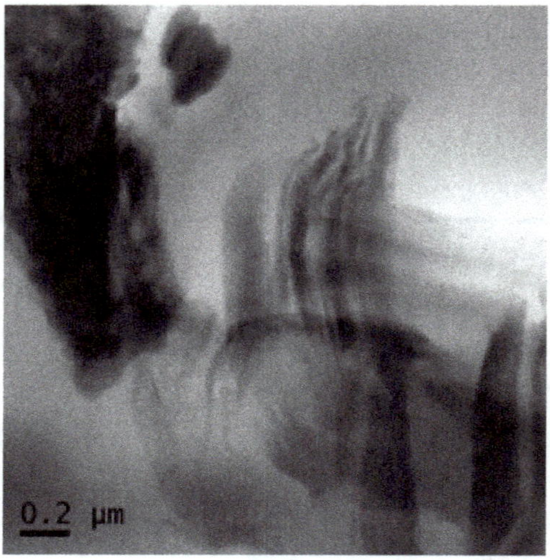

Figure 12.12 TEM image of PP/PP-g-DA6/5.0% GO composite.

planes of PP, respectively [72]. For the nanocomposites, only crystalline diffraction peaks of PP were observed, and the characteristic peak of filler was absent owing to its good degree of exfoliation in the matrix [73]. No β and γ phases were observed in the specimens, which was previously confirmed through DSC analysis.

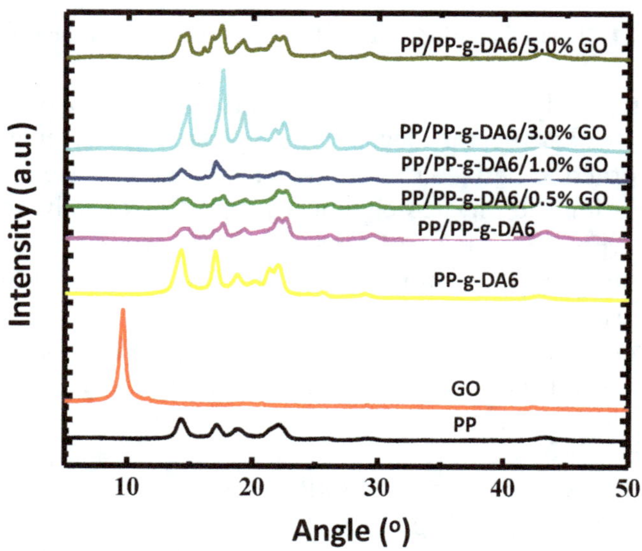

Figure 12.13 X-ray diffractograms of PP, GO, PP-g-DA6, PP/PP-g-DA6 and PP/PP-g-DA6/GO nanocomposites.

12.4 Conclusions

Nanocomposites of PP, GO and amine functionalized PP (PP-g-DA6) compatibilizer were prepared by melt mixing method. Addition of 10 wt% of compatibilizer (PP-g-DA6) and varying amounts of GO (0.5, 1.0, 3.0, 5.0 wt%) remarkably impacted the morphology as well as properties of PP. GO was reduced to r-GO owing to the chemical interaction with the PP-g-DA6 compatibilizer, as confirmed by FTIR and Raman analyses. The covalent interactions between the filler and compatibilizer phases resulted in enhanced filler distribution in the matrix, which was also confirmed through the microscopy analysis. As a result, the efficient load transfer to the filler owing to the effective polymer-filler interfacial contacts increased the tensile modulus in the nanocomposites. The stronger interaction between the polymer and filler also played a significant role in inhibiting the motility of the PP matrix and increasing the energy needed for crack initiation and propagation, thus, significantly enhancing the toughness of the nanocomposites. DSC results confirmed that the addition of amine functionalized PP compatibilizer to PP, along with GO, marginally increased the crystallinity of PP. The thermal stability of the nanocomposites increased with GO content due to the enhanced heat transfer to the filler phase at the polymer-filler interface. Further, the thermal conductivity was observed to significantly improve with increase in GO loading. Overall, owing to the optimal property profiles and performance, the materials developed in this study have high potential for various engineering applications.

Acknowledgement

The authors sincerely thank the PI Research Centre, The Petroleum Institute, UAE, for the financial support.

References

1. Kissel, W. J., Han, J. H., and Meyer, J. A. (2003) Polypropylene: Structure, properties, manufacturing processes, and applications. In: *Handbook of Polypropylene and Polypropylene Composites*, Karian, H. (ed.), Marcel Dekker AG, Switzerland, pp. 10-27.
2. Maier, C., and Calafut, T. (1998) *Polypropylene. The Definitive Data User's and Guidebook*, William Andrew Inc., USA.
3. *Polypropylene Handbook*, Pasquini, N. (ed.), 2nd edition, Carl Hanser

Verlag, Germany (2005).

4. Lin, Y., Chen, H., Chan, C. M., and Wu, J. (2008) High impact toughness polypropylene/$CaCO_3$ nanocomposites and the toughening mechanism. *Macromolecules*, **41**, 9204-9213.
5. Friedrich, K., Carlsson, L. A., Gillespie, J. W., and Karger-Kocsis, J. (1991) Fracture of thermoplastic composites. In: *Thermoplastic Composite Materials*, Carlsson, L. A. (ed.), Elsevier, The Netherlands, pp. 233-294.
6. Acton, Q. A. (2013) *Polyenes - Advances in Research and Application*, ScholarlyEditions, USA.
7. Scheirs, J. (2009) *A Guide to Polymeric Geomembranes*, John Willey & Sons, USA.
8. Mittal, V. (2011) *Advances in Polyolefin Nanocomposites*, Taylor and Francis, USA.
9. Pukánszky, B., Tüdos, F., and Kelen, T. (1986) Mechanical and rheological properties of multicomponent polypropylene blends. *Polymer Composites*, **7**, 106-115.
10. Jančář, J., and Dibenedetto, A. T. (1994) The mechanical properties of ternary composites of polypropylene with inorganic fillers and elastomer inclusions. *Journal of Material Science*, **29**, 4651-4658.
11. Stamhuis, J. E. (1988) Mechanical properties and morphology of polypropylene composites. 2. Effect of polar components in talc-filled polypropylene. *Polymer Composites*, **9**, 72-77.
12. Arencón, D., Velasco, J. I., Realinho, V., Sánchez-Soto, M. A., and Gordillo, A. (2007) Fracture toughness of glass microsphere filled polypropylene and polypropylene/poly(ethylene terephthalate-coisophthalate) blend-matrix composites. *Journal of Materials Science*, **42**, 19-29.
13. Chanda, M., and Roy, S. K. (2006) *Plastics Technology Handbook*, 4th edition, CTC Press, USA.
14. Novoselov, K. S., Geim, A. K., Morozov, S. V., Jiang, D., Zhang, Y., Dubonos, S. V., Grigorieva, I. V., and Firsov, A. A. (2004) Electric field effect in atomically thin carbon films, *Science*, **306**, 666-669.
15. Zhu, Y. W., Murali, S., Cai, W. W., Li, X. S., Suk, J. W., Potts, J. R., and Ruoff, R. S. (2010) Graphene and graphene oxide, synthesis, properties, and applications. *Advanced Materials*, **22**, 3906-3924.
16. Kuilla, T., Bhadra, S., Yao, D., Kim, N. H., Bose, S., and Lee, J. H. (2010) Recent advances in graphene based polymer composites. *Progress in Polymer Science*, **35**(11), 1350-1375.
17. Appel, A. K., Thomann, R., and Mülhaupt, R. (2012) Polyurethane nanocomposites prepared from solvent-free stable dispersions of functionalized graphene nanosheets in polyols. *Polymer*, **53**, 4931-4939.
18. Ahmad, S. R., Young, R. J., and Kinloch, I. A. (2015) Raman spectra and mechanical properties of graphene/polypropylene

nanocomposites. *International Journal of Chemical Engineering and Applications*, **6**(1), 1-5.

19. Song, P., Cao, Z., Cai, Y., Zhao, L., Fang, Z., and Fu, S. (2011) Fabrication of exfoliated graphene-based polypropylene nanocomposites with enhanced mechanical and thermal properties. *Polymer*, **52**(18), 4001-4010.
20. Mukhopadhyay, P., and Gupta, R. K. (2011) Trends and frontiers in graphene-based polymer nanocomposites. *Plastics Engineering*, **67**, 32-42.
21. Eda, G., Fanchini G., and Chhowalla, M. (2008) Large-area ultrathin films of reduced graphene oxide as a transparent and flexible electronic material. *Nature Nanotechnology*, **3**, 270-274.
22. Hummers, W. S., and Offeman, R. E. (1958) Preparation of graphitic oxide. *Journal of American Chemical Society*, **80**(6), 1339.
23. Dreyer, D. R., Park, S., Bielawski, C. W., and Ruoff, R. S. (2010) The chemistry of graphene oxide. *Chemical Society Reviews*, **39**, 228-240.
24. Eda, G., and Chhowalla, M. (2010) Chemically derived graphene oxide, Towards large-area thin-film electronics and optoelectronics. *Advanced Materials*, **22**, 2392-2415.
25. Kowalewski, T., Galeski, A., and Kryszewski, M. (1984) The structure and tensile properties of cold drawn modified chalk filled polypropylene. In: *Polymer Blends: Processing, Morphology and Properties*, Kryszewski, M., Galeski, A., and Martuscelli, E. (eds.), Plenum Press, USA, pp. 223-241.
26. Shin, K.-Y., Hong, J.-Y., Lee, S., and Jang, J. (2012) Evaluation of anti-scratch properties of graphene oxide/polypropylene nanocomposites. *Journal of Material Chemistry*, **22**, 7871-7879.
27. Ryu, S. H., and Shanmugharaj, A. M. (2014) Influence of hexamethylene diamine functionalized graphene oxide on the melt crystallization and properties of polypropylene nanocomposites. *Materials Chemistry and Physics*, **146**(3), 478-486.
28. Bao, R.-Y., Cao, J., Liu, Z.-Y., Yang, W., Xie, B.-H., and Bao, M.-B. Y. (2014) Towards balanced strength and toughness improvement of isotactic polypropylene nanocomposites by surface functionalized graphene oxide. *Journal of Materials Chemistry A*, **2**, 3190-3199.
29. Fuad, M. Y. A., Hanim, H., Zarina, R., Mohd. Ishak, Z. A., and Hassan, A. (2010) Polypropylene/calcium carbonate nanocomposites - effects of processing techniques and maleated polypropylene compatibilizer. *Polymer Letters*, **4**(10), 611-620.
30. Page, D. J. Y. S. and Gopakumar, T. G (2006) Properties and crystallization of maleated polypropylene/graphite flake nanocomposites. *Polymer Journal*, **38**, 920-929.
31. Hemmati, M., Shariatpanahi, H., Ahangari, M. G., Fereidoon, A., Narimani, A. (2012) Effect of polypropylene-grafted-maleic anhydride

compatibilizeron the physical properties of polypropylene/carhon nanotuhe composites. *Polymers and Polymer Composites*, **20**(6), 559-566.

32. Cui, L., and Paul, D. R. (2007) Evaluation of amine functionalized polypropylenes as compatibilizers for polypropylene nanocomposites. *Polymer*, **48**(6), 1632-1640.
33. Kobayashi, S., Song, J., Silvis, H. C., Macosko, C. W., and Hillmyer, M. A. (2011) Amino-functionalized polyethylene for enhancing the adhesion between polyolefins and polyurethanes. *Industrial Engineering and Chemistry Research*, **50**(6), 3274-3279.
34. Chaudhry, A. U., and Mittal, V. (2013) High density polyethylene nanocomposites using master batches of chlorinated polyethylene/graphene oxide. *Polymer Engineering and Science*, **53**(1), 78-88.
35. Tien, H. N., Luan, V. H., Lee, T. K., Kong, B. S., Chung, J. S., Kim, E. J., and Hur, S. H. (2012) Enhanced solvothermal reduction of graphene oxide in a mixed solution of sulfuric acid and organic solvent. *Chemical Engineering Journal*, **211**, 97-103.
36. Yuan, B., Bao, C., Song, L., Hong, N., Liew, K. M., and Hu, Y. (2014) Preparation of functionalized graphene oxide/polypropylene nanocomposite with significantly improved thermal stability and studies on the crystallization behavior and mechanical properties. *Chemical Engineering Journal*, **237**, 411-420.
37. Elias, D. C., Nair, R. R., Mohiuddin, T. M. G., Morozov, S. V., Blake, P., Halsall, M. P., Ferrari, A. C., Boukhvalov, D. W., Katsnelson, M. I., Geim, A. K., and Novoselovet, K. S. (2009) Control of graphene's properties by reversible hydrogenation, evidence for graphane. *Science*, **323**(5914), 610-613.
38. Rao, C. N. R., Sood, A. K., Subrahmanyam, K. S., and Govindaraj, A. (2009) Graphene, the new two-dimensional nanomaterial. *Angewandte Chemie, International Edition*, **48**, 7752-7777.
39. Mittal, V., Luckachan, G. E., and Matsko, N. B. (2014) PE/chlorinated-PE blends and PE/chlorinated-PE/graphene oxide nanocomposites: Morphology, phase miscibility, and interfacial interactions. *Macromolecular Chemistry and Physics*, **215**, 255-268.
40. Xu, Z., Bando, Y., Liu, L., Wang, W., Bai, X., and Golberg, D. (2011) Electrical conductivity, chemistry, and bonding alternations under graphene oxide to graphene transition as revealed by in situ TEM. *ACS Nano*, **5**(6), 4401-4406.
41. Jiao, L., Wang, X., Diankov, G., Wang, H., and H. Dai (2010) Facile synthesis of high-quality graphene nanoribbons., *Nature Nanotechnology*, **5**, 321-325.
42. Mittal, V., and Chaudhry, A. U. (2015) Polymer-graphene nanocomposites: Effect of polymer matrix and filler amount on properties. *Macromolecular Materials and Engineering*, **300**, 510-521.
43. Ryu, S. H., and Shanmugharaj, A. M. (2014) Influence of long-chain

alkylamine-modified graphene oxide on the crystallization, mechanical and electrical properties of isotactic polypropylene nanocomposites. *Chemical Engineering Journal*, **244**, 552-560.

44. Wypych, G. (2000) *Handbook of Fillers*, 4th edition, ChemTec Publishing, USA.
45. Shen, B., Zhai, W., Tao, M., Lu, D., and Zheng, W. (2013) Chemical functionalization of graphene oxide toward the tailoring of the interface in polymer composites. *Composites Science and Technology*, **77**, 87-94.
46. Mandhakini, M., Chandramohan, A., Vengatesan, M. R., Alagar, M. (2011) Synthesis and characterization of linseed vinyl ester fatty amide modified epoxy layered silicate nanocomposites. *High Performance Polymers*, **23**(5), 403-412.
47. Qian, D., Dickey, E. C., Andrews, R., and Rantell, T. (2000) Load transfer and deformation mechanisms in carbon nanotube polystyrene composites. *Applied Physics. Letters*, **76**(20), 2868.
48. Cotterell, B., Chia, J.Y.H., and Hbaieb, K. (2007) Fracture mechanisms and fracture toughness in semicrystalline polymer nanocomposites. *Engineering Fracture Mechanics*, **74**(7), 1054-1078.
49. Herrera-Ramírez, L. C., Castell, P., Fernández-Blázquez, J. P., Fernández, Á., and de Villoria, R. G. (2015) How do graphite nanoplates affect the fracture toughness of polypropylene composites? *Composites Science and Technology*, **111**(6), 9-16.
50. Arends, C. B. (1996) *Polymer Toughening*, Marcel Dekker Inc., USA.
51. Arencón, D., and Velasco, J. I. (2009) Fracture toughness of polypropylene-based particulate composites. *Materials*, **2**, 2046-2094.
52. Liang, J. Z., and Li, R. K. Y. (2000) Rubber toughening in polypropylene, A Review. *Journal of Applied Polymer Science*, **77**, 409-417.
53. Albdiry, M., Yousif, B., Ku, H., and Lau, K. (2012) A critical review on the manufacturing processes in relation to the properties of nanoclay/polymer composites. *Journal of Composites Materials*, **47**(9), 1093-1115.
54. De Villoria, R. G., and Miravete, A. (2007) Mechanical model to evaluate the effect of the dispersion in nanocomposites. *Acta Materialia*, **55**, 3025-3031.
55. Woo, D. K., Kim, B. C., and Lee, S. J. (2009) Preparation and rheological behavior of polystyrene/multi-walled carbon nanotube composites by latex technology. *Korea-Australia Rheology Journal*, **21**(3), 185-191.
56. Chaudhry, A. U., and Mittal, V. (2014) Blends of high-density polyethylene with chlorinated polyethylene, Morphology, thermal, rheological, and mechanical properties. *Polymer Engineering & Science*, **54**, 85-95.
57. Ferry, J. D. (1980) *Viscoelastic Properties of Polymers*, 3rd edition, Wiley, USA.

58. Poveda, R. L., and Gupta, N. (2016) *Carbon Nanofiber Reinforced Polymer Composites*, Springer, Germany.
59. Li, Y., Zhu, J., Wei, S., Ryu, J., Sun, L., and Guo Z. (2011) Poly(propylene)/graphene nanoplatelet nanocomposites, melt rheological behavior and thermal, electrical, and electronic properties. *Journal of Macromolecular Chemistry and Physics*, **212**(18), 1951-1959.
60. a) Chen, X., Wei, S., Yadav, A., Patil, R., Zhu, J., Ximenes, R., Sun, L., and Guo, Z. (2011) Poly (propylene)/carbon nanofiber nanocomposites: Ex situ solvent-assisted preparation and analysis of electrical and electronic properties. *Macromolecular Materials and Engineering*, **296**(6), 434-443; b) Wei, S., Patil, R., Sun, L., Haldolaarachchige, N., Chen, X., Young, D. P., and Guo, Z. (2011) Ex situ solvent-assisted preparation of magnetic poly(propylene) nanocomposites filled with Fe@FeO nanoparticles. *Macromolecular Materials and Engineering*, **296**(9), 850-857.
61. King, J. A., Via, M. D., Keith, J. M., and Morrison, F. A. (2009) Effects of carbon fillers on rheology of polypropylene based resins. *Journal of Composite Materials*, **43**(25), 3073-3089.
62. Sugimoto, M., Tanaka, T., Masubuchi, Y., Takimoto, J. I., and Koyama, K. (1999) Effect of chain structure on the melt rheology of modified polypropylene. *Journal of Applied Polymer Science*, **73**, 1493-1500.
63. *Polymer Handbook*, Brandup, J., Immergut, E. H., and Grulke, E. A. (eds.), 4th edition, John Wiley & Sons, USA (1999).
64. Lin, Z., Liu, Y., and Wong, C. P. (2010) Facile fabrication of superhydrophobic octadecylamine-functionalized graphite oxide film. *Langmuir*, **26**, 16110-16114.
65. Yang, B. X., Pramoda, K. P., Xu, G. Q., and Goh, S. H. (2007) Mechanical reinforcement of polyethylene using polyethylene-grafted multiwalled carbon nanotubes. *Advanced Functional Materials*, **17**, 2062-2069.
66. Vengatesan, M. R., Singh, S., Pillai, V. V., and Mittal, V. (2016) Crystallization, mechanical, and fracture behavior of mullite fiber-reinforced polypropylene nanocomposites. *Journal of Applied Science*, **133**(30), doi: 10.1002/app.43725.
67. Cheng, H. K., Sahoo, N. G., Tan, Y. P., Pan, Y., Bao, H., Li, L., Chan, S. H., and Zhao, J. (2012) Poly(vinyl alcohol) nanocomposites filled with poly(vinyl alcohol)-grafted graphene oxide. *ACS Applied Materials & Interfaces*, **4**, 2387-2394.
68. Li, B., and Zhong, W.-H. (2011) Review on polymer/graphite nanoplatelet nanocomposites. *Journal of Materials Science*, **46**, 5595-5614.
69. Prasher, R. (2010) Graphene spreads the heat. *Science*, **328**, 185-186.
70. Kalaitzidou, K., Fukushima, H., and Drzal, L. T. (2007) Multifunctional polypropylene composites produced by incorporation of exf-

oliated graphite nanoplatelets. *Carbon*, **45**(7), 1446-1452.

71. Yang, L., Phua, S. L., Teo, J. K., Toh, C. L., Lau, S. K., Ma, J., and Lu, X. (2011) A biomimetic approach to enhancing interfacial interactions, polydopamine-coated clay as reinforcement for epoxy resin. *ACS Applied Materials & Interfaces*, **3**, 3026-3032.
72. Nishino, T., Matsumoto, T., and Nakamae, K. (2000) Surface structure of isotactic polypropylene by X-ray diffraction. *Polymer Engineering & Science*, **40**, 336-343.
73. Cao, N., and Zhang, Y. (2015) Study of reduced graphene oxide preparation by hummers' method and related characterization. *Journal of Nanomaterials*, **2015**, Article ID 168125.

Index

A

B

C

D

E

F

G

N

P

R

S

T

U

V

W

X

Y

www.ingramcontent.com/pod-product-compliance
Lightning Source LLC
LaVergne TN
LVHW011516200926
848484LV00001B/53

* 9 7 8 1 9 2 2 6 1 7 1 2 5 *